# METHODS IN MOLECULAR BIOLOGY™



*Series Editor*
**John M. Walker**
**School of Life Sciences**
**University of Hertfordshire**
**Hatfield, Hertfordshire, AL10 9AB, UK**

For further volumes:
http://www.springer.com/series/7651

# Gene Synthesis

## Methods and Protocols

Edited by

**Jean Peccoud**

*Virginia Bioinformatics Institute, Virginia Tech, Blacksburg, VA, USA*

Humana Press

*Editor*
Jean Peccoud, Ph.D.
Virginia Bioinformatics Institute
Virginia Tech
Blacksburg, VA, USA
jpeccoud@vbi.vt.edu

ISSN 1064-3745 e-ISSN 1940-6029
ISBN 978-1-4939-6200-6 ISBN 978-1-61779-564-0 (eBook)
DOI 10.1007/978-1-61779-564-0
Springer New York Dordrecht Heidelberg London


Softcover re-print of the Hardcover 1st edition 2012


Printed on acid-free paper

Humana Press is part of Springer Science+Business Media (www.springer.com)

# Preface

The de novo fabrication of custom DNA molecules is a transformative technology that significantly affects the biotechnology industry. Basic genetic engineering techniques for manipulating DNA in vitro opened an incredible field of opportunity in the life sciences. However, genetic engineering has now moved beyond the introduction of single genes into cells to multigene cassettes, and is rapidly progressing toward whole genome engineering. In this new context, the synthesis of DNA molecules has resurged as the time and cost-limiting step in genetic engineering.

Today, most multigene engineering projects involve ad hoc methods of DNA assembly. A variety of PCR-based methods are in common use alongside more traditional restriction enzyme-based assembly methods. Their essential feature is the piecing together of existing DNAs that are cloned from natural sources. These techniques present a number of limitations. The use of restriction sites within natural sequences necessitates a labor intensive custom cloning strategy that is difficult to automate. As a result, molecular biologists often reach a tacit compromise between obtaining a desired sequence and the number of steps in the cloning process they are willing or able to undertake in constructing it.

Theoretically, DNA fabrication methods that are rooted in chemical synthesis could transform synthesis into a generic, predictable, and scalable process allowing the generation of any user-defined DNA sequence. By liberating the process from the confines of preexisting sequences, the problem of composition design becomes orthogonal to the problem of physical construction. Therefore, as gene synthesis becomes a commodity, biologists will spend more time designing custom DNA molecules and characterizing their performance, and less time constructing them.

One day, DNA may be fabricated using a purely chemical process. Today, however, DNA fabrication still involves sophisticated cloning techniques, but nevertheless a transition period has already emerged. Academic and commercial operators experiment with complex processes that combine the assembly of chemically synthesized oligos with cloning steps in attempts to construct long DNA molecules. Even though a number of companies have rushed to and sometimes later walked away from the gene synthesis market, DNA fabrication is not a black box that would involve radically different techniques than those commonly used in a molecular biology laboratory, nor does it require expensive equipment. Depending on the context it might make sense to outsource DNA fabrication to an external vendor, but in other cases there might be value in performing part of the process in house. In fact, gene synthesis projects are approachable by undergraduate students enabled by straightforward protocols and training in a relatively small set of molecular biology skills. In any case, it is important to understand that the fabrication of small DNA fragment (less than 1 kb) is often very straightforward, but the assembly of longer DNA molecules raises a number of inherent technical difficulties that need to be understood.

This book provides step-by-step protocols for the different stages of a DNA fabrication process. Section I focuses on protocols used for the assembly of oligonucleotides in building blocks also called synthons. The cloning of synthons into larger fragments up to the size of bacterial genomes is the focus of Section II. Bioinformatics protocols and software applications necessary to design gene synthesis protocols are described in Section III. Finally, Section IV describes the educational and biosecurity impacts of gene synthesis.

Any laboratory relying on recombinant DNA technology for its research is a potential user of gene synthesis. Few laboratories will develop a completely home grown gene synthesis process. Oligonucleotide synthesis or sequencing will most likely be outsourced to a core facility or a commercial operator. In other cases, the synthesis of longer fragments may also be outsourced. By providing step-by-step descriptions of all the different stages of a complex gene synthesis process, this book will help readers refine their understanding of gene synthesis and determine what part of the process they can or should do in their laboratory and what parts should be contracted to a specialized service provider.

*Blacksburg, VA, USA* *Jean Peccoud, Ph.D.*

# Contents

# Contributors


RINJI AKADA • *Department of Applied Molecular Bioscience, Yamaguchi University Graduate School of Medicine, Ube, Japan*

NARAYANA ANNALURU • *Department of Environmental Health Sciences, Johns Hopkins University School of Public Health, Baltimore, MD, USA*

JOEL S. BADER • *High Throughput Biology Center, Johns Hopkins University School of Medicine, Baltimore MD, USA; Whiting School of Engineering, Johns Hopkins University, Baltimore, MD, USA*

GWYNEDD A. BENDERS • *J. Craig Venter Institute Inc., San Diego, CA, USA*

KAJAL BISWAS • *Mouse Cancer Genetics Program, Center for Cancer Research, National Cancer Institute at Frederick, Frederick, MD, USA*

MARCUS BODE • *Institute of Bioengineering and Nanotechnology, The Nanos, Singapore*

JEF D. BOEKE • *Department of Molecular Biology and Genetics, High Throughput Biology Center, Johns Hopkins University School of Medicine, Baltimore, MD, USA*

A. MALCOLM CAMPBELL • *Department of Biology, Davidson College, Davidson, NC, USA; Genome Consortium for Active Teaching, Davidson, NC, USA*

JERONIMO CELLO • *Department of Molecular Genetics and Microbiology, Stony Brook University, Stony Brook, NY, USA*

KAMONCHAI CHA-AIM • *Department of Applied Molecular Bioscience, Yamaguchi University Graduate School of Medicine, Ube, Japan*

SRINIVASAN CHANDRASEGARAN • *Department of Environmental Health Sciences, Johns Hopkins University School of Public Health, Baltimore, MD, USA*

SUHWAN CHANG • *Mouse Cancer Genetics Program, Center for Cancer Research, National Cancer Institute at Frederick, Frederick, MD, USA*

WAI CHYE CHEONG • *Institute of Bioengineering and Nanotechnology, The Nanos, Singapore*

CHRISTOPHER F. CHYBA • *Program on Science and Global Security, Woodrow Wilson School, Princeton University, Princeton, NJ, USA*

ERIC M. COOPER • *High Throughput Biology Center, Johns Hopkins University School of Medicine, Baltimore, MD, USA*

JESSICA S. DYMOND • *High Throughput Biology Center, Johns Hopkins University School of Medicine, Baltimore, MD, USA*

TODD T. ECKDAHL • *Department of Biology, Missouri Western State University and Genome Consortium for Active Teaching., St. Joseph, MO, USA; Genome Consortium for Active Teaching, Davidson, NC, USA*

STEPHEN J. ELLEDGE • *Department of Genetics, Howard Hughes Medical Institute, Harvard Medical School, Boston, MA, USA; Division of Genetics, Department of Medicine, Brigham and Women's Hospital, Boston, MA, USA*

TOMOAKI FUKUNAGA • *Department of Applied Molecular Bioscience, Yamaguchi University Graduate School of Medicine, Ube, Japan*

DANIEL G. GIBSON • *Department of Synthetic Biology, J. Craig Venter Institute, Inc., Rockville, MD, USA*
DAVID HOOVER • *Scientific Computing Branch, Center for Information Technology, National Institutes of Health, Bethesda, MD, USA*
HISASHI HOSHIDA • *Department of Applied Molecular Bioscience, Yamaguchi University Graduate School of Medicine, Ube, Japan*
OLIVIA HO-SHING • *Department of Biology, Davidson College, Davidson, NC, USA*
MO CHAO HUANG • *Institute of Bioengineering and Nanotechnology, The Nanos, Singapore*
KARTHIKEYAN KANDAVELOU • *Pondicherry Biotech Private Limited, IT Park, Pondy Technopolis, Pillaichavady, Puducherry, India*
KIN H. LAU • *Department of Biology, Davidson College, Davidson, NC, USA*
MAMIE Z. LI • *Department of Genetics, Howard Hughes Medical Institute, Harvard Medical School, Boston, MA, USA; Division of Genetics, Department of Medicine, Brigham and Women's Hospital, Boston, MA, USA*
MO-HUANG LI • *Institute of Bioengineering and Nanotechnology, The Nanos, Singapore*
LI SHI LIM • *Institute of Bioengineering and Nanotechnology, The Nanos, Singapore*
GREGORY LINSHIZ • *Department of Computer Science and Applied Mathematics, Weizmann Institute of Science, Rehovot, Israel; Department of Biological Chemistry, Weizmann Institute of Science, Rehovot, Israel*
STEFFI LIU • *Department of Biomedical Engineering, Johns Hopkins University, Baltimore, MD, USA*
VIKTORIYA LONDON • *Department of Environmental Health Sciences, Johns Hopkins University School of Public Health, Baltimore, MD, USA*
JULIE A. MARCHAND • *Virginia Bioinformatics Institute, Virginia Tech, Blacksburg, VA, USA*
WAYNE MATERI • *Carbonitum Energy Corporation, Edmonton, AB, Canada*
JEREMY MINSHULL • *DNA2.0, Inc., Menlo Park, CA, USA*
STEFFEN MUELLER • *Department of Molecular Genetics and Microbiology, Stony Brook University, Stony Brook, NY, USA*
HÉLOÏSE MULLER • *Department of Environmental Health Sciences, Johns Hopkins University School of Public Health, Baltimore, MD, USA*
ALI NOURI • *Program on Science and Global Security Woodrow Wilson School, Princeton University, Washington DC, USA*
JEAN PECCOUD • *Virginia Bioinformatics Institute, Virginia Tech, Blacksburg, VA, USA*
SIVAPRAKASH RAMALINGAM • *Department of Environmental Health Sciences, Johns Hopkins University School of Public Health, Baltimore, MD, USA*
SARAH M. RICHARDSON • *High Throughput Biology Center, Johns Hopkins University School of Public Health, Baltimore, MD, USA*
JOY WU SCHWERZMANN • *Department of Environmental Health Sciences, Johns Hopkins University School of Public Health, Baltimore, MD, USA*
EHUD SHAPIRO • *Department of Biological Chemistry, Weizman Institute of Science, Rehovot, Israel; Department of Computer Science and Applied Mathematics, Weizman Institute of Science, Rehovot, Israel*
SHYAM K. SHARAN • *Mouse Cancer Genetics Program, Center for Cancer Research, National Cancer Institute at Frederick, Frederick, MD, USA*
STACEY STAUFFER • *Mouse Cancer Genetics Program, Center for Cancer Research, National Cancer Institute at Frederick, Frederick, MD, USA*

WILLIAM VERNON • *Department of Biology, Missouri Western State University, St. Joseph, MO, USA*
ALAN VILLALOBOS • *DNA2.0, Inc., Menlo Park, CA, USA*
MARK WELCH • *DNA2.0, Inc., Menlo Park, CA, USA*
HONGYE YE • *Institute of Bioengineering and Nanotechnology, The Nanos, Singapore*
TUVAL BEN YEHEZKEL • *Department of Biological Chemistry, Weizman Institute of Science, Rehovot, Israel*

# Part I

## Assembly of Oligonucleotides in Synthons

# Chapter 1

# Building Block Synthesis Using the Polymerase Chain Assembly Method

**Julie A. Marchand and Jean Peccoud**

## Abstract

De novo gene synthesis allows the creation of custom DNA molecules without the typical constraints of traditional cloning assembly: scars, restriction site incompatibility, and the quest to find all the desired parts to name a few. Moreover, with the help of computer-assisted design, the perfect DNA molecule can be created along with its matching sequence ready to download. The challenge is to build the physical DNA molecules that have been designed with the software. Although there are several DNA assembly methods, this section presents and describes a method using the polymerase chain assembly (PCA).

**Key words:** Gene synthesis, Polymerase chain assembly, Building blocks, DNA fabrication, Computer-assisted design

## 1. Introduction

At its core, DNA fabrication relies on the synthesis of DNA oligomers at base level. The essential feature of DNA fabrication is that no naturally isolated DNA is used. Although clonal plasmid-based intermediates might exist during the assembly of a target DNA, every base originated as a phosphoramidite molecule at the beginning of the process. Today, all fabrication methods begin with solid-phase phosphoramidite chemistry to construct single-stranded DNA molecules that are between 10 and 100 base pairs (bp) long, which are enzymatically assembled into larger molecules. This process is commonly referred to as "gene synthesis" and can be used to synthesize sequences up to 1 kilobase (kb) long. Still larger target DNA sequences require investigators to assemble partial products into the desired full-length construct. DNA of several kb in length can be enzymatically assembled from 1-kb DNA segments, whereas DNA of megabase length requires in vivo

Jean Peccoud (ed.), *Gene Synthesis: Methods and Protocols*, Methods in Molecular Biology, vol. 852, DOI 10.1007/978-1-61779-564-0_1, 

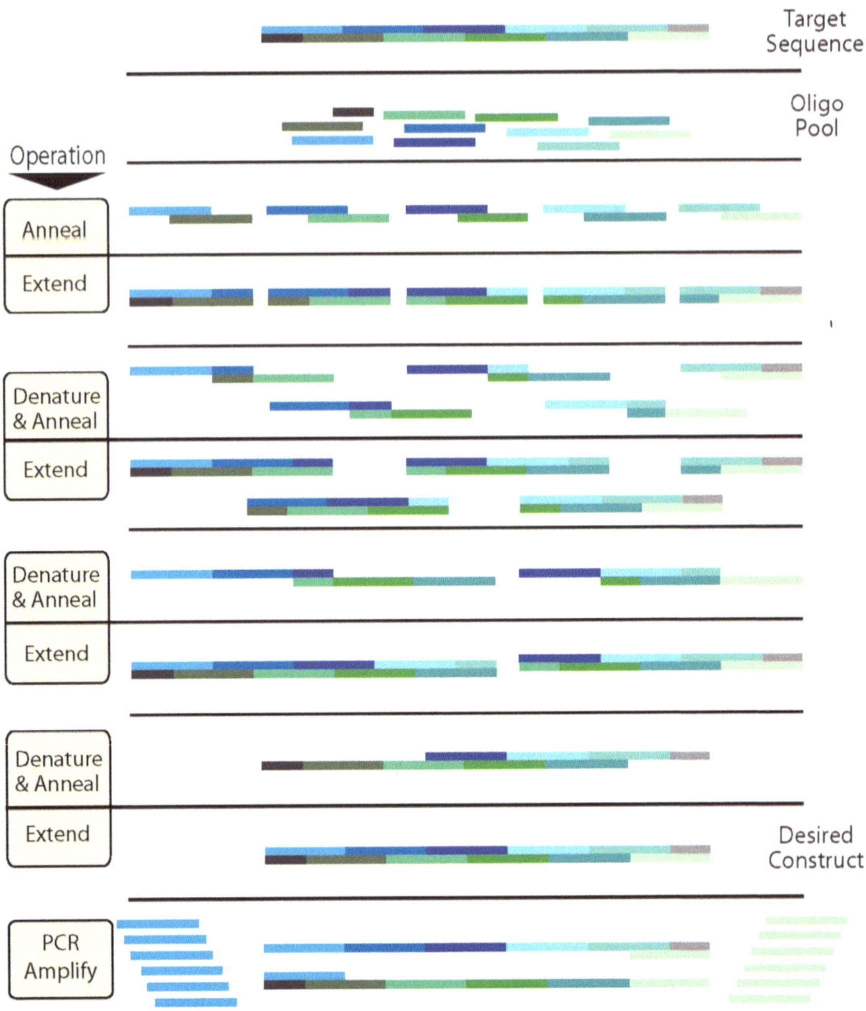


Fig. 1. PCA assembly of a DNA construct. A target sequence is shown in the *top panel* of this figure. The *different color segments* represent the oligos that are synthesized to build the construct. The pool of oligos is assembled in equimolar amounts and allowed to anneal. The annealed oligos are extended in the 3′ direction until the end of their partner oligo is reached. The double-stranded DNA is melted and reannealed with extension products and any remaining oligos. Each extension reaction results in progressively longer products, and full-length products are eventually synthesized. At this step, the terminal oligos are added to the reaction, and full-length products from the previous reactions are amplified by PCR and subsequently cloned and sequenced. Figure reproduced with permission from ref. 1.

recombination methods (Fig. 1). The details of DNA fabrication are therefore not monolithic and are distinct based on the size of the target DNA (1).

Gene synthesis is the step during which oligonucleotides (oligos) are combined into DNA fragments of several hundred bases in length. Numerous protocols have been described and extensively reviewed, including polymerase chain assembly (PCA), thermodynamically balanced inside-out synthesis (TBIO) (2), and ligase chain reaction (LCR) (1, 3). Here the PCA method (4) is used to synthesize 750-bp building blocks from the right arm of yeast chromosome 6. The right chromosome arm was first subdivided in 12 segments, each on average 12 kb, and then each segment was further subdivided in 15 building blocks of 750 bp each. The building blocks were further dissected into a set of about 12–13 oligonucleotides of about 60 bp in length. The building blocks were further synthesized from the oligonucleotides using the PCA method, cloned, and sent for sequencing.

## 2. Materials

### *2.1. Software*

The primers are designed using the software GeneDesign, a web-based program for the design of synthetic genes (5). It consists of several modules that automate the tasks associated with the manipulation of synthetic sequences. The source code is from http://github.com/notadoctor/GeneDesign/.

### *2.2. Gene Synthesis and Analysis*

1. The primers were purchased from Integrated DNA Technologies, Inc. (IDT, Coralville, IA). They were ordered wet frozen (in water) in 96-well plates at a concentration of 60 μmol/L and a volume of 83.33 μl/well.
2. HotStarTaq master mix kit.
3. Nuclease-free water (not DEPC-treated).
4. 96-well PCR plates.
5. DNA 12000 kit (catalogue number 5067-1508 Agilent Technologies, Inc., Wilmington, DE).
6. Agilent Bioanalyzer (Agilent Technologies, Inc., Wilmington, DE).
7. 96-well block, 1 ml volume/well.

### *2.3. Cloning of Building Blocks*

1. TOPO TA cloning® kit for sequencing (catalogue number K457501, Life technologies, Carlsbad, CA).
2. TOP10 chemically competent cells (catalogue number C404003, Life technologies, Carlsbad, CA).
3. Terrific agar plates supplemented with 100 μg/ml carbenicillin.

### *2.4. Bacterial Culture*

1. Luria broth (LB) medium supplemented with 10% glycerol and 50 μg/ml carbenicillin.
2. SOC medium: complement 1 L of LB medium with 10 ml of 1 M $MgSO_4$ (10 mM final), 10 ml of 1 M $MgCl_2$ (10 mM final), and 18 ml of 20% dextrose (0.36% w/vol).
3. Air-permeable sealing membrane.

4. Aluminum seal film.
5. 96-well culture plates.
6. Carbenicillin.

## 3. Methods

The GeneDesign software (5) allows for breaking large sequences into several building blocks of about 750 bp, which, in turn, are further dissected into about 12–13 60-bp oligonucleotides used for the subsequent synthesis. The building blocks are synthesized manually here, but this can also be performed with a high-throughput liquid handling system. To create the DNA template for the building blocks, this protocol uses the polymerase chain assembly (PCA) method in which oligonucleotides, present in equimolar quantity, span both strands of a DNA sequence, anneal through partial overlap, and are extended in such a way that each are increased in length and can be extended further by hybridizing to other oligonucleotides or products of subsequent extensions (1). Upon its creation, the full-length DNA template corresponding to a specific building block is amplified by polymerase chain reaction (PCR) using the two outermost oligonucleotides from the PCA step. Following the completion of the final amplification step, analysis of the PCR reaction by electrophoresis is used to visualize the presence and size of the generated building block. This final amplicon is then cloned in a TOPO®TA cloning vector and transformed. For each synthesized building block, an average of 12 clones are sent for sequencing in the form of liquid culture. To avoid template degradation, it is important to clone the amplicons immediately after the electrophoresis.

### 3.1. Design of the Building Blocks

1. Using the GeneDesign software (see Subheading 2.1), the 750-bp building block sequences are entered in the program using option X (calibrate the Tm, length restriction sites added). This function will subdivide the sequences in several primers (usually 12–13 primers, but can be up to 20 for a longer building block), each with an average length of 60 bp, and will also deliver a list of primers.
2. Using the primer list, the primers are ordered from commercial suppliers as described in Subheading 2.2.

### 3.2. Building Block Synthesis

1. Using a multichannel pipette, the primers are diluted to 6 μM in nuclease-free water to a final volume of 100 μl in a second 96-well block to produce the working stock. From this 6 μM primer working stock, a template primer mix (TPM) and outer primer mix (OPM) are prepared.

2. In the TPM, all primers must be present at a concentration of 300 nM (the primers must all be diluted by 1/20). These dilutions are prepared in a 0.5-ml reaction tube, and primer mixes are stored at –20°C when not in use. To prepare the TPM for a building block that consists of up to 20 primers, add 10 μl of each primer and if less than 20 primers are used, add nuclease-free water instead to yield a final volume of 200 μl; mix thoroughly. If more than 20 primers are used, add 10 μl of each primer, but no additional water. The primer concentration should be around 250–275 nM.
3. In the OPM, the outer primers must be present at a concentration of 3 μM (i.e., both primers are diluted by 1/2). Again, these dilutions are prepared in a 0.5-ml reaction tube in nuclease-free water, and primer mixes are stored at –20°C when not in use. To obtain the OPM, add 25 μl of the first and last primer and mix thoroughly to give a total volume of 50 μl.
4. The building block template synthesis is performed by polymerase chain assembly (PCA) in a thermocycler. This reaction is also called templateless PCR; it has a final volume of 25 μl and contains the following reagents: 12.5 μl of HotStarTaq master mix 2× (containing the buffer, the dNTPS, and the Taq polymerase), 2.5 μl of TPM, and 10 μl of nuclease-free water. The assembly is performed using the following program: 95°C for 15 min, 55°C for 30 s, and 72°C for 1 min; then 25 cycles of 95°C for 30 s, 55°C for 30 s, and 72°C for 1 min; followed by 72°C for 3 min and 10°C forever (see Note 1).
5. The building block templates are then amplified by PCR using the OPM to create the finished PCR and thus the building blocks. This reaction has a final volume of 25 μl and contains the following reagents: 12.5 μl of HotStarTaq master mix 2× (the buffer, the dNTPS, and the Taq polymerase), 2.5 μl of templateless PCR reaction from step 4 above (diluted 1:5), 2 μl of OPM, and 10 μl of nuclease-free water. The amplification is performed with the following program: 95°C for 15 min, 55°C for 30 s, and 72°C for 1 min; 25 cycles of 95°C for 30 s, 55°C for 30 s, and 72°C for 1 min; followed by 72°C for 3 min and 10°C forever (see Note 1).
6. The building block synthesis is monitored with a microfluidic electrophoresis using the DNA 12000 kit and the Bioanalyzer instrument from Agilent according to the manufacturer's instructions (Fig. 2). This analysis can also be performed with an agarose gel.

### 3.3. Cloning of Building Blocks

1. The building blocks are cloned in the TOPO®TA cloning vector pCR4® (see Subheading 2.3). All reagents are provided in the kit. The ligation reaction is composed of 2 μl of the PCR reaction from step 6 above (see Note 2), 1 μl of ultra salt solution,

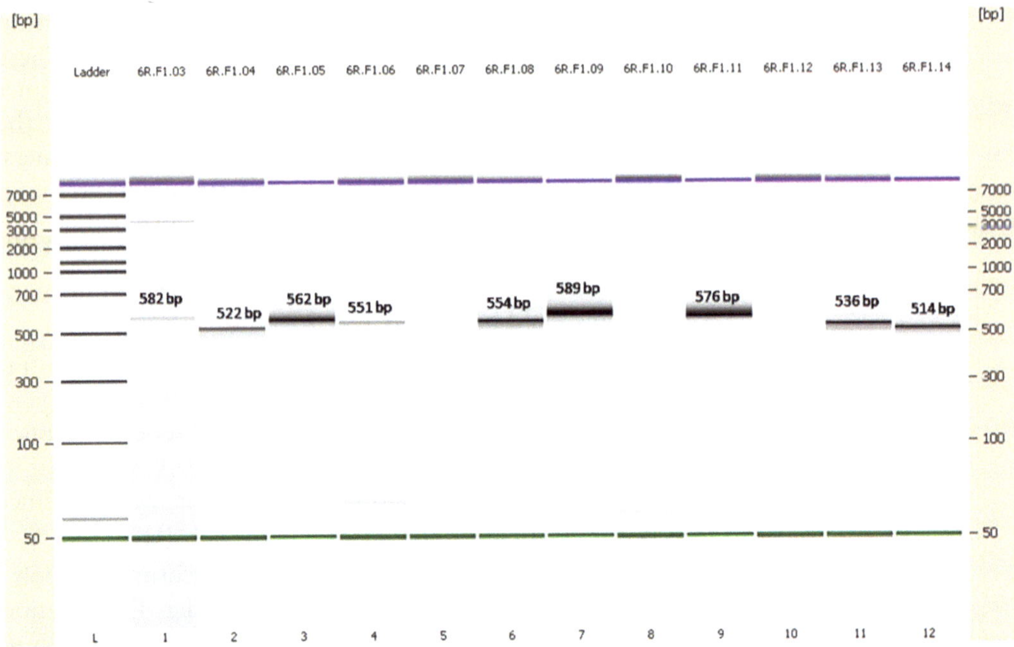


Fig. 2. Gel electrophoresis for a selection of synthesized building blocks from segment 15 of the right arm of yeast chromosome 6. These building blocks were synthesized using the PCA method, and a microfluidic electrophoresis was performed to monitor the success of the synthesis. For each building block, positive synthesis is demonstrated by the presence of a band with a molecular weight that corresponds to the associated building block.

2 μl of ultra pure water, and 1 μl of pCR4® TOPO® vector, given a final volume of 6 μl. The reagents are added in the specified order, and the reaction is incubated at room temperature for 12 min (see Note 3).

2. The cloned building blocks are transformed into chemically competent bacterial cells *E. coli* TOP10 (see Note 4). The entire ligation reaction (6 μl) is added to 50 μl of chemically competent *E. coli* TOP10 cells and incubated on ice for 30 min, before being heat-shocked at 42°C for 30 s and placed back onto ice for 2 min. Then 400 μl of SOC medium without antibiotics is added, and the cells are incubated at 37°C for 1 h. Finally, the cells are spread on terrific agar plates supplemented with carbenicillin at 100 μg/ml and incubated at 37°C for 18 h (see Note 6). The vector has a negative selection with *E. coli* lethal gene ccdB fused to LacZα fragment. Upon ligation of an insert, the LacZα-ccdB gene fusion expression is disrupted, and thus the cell survives. The host must not express the ccdA gene (see Note 5).

### 3.4. Culturing of Clones for Sequencing

1. For each construct, 12 colonies are manually picked from the 18-h agar plates and inoculated in a 96-well culture plate containing 200 μl of LB broth supplemented with 10% glycerol and 50 μg/ml carbenicillin (see Note 6). The plate is sealed

with an air-permeable membrane and incubated without agitation at 37°C for 18 h.

2. The 18-h plate is then used to inoculate two replicate 96-well culture plates. Using a multichannel pipette, 10 μl of the 18-h culture plate is transferred to each of the 96-well plates containing 190 μl of LB broth supplemented with 10% glycerol and 50 μg/ml carbenicillin. The plates are sealed with an air-permeable membrane and incubated without agitation at 37°C for exactly 12 h. The 18-h plate is sealed with aluminum seal and stored at −80°C.
3. Upon completion of the incubation time, the two 12-h plates are sealed with aluminum seal and stored at −80°C. One of the 12-h plates will be sent to an external company for sequencing, and the other is kept.
4. After the sequencing results are analyzed, the selected perfect clones are picked from their respective culture plates and grown in tubes containing 3 ml of LB broth supplemented with 10% glycerol and 100 μg/ml carbenicillin and incubated in an incubator shaker at 37°C and 250 RPM for exactly 12 h. The culture is then frozen as a glycerol stock in a cryogenic vial. The 96-well plates containing the incorrect clones are discarded.

## 4. Notes

1. The annealing temperature for PCA or finishing PCR reaction must be adjusted according to the Tm of the primers.
2. The volume of PCR reactions used for the ligation can range from 0.5 μl to 4 μl depending on the yield of the amplicons.
3. The building block can also be cloned by traditional ligation provided that restriction sites are added on the outermost primers, or by cloning systems such as Gateway® that has recombination sites added to the outmost primers. If a uracil-specific excision reagent (USER) fusion system is used, it is important to note that wild-type archaeal DNA polymerases are inhibited by the deoxyuracil.
4. The selection of the bacterial strain should be made according to both the strain genotype and the type of insert. For the expression of yeast parts, the *E. coli* TOP10 cells appear more suitable, and we have observed increased number of colonies and good growth with this strain.
5. Despite the presence of the ccdB lethal gene, we have observed some negative colonies that do not appear to be satellite colonies. Random screening or sending more clones for sequencing might be advisable.

6. This growth medium is recommended by Beckman Coulter Genomics where our group receives the sequencing service. The selective antibiotic is based on the resistance encoded by the vector of choice.

## References


1. Czar MJ, Anderson JC, Bader JS and Peccoud J. (2009) Gene synthesis demystified. *Trends Biotechnol* 27:63–72.
2. Xiong AS, Peng RH, Zhuang J, Gao F, Li Y, Cheng Z M and Yao, QH (2008) Chemical gene synthesis: strategies, softwares, error corrections, and applications. *FEMS Microbiol* 32:522–540.
3. Cello J, Paul AV and Wimmer E (2002) Chemical Synthesis of Poliovirus cDNA: Generation of Infectious Virus in the Absence of Natural Template. Science 297:1016–1018.
4. Dymond J, Scheifele L, Richardson S, Lee P, Chandrasegaran S, Bader J and Boeke JD (2009) Teaching Synthetic Biology, Bioinformatics, and Engineering to Undergraduates: The Interdisciplinary Build-a-Genome Course. *Genetics* 18:13–21.
5. Richardson SM, Wheelan SJ, Yarrington, RM and Boeke, JD (2006) GeneDesign: rapid, automated design of multikilobase synthetic genes. *Genome Res* 16:550–6.

# Chapter 2

# Oligonucleotide Assembly in Yeast to Produce Synthetic DNA Fragments

**Daniel G. Gibson**

## Abstract

The yeast *Saccharomyces cerevisiae* can take up and assemble at least 38 overlapping single-stranded oligonucleotides and a linear double-stranded vector in one transformation event. These oligonucleotides can overlap by as few as 20 bp and can be as long as 200 nucleotides in length to produce kilobase-sized synthetic DNA molecules. A protocol for designing the oligonucleotides to be assembled, transforming them into yeast, and confirming their assembly is described here. This straightforward scheme for assembling chemically synthesized oligonucleotides can be a useful tool for building synthetic DNA molecules.

**Key words:** In vivo DNA assembly, Yeast transformation, Gene synthesis, Oligonucleotides, Synthetic biology

## 1. Introduction

Chemically synthesized oligonucleotides (oligos) are often joined into larger DNA fragments containing full-length genes. This was first demonstrated in 1970 when Khorana and colleagues synthesized the 77-nucleotide gene encoding a yeast alanine transfer RNA from 17 overlapping oligonucleotides (1). Since then, chemical oligonucleotide synthesis has improved tremendously (2), and a number of in vitro enzymatic strategies are available for the assembly of oligos into larger constructs (3–5). It is now possible to produce genes, biosynthetic pathways, and even entire chromosomes from chemically synthesized DNA (6, 7). Because absolute control can be exerted over the sequence of chemically derived DNA molecules, genetic components can be exhaustively optimized.

The capacity of the yeast *Saccharomyces cerevisiae* to take up and recombine DNA fragments has made it a model eukaryote for studying numerous cellular processes. This is mainly because DNA sequences can be genetically altered by transforming yeast with

Jean Peccoud (ed.), *Gene Synthesis: Methods and Protocols*, Methods in Molecular Biology, vol. 852,
DOI 10.1007/978-1-61779-564-0_2, © Springer Science+Business Media, LLC 2012

either double-stranded (ds) DNA fragments (8) or single-stranded (ss) oligos (9). In addition, homologous recombination in yeast can be used to build DNA fragments from overlapping constituent parts. This was first demonstrated when a plasmid was constructed from two dsDNA fragments containing homologous ends (10). Two nonhomologous dsDNA fragments can also be bridged by single-stranded oligonucleotides that join the ends of the two fragments (11). Previously, we showed that six overlapping dsDNA fragments could be assembled by yeast into an entire *Mycoplasma genitalium* genome (6). Subsequently, this process was improved, and 25 overlapping fragments, between 17 kb and 35 kb in length, were assembled at once into this genome (12). More recently, we reported on the synthesis of a 1.08-Mbp *Mycoplasma mycoides* genome, which was used to produce a cell controlled only by this synthetic genome (7). Using yeast recombination, the synthetic *M. mycoides* genome was assembled in three stages from 1,078 overlapping 1,080-bp DNA fragments that were each chemically synthesized.

To exclusively use yeast in the production of whole genomes and large constructs of any reasonable sequence, what remained was the demonstration of the assembly of chemically synthesized oligonucleotides into appropriate dsDNA molecules, which we reported in 2009 (13). There we showed that yeast could take up and assemble at least 38 overlapping single-stranded oligonucleotides and a linear double-stranded vector in one transformation event to produce ~1.2-kb dsDNA fragments. These oligonucleotides can overlap by as few as 20 bp and can be as long as 200 nucleotides in length. A protocol for synthesizing kilobase-sized DNA fragments in yeast from a series of overlapping oligos is described here.

## 2. Materials

### 2.1. Design and Preparation of the Oligonucleotides and Assembly Vector

1. Yeast/*E. coli* shuttle vector [e.g., pRS313 (ATCC 77142), pRS314 (ATCC 77143), pRS315 (ATCC 77144), and pRS316 (ATCC 77145)].
2. Primers for assembly vector amplification.
3. Overlapping synthetic oligonucleotides to be assembled.
4. High-fidelity polymerase chain reaction (PCR) amplification kit (e.g., Phusion® polymerase (New England BioLabs®, Inc. [NEB])).
5. Gel extraction kit (e.g., QIAquick Gel Extraction Kit, Qiagen).
6. Tris–EDTA buffer pH 8.0 (TE buffer).
7. DNA analysis software (e.g., Vector NTI® [Invitrogen], Clone Manager [Sci-Ed], and CLC Genomics Workbench [CLC bio]).

#### 2.2. Yeast Transformation

1. 100× adenine hemisulfate solution: 1% (w/v) adenine hemisulfate. Autoclave or filter sterilize, and store at room temperature.
2. $YPAD^{100}$ liquid medium: 2% (w/v) bacto peptone, 1% (w/v) bacto yeast extract, 2% (w/v) dextrose, 1× adenine hemisulfate solution. Autoclave or filter sterilize and store at 4°C.
3. $YPAD^{100}$ agar plates: $YPAD^{100}$ liquid medium plus 2% (w/v) bacto agar. Autoclave and store plates at 4°C.
4. Yeast strain to be transformed (e.g., VL6-48, ATCC Number MYA-3666).
5. Sterile water.
6. 1 M sorbitol.
7. Sorbitol/sodium phosphate/EDTA (SPE) solution: 1 M sorbitol, 0.01 M sodium phosphate, 0.01 M $Na_2EDTA$ (pH 7.5). Autoclave or filter sterilize and store at room temperature.
8. Beta-mercaptoethanol (BME), 14 M.
9. Zymolyase-20T solution: 10 mg/ml Zymolyase-20T (ICN Biochemicals, cat. no. 320921), 25% (w/v) glycerol, 50 mM Tris–HCl, pH 7.5. Aliquot 500 μl portions and store at −20°C.
10. Sorbitol/Tris–Cl/$CaCl_2$ (STC) solution: 1 M sorbitol, 0.01 M Tris–HCl, pH 7.5, 0.01 M $CaCl_2$. Autoclave or filter sterilize and store at room temperature.
11. Transforming DNA.
12. PEG/$CaCl_2$ solution: 20% (w/v) PEG 8000 (US Biological, cat. no. 19966), 10 mM $CaCl_2$, 10 mM Tris–HCl, pH 7.5. Store at room temperature for up to 2 weeks.
13. SOS solution: 1 M sorbitol, 6.5 mM $CaCl_2$, 0.25% bacto yeast extract, 0.5% bacto peptone. Autoclave or filter sterilize and store at room temperature.
14. Selective regeneration bottom plates: Supplement complete minimal (CM) dropout plates (see below) with 1 M sorbitol and 1× adenine hemisulfate solution. Autoclave and store plates at 4°C.
15. Selective regeneration top agar: Supplement CM dropout plates with 1 M sorbitol, 1× adenine hemisulfate solution, and bacto agar up to 3% (w/v). Autoclave and store at room temperature.
16. Sorbitol/DMSO solution (optional): 1 M sorbitol, 15% DMSO. Prepare fresh from sterile solutions.

#### 2.3. Identifying Yeast Clones Containing the Assembled Products

1. Complete minimal (CM) dropout plates: 0.17% (w/v) yeast nitrogen base, 0.5% (w/v) ammonium sulfate, 2% (w/v) dextrose, 2% (w/v) bacto agar, complete supplemental mixture.
2. Cell resuspension buffer (Qiagen buffer P1): 50 mM Tris–Cl (pH 8.0), 10 mM EDTA. Autoclave or filter sterilize and store at room temperature.

3. Zymolyase-100T solution: 20 mg/ml Zymolyase-100T (US Biological, cat. no. Z1004), 50% (w/v) glycerol, 2.5% (w/v) glucose, 50 mM Tris–HCl (pH 7.5). Prepare from sterilized solutions and store at −20°C.
4. Beta-mercaptoethanol (BME), 14 M.
5. Alkaline-lysis solution (Qiagen solution P2): 200 mM NaOH, 1% SDS (w/v). Filter sterilize and store at room temperature.
6. Neutralization solution (Qiagen solution P3): 3 M potassium acetate, pH 5.5. Autoclave or filter sterilize and store at room temperature.
7. QIAprep Spin Miniprep Kit, Qiagen (optional).
8. Isopropanol.
9. Multiplex PCR screening kit (e.g., Qiagen Multiplex PCR kit).
10. PCR kit for screening yeast clones (e.g., Hot Start Phusion® polymerase, NEB).
11. Diagnostic primers to confirm the assembled product.
12. 70% ethanol.
13. TE buffer, pH 8.0.
14. Electrocompetent *E. coli* cells.
15. Electrocuvettes.
16. 14-ml round-bottom tubes.
17. SOC.
18. LB plates containing antibiotic.

## 3. Methods

### 3.1. Design and Preparation of the Oligonucleotides and Assembly Vector

Yeast can take up and assemble at least 38 overlapping single-stranded oligonucleotides and a linear double-stranded vector in one transformation event to produce gene-sized fragments. These oligos can overlap by as few as 20 bp and can be as long as 200 nucleotides in length. Thus, gaps as long as 160 nucleotides can be filled by yeast. In this method, the oligos are assembled with a vector to form a circular product. The terminal oligos in the set contain 20 bp overlapping sequence to the ends of a yeast/*E. coli* shuttle vector and restriction sites to release the synthesized dsDNA fragment from the vector:

1. PCR amplify a yeast/*E. coli* shuttle vector (see Note 1).
2. Purify the PCR-amplified vector from an agarose gel following electrophoresis with a commercially available kit (e.g., QIAquick Gel Extraction Kit, Qiagen).

**a**

*5'caggtcgactctagaggatc**gcggccgc**GAAAAAAAGAACCTTTCGGCTATATAGGAATAGTATGAGCAATAATGTCTATTGGCTTTCTA*
*GGCTTTATTGTATGAGCCCACCACATATTCACAGTAGGATTAGATGTAGACACACGAGCTTACTTTACATCAGCCACTATAA*
*TTATCGCAATTCCTACCGGTGTCAAAGTATTTAGCTGACTTGCAACCCTACACGGAGGTAATATTAAATGATCTCCAGCTATA*
*CTATGAGCCTTAGGCTTTATTTTCTTATTTACAGTTGGTGGTCTAACCGGAATTGTT**gcggccgc**cgggtaccgagctcgaattc3'*

**b**

*Oligo 1 = 5'-caggtcgactctagaggatc**gcggccgc**GAAAAAAAGAACCTTTCGGCTATATAGGAATA-3'*
*Oligo 2 = 5'-CAATAAAGCCTAGAAAGCCAATAGACATTATTGCTCATACTATTCCTATATAGCCGAAAG-3'*
*Oligo 3 = 5'-TGGCTTTCTAGGCTTTATTGTATGAGCCCACCACATATTCACAGTAGGATTAGATGTAGA-3'*
*Oligo 4 = 5'-TGCGATAATTATAGTGGCTGATGTAAAGTAAGCTCGTGTGTCTACATCTAATCCTACTGT-3'*
*Oligo 5 = 5'-CAGCCACTATAATTATCGCAATTCCTACCGGTGTCAAAGTATTTAGCTGACTTGCAACCC-3'*
*Oligo 6 = 5'-CATAGTATAGCTGGAGATCATTTAATATTACCTCCGTGTAGGGTTGCAAGTCAGCTAAAT-3'*
*Oligo 7 = 5'-TGATCTCCAGCTATACTATGAGCCTTAGGCTTTATTTTCTTATTTACAGTTGGTGGTCTA-3'*
*Oligo 8 = 5'-gaattcgagctcggtacccg**gcggccgc**AACAATTCCGGTTAGACCACCAACTGTAAATA-3'*

Fig. 1. Overlapping oligonucleotide design for assembly into a yeast vector. (**a**) A 340-bp sequence, which includes 20 bp overlapping sequence to PCR-amplified pRS313 (*nonbolded lowercase*) and NotI restriction sites (*bolded and underlined*). Because 56 bp is used for assembly into and release from the vector, only 284 bp of unique sequence (*uppercase*) is synthesized. (**b**) The sequence shown in (**a**) can be synthesized from the eight 60-mer oligos shown, which contain 20-bp overlaps.

3. Quantify the PCR product and dilute to 100 ng/μl in TE buffer.
4. Synthesize or purchase oligonucleotides (see Note 2). Oligonucleotides can range from 40 to 200 bases and overlap neighboring oligos by 20–30 bases (see Note 3). Terminal oligos should have 20 bases of sequence that overlap with the PCR-amplified assembly vector. If PCR-amplified pRS313 (described in Note 1) is chosen as the assembly vector, these terminal oligo sequences would be 5′-caggtcgactctagaggatcx—xW—W-3′ for the first oligo in the series and 5′-gaattcgagctcg gtacccg x—xW—W-3′ for the last oligo in the series, where x—x are restriction sites to release the assembled insert from the vector (e.g., NotI restriction site, gcggccgc), and W—W is new DNA sequence that is synthesized. See Fig. 1 for an example of how to design the overlapping oligos.
5. Adjust each oligo to 50 μM with TE buffer.
6. Combine equal volumes of oligonucleotides and dilute to a per-oligonucleotide concentration of 60–240 nM in TE buffer (see Note 4).
7. Combine 20 μl of the oligo pool with 2 μl of PCR-amplified vector (step 3) and use as the transforming DNA described in the transformation procedure below.

### 3.2. Yeast Transformation

To assemble genes and genome-sized fragments from overlapping DNA molecules, the yeast spheroplast transformation procedure is carried out. In this method, cells are treated with Zymolyase® to weaken the cell wall. These yeast spheroplasts are then made competent to take up the overlapping DNA fragments by treatment with polyethylene glycol (PEG) and $CaCl_2$. A slightly modified protocol described by Kouprina and Larionov (14) is carried out. This procedure is optimized for use with the VL6-48 yeast strain (ATCC Number MYA-3666):

1. Streak a frozen glycerol stock containing the yeast strain onto a $YPAD^{100}$ agar plate. Incubate the plate at 30°C for 2–3 days or until individual colonies appear. Store the plate at 4°C.
2. Inoculate a single colony into 50 ml $YPAD^{100}$ medium (see Note 5).
3. Harvest the cells in a 50-ml tube at 1,600 × *g* for 3 min once the cells reach an $OD_{600}$ of 0.5–0.6 (~$10^7$ cells/ml) (see Note 6).
4. Resuspend cell pellets in 50 ml sterile water. Harvest the cells as in step 3.
5. Resuspend the cells in 20 ml 1 M sorbitol. Leave the cells on ice in a covered bucket at 4°C for 4 h. Alternatively, the cells can remain on ice for up to 24 h, and the transformation procedure can be continued from step 6 on the following day.
6. Invert the tube several times to resuspend the cells that have settled. Harvest the cells as in step 3.
7. Resuspend the cells in 20 ml SPE solution. Add 40 μl BME and invert to mix. Add 40 μl Zymolyase-20T solution and invert to mix (see Note 7).
8. Incubate for 40 min in a 30°C air incubator at 50 RPM with the tube on its side. Invert the tube three to four times halfway through the incubation (see Note 8).
9. Add 1 M sorbitol up to 50 ml. Invert to mix.
10. Harvest cells at 1,600 × *g* for 5 min. Pour off the supernatant.
11. Resuspend the spheroplasts in 20 ml of 1 M sorbitol by pipetting up and down with a 25-ml pipette (see Note 9). Add 1 M sorbitol up to 50 ml. Invert to mix.
12. Harvest the yeast spheroplasts as in step 10 (see Note 10).
13. Resuspend the spheroplasts in 2.8 ml STC solution by pipetting up and down with a 5-ml pipette (see Note 11).
14. Incubate the spheroplasts at room temperature for 10 min.
15. Add 200 μl spheroplasts to the transforming DNA solution already contained in a microfuge tube (see Note 12). Mix the spheroplasts with the DNA by slowly adding them to the DNA while stirring at the same time.
16. Incubate the spheroplasts/DNA mixture at room temperature for 10 min.
17. Add 1 ml PEG/$CaCl_2$ solution. Mix by inverting the tube ten times.
18. Incubate the tube at room temperature for 20 min.
19. Harvest the cells at 1,500 × *g* for 8 min in a microfuge.

20. Remove supernatant with a 1-ml pipette.
21. Add 800 μl SOS solution. Resuspend by pipetting up and down with a wide-bore 1-ml pipette tip.
22. Incubate the tube in a 30°C water bath for 30 min.
23. During the incubation in step 22, add 8 ml equilibrated selective regeneration top agar to a 15-ml tube. Keep tube in a 55°C water bath.
24. Add cells to the 8-ml selective regeneration top agar, invert three times, and then pour onto a selective regeneration bottom plate (see Note 13).
25. Incubate the plate at 30°C for 3–4 days.

### 3.3. Identifying Yeast Clones Containing the Assembled Products

Yeast clones containing full-length assemblies can be screened by PCR or restriction digestion following its transfer to *E. coli*. Because of the error rates associated with oligo synthesis, assembled inserts should also be sequenced. PCR screening and DNA sequencing reactions can be carried out with primers that anneal to the vector and point inward toward the insert. The primers M13F (5′-tgtaaaac gacggccagt-3′) and M13R (5′-caggaaacagctatgacc-3′) will anneal to many commonly used vectors, including the pRS313-316 vector series described above. Alternatively, the plasmid DNA can be transferred from yeast to *E. coli*, where it can be extracted and analyzed following restriction digestion or DNA sequencing. The protocol described here will provide DNA of sufficient quality and quantity for PCR analysis and *E. coli* transformation. In this method, primary transformants are transferred and grown on selective plates as small patches. The cells are first treated with Zymolyase® to remove the cell wall, and then a standard alkaline-lysis procedure, as performed with *E. coli*, is carried out:

1. Use thin pipette tips (e.g., 10-μl tips) to transfer individual colonies to CM dropout plates in ~0.5 $cm^2$ patches.
2. Incubate the plates overnight (16–24 h) at 30°C (see Note 14).
3. Add 250 μl cell resuspension buffer containing 0.25 μl BME and 2.5 μl Zymolyase-100T solution to a microfuge tube.
4. Use a 1-ml pipette tip to scrape cells from the yeast patch and combine them with the 250-μl buffer from step 3.
5. Vortex to resuspend the cells in this mixture.
6. Incubate at 37°C for 1 h.
7. Add 250 μl alkaline-lysis solution and invert the tube seven times.
8. Incubate the tube for 5 min at room temperature.
9. Add 250 μl cold neutralization solution and invert the tube seven times (see Note 15).
10. Incubate the tube on ice for 10 min.

11. Centrifuge the sample at 4°C for 10 min at 16,500×*g* in a microfuge.
12. Pour supernatant into a fresh tube containing 700 μl isopropanol.
13. Invert the tube ten times to mix.
14. Incubate the sample at room temperature for 10 min.
15. Centrifuge at 16,500×*g* for 10 min.
16. Pour off the isopropanol.
17. Wash the DNA pellet with 1 ml 70% ethanol.
18. Centrifuge at 16,500×*g* for 5 min.
19. Pour off the 70% ethanol.
20. Spin again briefly to bring the ethanol to the bottom of the tube.
21. Remove the excess ethanol by pipetting or aspirating.
22. Allow the DNA pellet to air dry for 5 min.
23. Resuspend the DNA pellet in 50 μl TE buffer (see Note 16). If an *E. coli* clone is not used, proceed to step 31.
24. Electroporate 3 μl DNA from step 23 into *E. coli* cells (see Note 17).
25. Recover cells at 37°C for 1.5 h in 1 ml SOC medium in 14-ml round-bottom tubes.
26. Plate cells onto LB medium containing the appropriate antibiotic (for the pRS313-316 vector series, use 100 μg/ml carbenicillin or ampicillin).
27. Incubate the plates at 37°C for 12–18 h.
28. Grow individual colonies in 1 ml LB medium + 100 μg/ml carbenicillin (see Note 18).
29. Extract plasmid DNA using a commercially available miniprep kit (e.g., QIAprep Spin Miniprep Kit, Qiagen) (see Note 19).
30. Analyze the restriction patterns of the plasmid DNA on an agarose gel following electrophoresis (see Note 20).
31. Sequence both strands of the insert DNA. For the pRS313-316 vector series, the M13F and M13R primers can be used. Standard Sanger sequencing reactions can be carried out on a 3100 sequencer (Applied Biosystems).
32. Align trace files with the reference sequence (see Note 21).

## 4. Notes

1. The pRS313 vector has been demonstrated to work well for assembling oligonucleotides in yeast. This vector can be linearized by restriction digestion with Bam HI then extracted from

an agarose gel following electrophoresis. This linearized vector can then be PCR-amplified with a forward primer having the sequence 5′-gatcctctagagtcgacctg**caggaattcgatatcaagcttatcg**-3′ and a reverse primer having the sequence 5′-cgggtaccgagctcga-attc**ggagctccaattcgccctat**-3′ where pRS313-specific sequence is bolded. The gel purification of the Bam HI restriction fragments and its PCR amplification help reduce the background of undesired vector-only clones following yeast transformation.

2. The oligos can be synthesized without modifications and with standard desalting.
3. Shorter oligonucleotides, such as 60-mers, can be used to avoid secondary structures and increased error rates that may occur with longer oligos.
4. A 60-nM concentration for each oligo typically works well for oligos up to 90 bases. However, for oligos that are more than 90 bases, a concentration of 200 nM for each oligo is recommended. If longer oligos are to be used (e.g., ≥90-mers), a high-fidelity synthesis process and/or additional oligo purification (e.g., polyacrylamide gel electrophoresis purification, PAGE purification) should be considered to reduce the errors commonly associated with longer oligos.
5. This amount of culture can be used for up to 14 transformations. Inoculate a larger culture volume if more transformations are desired. To ensure that a logarithmic phase culture is ready to be processed the following day, a second culture that is a 1/5 dilution of the first can also be inoculated.
6. Cultures from a freshly streaked plate of yeast will typically be ready within 12–16 h. However, cultures from plates that are more than 1 month old may take as long as 24 h to reach an $OD_{600}$ of 0.5.
7. Yeast spheroplasts are more fragile than cells with an intact cell wall. To avoid a reduction in transformation efficiency, the yeast should be handled with care once the Zymolyase® solution is added. For example, yeast spheroplasts should not be vortexed, and pipetting should be carried out with wide-bore pipette tips.
8. During the 40-min incubation (step 8), prewarm the selective regeneration bottom plates at 37°C, melt (by microwave), and then equilibrate the selective regeneration top agar to 55°C.
9. It is normal for it to take 2–3 min to completely resuspend the yeast spheroplasts. Yeast spheroplasts are more difficult to resuspend than yeast cells with intact cell walls.
10. Optional: At this point, yeast spheroplasts can be resuspended in sorbitol/DMSO solution and stored at −80°C for later use if a seven to ten times reduction in transformation efficiency is acceptable. If this route is chosen, resuspend the yeast

spheroplasts in 2.8 ml sorbitol/DMSO solution, aliquot 200 μl samples to 14 microfuge tubes, and then freeze yeast spheroplasts in a dry ice/ethanol bath and store aliquots at –80°C.

11. Optional: The yeast transformation procedure can be carried out from this point using previously frozen yeast spheroplasts (see Note 10). If this route is chosen, thaw spheroplasts on ice, harvest them at 1,500×*g* for 8 min in a microfuge, and then resuspend them in 200 μl STC.
12. The volume of the transforming DNA solution should not exceed 40 μl.
13. This step must be done quickly to ensure that the top agar does not solidify prior to being poured onto the plate.
14. Alternatively, single colonies can be inoculated into 0.5 ml CM dropout liquid medium and grown overnight with agitation at 30°C. If this route is chosen, harvest the cells in a microfuge tube by centrifugation at 16,500×*g* for 30 s, remove the supernatant, wash the cells with 1 ml sterile water, harvest the cells as above, and then proceed with step 3.
15. Alternatively, at this step, the QIAprep Spin Miniprep Kit (Qiagen) can be used. In this case, 350 μl buffer N3 (Qiagen) is added to the sample, and the procedure is carried out as described in the instructions provided in the kit.
16. If an *E. coli* clone is not used, this DNA may be used as template in PCR reactions in order to screen for full-length assemblies (e.g., with the M13 F and M13 R primer set). These PCR products may then be sequenced.
17. The EPI300™ (Epicentre) electrocompetent *E. coli* cells work well with this procedure. Combine 3 μl DNA with 30 μl of these cells in a 1-mm cuvette (BioRad) and electroporate the cells at 1,200 V, 25 μF, and 200 Ω using a Gene Pulser Xcell electroporation system (BioRad).
18. This transformation usually results in hundreds to thousands of *E. coli* clones. It is usually only necessary to pick one or two *E. coli* clones because the DNA was derived from a single yeast clone, and thus, most colonies will contain the same plasmid DNA sequence.
19. Alternatively, the method described in steps 3–23 can be carried out with *E. coli* cell pellets, with two exceptions: (1) The Zymolyase® solution and BME do not need to be added to the resuspension buffer, and the 1-h incubation step does not need to be carried out; and (2) the DNA pellet (step 23) should be suspended in TE buffer containing 0.1 μg/ml RNAse A and incubated at 37°C for 30 min.
20. If NotI restriction sites were designed into the assembly strategy, the NotI restriction enzyme can be used to release the insert to determine if a full-length assembly is present.

21. ClustalW Multiple alignment (15), contained within the BioEdit Sequence Alignment Editor software can be used for this purpose.

## Acknowledgments

The author would like to thank the Synthetic Biology Group at JCVI for the helpful discussions and Synthetic Genomics, Inc. for funding this work.

### References

1. Agarwal, K. L., Buchi, H., Caruthers, M. H., Gupta, N., Khorana, H. G., Kleppe, K., Kumar, A., Ohtsuka, E., Rajbhandary, U. L., Van de Sande, J. H., Sgaramella, V., Weber, H., and Yamada, T. (1970) Total synthesis of the gene for an alanine transfer ribonucleic acid from yeast, *Nature 227*, 27–34.
2. Reese, C. B. (2005) Oligo- and poly-nucleotides: 50 years of chemical synthesis, *Org. Biomol. Chem. 3*, 3851–3868.
3. Xiong, A. S., Peng, R. H., Zhuang, J., Gao, F., Li, Y., Cheng, Z. M., and Yao, Q. H. (2008) Chemical gene synthesis: strategies, softwares, error corrections, and applications, *FEMS Microbiol. Rev. 32*, 522–540.
4. Xiong, A. S., Peng, R. H., Zhuang, J., Liu, J. G., Gao, F., Chen, J. M., Cheng, Z. M., and Yao, Q. H. (2008) Non-polymerase-cycling-assembly-based chemical gene synthesis: strategies, methods, and progress, *Biotechnol. Adv. 26*, 121–134.
5. Czar, M. J., Anderson, J. C., Bader, J. S., and Peccoud, J. (2009) Gene synthesis demystified, *Trends. Biotechnol.*, 27(2):63–72.
6. Gibson, D. G., Benders, G. A., Andrews-Pfannkoch, C., Denisova, E. A., Baden-Tillson, H., Zaveri, J., Stockwell, T. B., Brownley, A., Thomas, D. W., Algire, M. A., Merryman, C., Young, L., Noskov, V. N., Glass, J. I., Venter, J. C., Hutchison, C. A., 3rd, and Smith, H. O. (2008) Complete chemical synthesis, assembly, and cloning of a *Mycoplasma genitalium* genome, *Science 319*, 1215–1220.
7. Gibson, D. G., Glass, J. I., Lartigue, C., Noskov, V. N., Chuang, R. Y., Algire, M. A., Benders, G. A., Montague, M. G., Ma, L., Moodie, M. M., Merryman, C., Vashee, S., Krishnakumar, R., Assad-Garcia, N., Andrews-Pfannkoch, C., Denisova, E. A., Young, L., Qi, Z. Q., Segall-Shapiro, T. H., Calvey, C. H., Parmar, P. P., Hutchison, C. A., 3rd, Smith, H. O., and Venter, J. C. (2010) Creation of a bacterial cell controlled by a chemically synthesized genome, *Science 329*, 52–56.
8. Orr-Weaver, T. L., Szostak, J. W., and Rothstein, R. J. (1981) Yeast transformation: a model system for the study of recombination, *Proc. Natl. Acad. Sci. USA 78*, 6354–6358.
9. Moerschell, R. P., Tsunasawa, S., and Sherman, F. (1988) Transformation of yeast with synthetic oligonucleotides, *Proc. Natl. Acad. Sci. USA 85*, 524–528.
10. Ma, H., Kunes, S., Schatz, P. J., and Botstein, D. (1987) Plasmid construction by homologous recombination in yeast, *Gene 58*, 201–216.
11. Raymond, C. K., Sims, E. H., and Olson, M. V. (2002) Linker-mediated recombinational subcloning of large DNA fragments using yeast, *Genome Res. 12*, 190–197.
12. Gibson, D. G., Benders, G. A., Axelrod, K. C., Zaveri, J., Algire, M. A., Moodie, M., Montague, M. G., Venter, J. C., Smith, H. O., and Hutchison, C. A., 3rd. (2008) One-step assembly in yeast of 25 overlapping DNA fragments to form a complete synthetic *Mycoplasma genitalium* genome, *Proc. Natl. Acad. Sci. USA 105*, 20404–20409.
13. Gibson, D. G. (2009) Synthesis of DNA fragments in yeast by one-step assembly of overlapping oligonucleotides, *Nucleic Acids Res. 37*, 6984–6990.
14. Kouprina, N., and Larionov, V. (2008) Selective isolation of genomic loci from complex genomes by transformation-associated recombination cloning in the yeast *Saccharomyces cerevisiae*, *Nat Protoc. 3*, 371–377.
15. Thompson, J. D., Higgins, D. G., and Gibson, T. J. (1994) CLUSTAL W: improving the sensitivity of progressive multiple sequence alignment through sequence weighting, position-specific gap penalties and weight matrix choice, *Nucleic Acids Res. 22*, 4673–4680.

# Chapter 3

# TopDown Real-Time Gene Synthesis


## Mo Chao Huang, Wai Chye Cheong, Hongye Ye, and Mo-Huang Li



## Abstract

This chapter introduces a simple, cost-effective TopDown one-step gene synthesis method, which is suitable for the sequence assembly of fairly long DNA. This method can be distinguished from conventional gene synthesis methods by two key features: (1) the melting temperature of the outer primers is designed to be ~8°C lower than that of the assembly oligonucleotides, and (2) different annealing temperatures are utilized to selectively control the efficiencies of oligonucleotide assembly and full-length template amplification. This method eliminates the interference between polymerase chain reactions (PCR) assembly and amplification in one-step gene synthesis. Additionally, the TopDown gene synthesis has been combined with the LCGreen I DNA fluorescence dye in a real-time gene synthesis approach for investigating the stepwise efficiency and kinetics of PCR-based gene synthesis. The obtained real-time fluorescence signals are compared with gel electrophoresis results to optimize gene synthesis conditions.



**Key words:** TopDown gene synthesis, PCR, Real-time gene synthesis, De novo gene synthesis, LCGreen I, Assembly efficiency


## 1. Introduction

De novo gene synthesis is a powerful molecular tool for creating man-made DNA sequences. This technology has broad applications for protein engineering (1, 2), development of artificial gene networks (3, 4), and creation of synthetic genomes (5, 6). Current gene synthesis methods include ligase chain reaction (LCR) (7) and polymerase chain reaction (PCR) assembly (8), which both rely on the use of overlapping oligonucleotides to construct genes. Various PCR-based methods that have been reported include the thermodynamically balanced inside-out (TBIO) method (9), successive PCR (10), dual asymmetrical PCR (DA-PCR) (11), overlap extension PCR (OE-PCR) (12, 13), PCR-based two-step DNA synthesis (8, 10, 14), and one-step gene synthesis (15).

Although the PCR assembly method has been commonly used for de novo gene synthesis, there is a lack of a universal synthesis


Jean Peccoud (ed.), *Gene Synthesis: Methods and Protocols*, Methods in Molecular Biology, vol. 852,
DOI 10.1007/978-1-61779-564-0_3, © Springer Science+Business Media, LLC 2012

method and capability in accurately predicting the gene synthesis. Herein, we present a simple, cost-effective TopDown one-step gene synthesis approach (16) which can selectively control the efficiencies of oligonucleotide assembly and full-length template amplification of relatively long genes. This method utilizes a computer program to design the outer primers with the melting temperature ~8°C lower than that of the assembly oligonucleotides to minimize the interference between the PCR assembly and amplification in the one-step gene synthesis. The overlapping gene synthesis is performed in one PCR mixture with two annealing temperature segments for oligonucleotide assembly and full-length template amplification, respectively. The outer primers are subjected to an elevated annealing condition during the assembly process, which prevents mispairing among primers and oligonucleotides (Fig. 1). The assembly process automatically switches to a preferential full-length amplification as the full-length template emerges. This greatly improves the assembly efficiency of the PCR process as

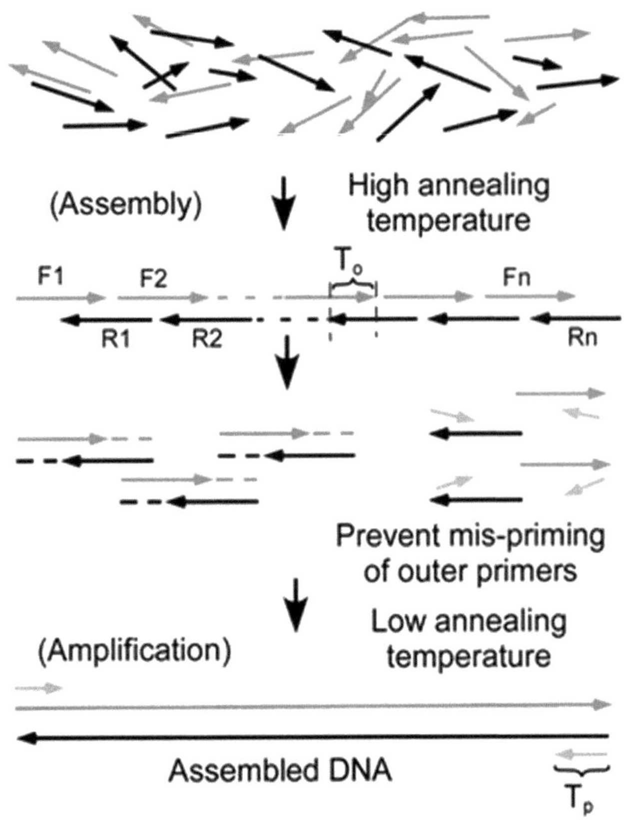


Fig. 1. Schematic illustration of TopDown one-step gene synthesis. This approach combines PCR assembly and amplification into a single stage by employing different annealing temperatures for assembly and amplification. The melting temperatures of the inner oligos ($T_o$) and outer primers ($T_p$) are designed so that $T_o - T_p \geq 8°C$ to minimize potential interference during PCR.

compared to the conventional one-step and two-step gene synthesis processes.

Furthermore, the TD one-step method is combined with real-time fluorescence analysis to investigate the gene synthesis process. Comparing the real-time fluorescence signals with gel electrophoresis results allows to optimize the gene synthesis conditions. The effects of the concentrations of oligonucleotides and of outer primer, the stringency of annealing temperature, and the number of PCR cycles can so be analyzed and the PCR conditions optimized.

## 2. Materials

### 2.1. Reagents

1. 100 μM Oligonucleotides (desalted without additional purification, Research Biolabs, Singapore) (see Note 1).
2. 100 μM Forward and reverse primers (Research Biolabs, Singapore) (see Note 1).
3. 25 mM $MgSO_4$ (see Note 2).
4. dNTP mixture (containing 25 mM dATP, 25 mM dGTP, 25 mM dCTP, and 25 mM dTTP) (see Note 2).
5. High-fidelity KOD Hot Start DNA polymerase (1.0 U/μl) and 10× KOD buffer (Novagen) (see Note 3).
6. 10 mg/ml Bovine serum albumin (BSA).
7. 10× LCGreen I (Idaho Technology Inc.) (see Note 4).
8. Deionized distilled water.
9. Agarose gel powder.
10. 10× TBE buffer.
11. 50 ng/μl 100 bp DNA ladder.
12. 6× DNA loading dye.

### 2.2. Equipment

1. Computer software to design oligonucleotides (e.g., TmPrime and DNAWorks).
2. Vortex mixer.
3. Gel electrophoresis apparatus, digital electrophoresis power supply.
4. Real-time PCR thermocycler, such as LightCycler® 1.5 (Roche), CFX96 (Bio-Rad), or ABI 7300/7500 (Applied Biosystems).
5. Gel imaging system such as Typhoon 9200 imager or Gel Doc XR.
6. Microcentrifuge.
7. LightCycler® centrifuge adapters (Roche).

## 3. Methods

### 3.1. Reagent Setup: Designing the DNA of Interest

The DNA sequence to be synthesized can be designed manually or using computer software. We strongly recommend using computer software [e.g., DNAWorks (http://helixweb.nih.gov/dnaworks) (17) or TmPrime (http://prime.ibn.a-star.edu.sg/) (18)] to design the gene sequence. These computer programs allow for the construction of oligonucleotides with a uniform melting temperature which increases the yield of the assembled full-length DNA product of the PCR gene assembly. Additionally, these programs also analyze the potential for mishybridization and secondary structures among the oligonucleotides, which you may want to check before conducting the gene assembly. For DNA with high sequence repeats, the PCR-based gene synthesis may not be the best choice, and the LCR-based approach is more effective for these challenging DNAs (19).

#### 3.1.1. Design of Oligonucleotides and Outer Primers

We use TmPrime to design the gene sequence. Figure 2 illustrates all the parameters needed to generate the oligonucleotide set when using TmPrime. Most of the parameters are self-explanatory. For instance, the user is asked to provide gene information, gene assembly buffer condition, oligonucleotide and outer primer concentrations, optional parameters for long DNA assembly, and parameters for mispriming analysis. The software will report the melting temperatures, oligonucleotide sequences, potential formation of secondary structures, and statistical information for the oligo sets of each pool in a PDF file (see Note 5). To ensure successful TopDown gene synthesis, oligos are designed to have a melting temperature that is ~8°C higher than that of the outer primers to minimize the competition between PCR assembly and PCR amplification of the assembled product in the one-step gene synthesis (see Note 6). The sequence of the human calcium-binding protein A4 promoter (S100A4, 752 bp; chr1:1503312036–1503311284) has been selected here as the target DNA for demonstration. The average melting temperature of the designed oligonucleotides is 66°C and consists of a pool of 30 oligos ranging in length from 41 nucleotides (nt) to 66 nt.

#### 3.1.2. Designing Gene-Specific Primers

The primers that anneal to the target gene can be designed using IDT SciTools (http://www.idtdna.com/SciTools/SciTools.aspx) (20) and TmPrime with an outer primer concentration of 300–400 nM. It is important that the primers are designed to anneal only to the outermost 3′ and 5′ ends of the target gene.

### 3.2. Real-Time TopDown Gene Synthesis

1. Set up the master mix containing the inner oligos: add 2 μl (100 μM) of each of the oligos to a 600-μl microfuge tube, and add deionized distilled water to a final volume of 200 μl.

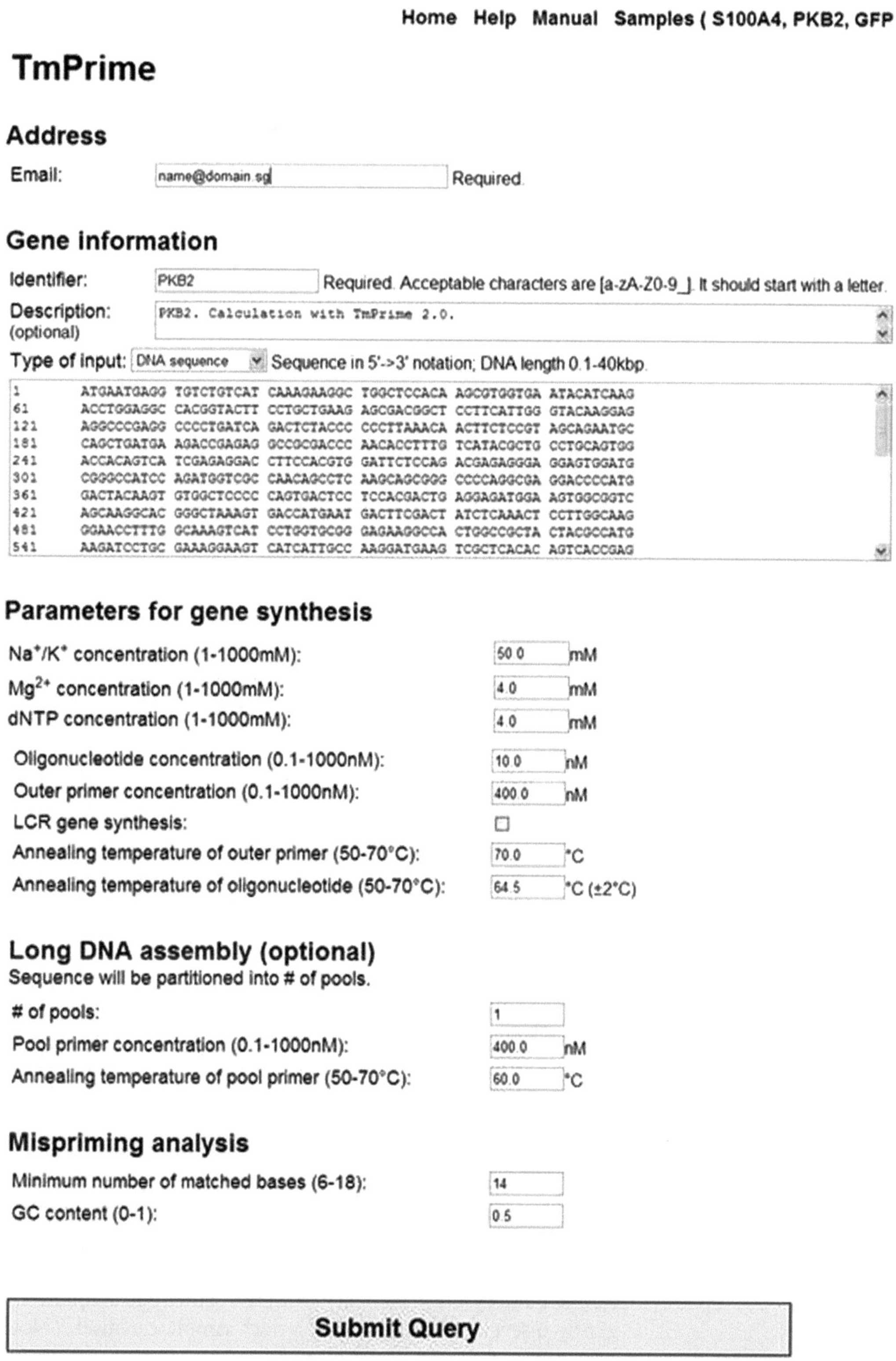


Fig. 2. Web interface for TmPrime. TmPrime is implemented in functional modules, each module reflecting a different aspect of the oligonucleotide design process with the interface elements organized in a coherently grouped fashion.

The final concentration of the oligo master mix is 1 μM. Then pipette 10 μl of the master mix into a 200-μl PCR tube, and add deionized distilled water to 40 μl. The concentration of the diluted master mix is 250 nM. The master mix should be stored at −20°C to prevent degradation of the oligos.

2. Set up the master mix of the gene-specific outer primers: add 5 μl (100 μM) of each outer primer into a 600-μl microfuge tube, and add deionized distilled water to 50 μl. Mix the contents of the tube by flicking and then pulse vortex in a vortex mixer. The final concentration of the primer master mix is 10 μM.

3. Prepare the TopDown gene synthesis mixture: add the PCR reaction components below to a thin-walled 200-μl PCR tube and mix the reaction mixture by flicking and spinning briefly. Pipette 20 μl of this reaction mixture into the reaction capillary of a LightCycler® 1.5 real-time thermal cycling machine. Cap the reaction capillary manually or using the LightCycler® capping tool. Throughput this procedure, all reaction solutions should be stored and handled on ice.

| Component | Amount | Final amount/concentration |
|---|---|---|
| dNTP (100 mM) | 4 μl | 4 mM |
| Oligo mix (250 nM) | 2 μl | 10 nM |
| Primer mix (10 μM) | 2 μl | 400 nM |
| $MgSO_4$ (25 mM) | 8 μl | 4 mM |
| 10× KOD buffer | 5 μl | 1× |
| KOD Hot Start polymerase | 1 μl | 1 U |
| BSA (10 mg/ml) | 2.5 μl | 0.5 mg/ml |
| 10× LCGreen I | 10 μl | 1× |
| $ddH_2O$ | 15.5 μl | |
| Total volume | 50 μl | |

4. Gently insert the capillary into LightCycler® centrifuge adapters. Transfer the centrifuge adapters into a standard microcentrifuge and briefly spin for 3–5 s at 500–1,000 rpm. Check and ensure that the reaction mix fills the capillary. (Note: repeat the centrifugation step if the reaction mix fails to fill the length of capillary or large air bubbles are found within the capillary as this will degrade signal detection during real-time PCR.) Remove the reaction capillary from the centrifuge adapters and gently insert it into the LightCycler® sample carousel. (Note: take note of the position of capillary on carousel.) You can use a different real-time PCR thermal cycler as long as the

thermal cycler can detect the LCGreen I (optimum excitation 440–470 nm, optimum emission 470–520 nm).

5. Carry out the real-time TopDown gene synthesis in a Roche LightCycler® 1.5 (or another real-time thermocycler) with the following thermal cycling conditions: 2 min initial denaturation at 95°C; 15 cycles of 95°C for 5 s, 65–70°C (according to the $T_m$ of inner oligos) for 60 s, 72°C for 30 s; followed by 15 cycles of 95°C for 5 s, 50–55°C (according the $T_m$ of outer primers) for 60 s, 72°C for 30 s; followed by a final extension step at 72°C for 10 min (see Note 7).

### 3.3. Agarose Gel Electrophoresis

1. Dilute the 10× TBE buffer with deionized water to the final concentration of 0.5×.
2. Prepare a 1.5% agarose gel solution by adding 3 g of agarose power to 200 ml of 0.5× TBE buffer using a 500-ml beaker or plastic bottle.
3. Heat the beaker in a microwave oven on medium power for 3–5 min until the solution is boiling. This step ensures that the agarose powder is fully dissolved.
4. Cool the solution to 50–60°C prior to gel casting. Cast the gel to the gel tray and wait for ~1 h until the gel is solidified.
5. Prepare DNA samples: mix 5 μl of the assembled product (from step 5, Subheading 3.2) with 1 μl of loading dye. For real-time gene synthesis, the assembly mixture already contains LCGreen I; thus, no additional LCGreen is needed.
6. Prepare 100 bp DNA ladder: mix 5 μl of 100 bp DNA ladder with 1 μl of LCGreen I.
7. Load the DNA ladder and DNA samples to the wells of cast gel.
8. Perform gel electrophoreses at 60 V for 60 min.
9. Scan the gel image with the Typhoon 9200 image scanner or any type of gel imaging system with emission filter for LCGreen I (optimum excitation 440–470 nm, optimum emission 470–520 nm).

## 4. Notes

1. The optimum oligo and outer primer concentrations are 10–20 nM and 0.3–0.4 μM, respectively for TopDown gene synthesis (Figs. 3 and 4).
2. dNTP and $Mg^{2+}$ concentration: the dNTPs concentration has been increased from 0.2 mM each as used in standard PCR to 1 mM each for TopDown gene synthesis to prevent the depletion of dNTPs. The $Mg^{2+}$ concentration has been empirically

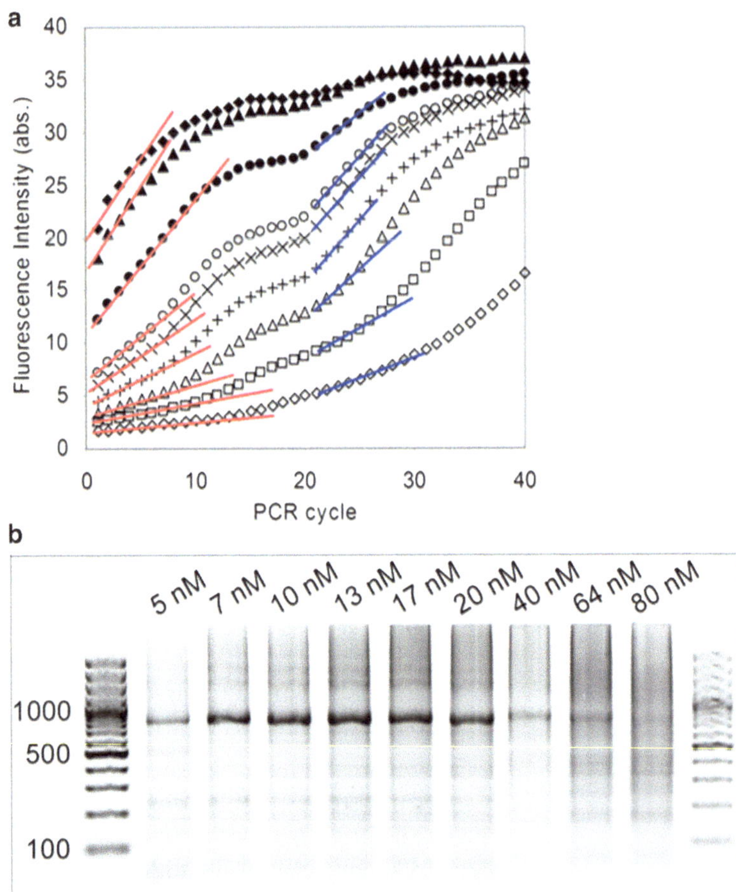


Fig. 3. Effect of oligo concentration of gene synthesis. The oligonucleotide concentration is critical for successful gene synthesis. S100A4 (752 bp) was synthesized using various oligonucleotide concentrations ranging from 5 to 80 nM and annealing temperatures of 67°C (first 20 cycles) and 49°C (next 20 cycles). (**a**) Fluorescence intensity versus cycle number plot for different oligonucleotide concentrations: 5 nM (*open diamond*), 7 nM (*open square*), 10 nM (*open triangle*), 13 nM (*plus sign*), 17 nM (*multiplication sign*), 20 nM (*open circle*), 40 nM (*filled circle*), 64 nM (*filled triangle*), and 80 nM (*filled square*). (**b**) Corresponding agarose gel (1.5%) electrophoresis results. The increasing slope of the fluorescence intensity during the early cycles and again around cycle number 21 indicates the efficiency of the assembly and amplification process, respectively.

optimized (with an optimum of 4 mM) based on the concentration of dNTPs that can chelate $Mg^{2+}$, thereby affecting polymerase activity (21, 22).

3. Choice of DNA polymerase: KOD Hot Start polymerase is recommended for TopDown gene synthesis as we have observed that this polymerase outperforms Taq and *Pfu* polymerases.
4. Choice of DNA fluorescence dye: LCGreen I, which has a similar fluorescence spectrum to SYBR Green I, which is commonly used in real-time PCR (23), is more suitable for studying

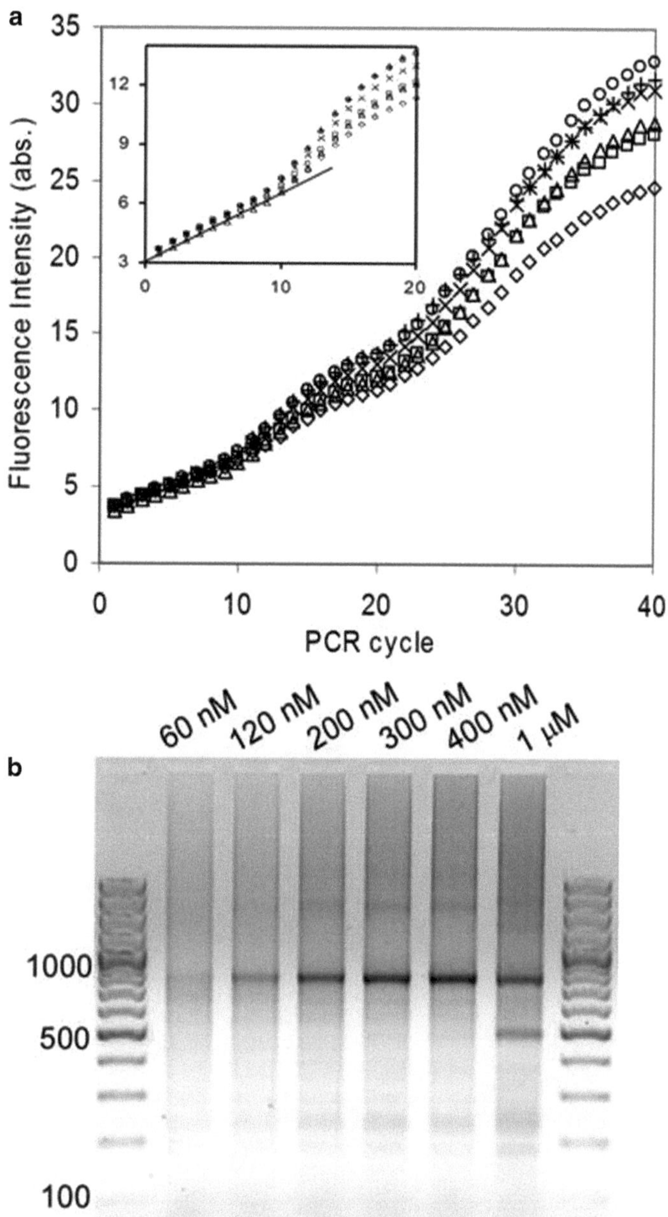


Fig. 4. Effect of outer primer concentration on gene synthesis. S100A4 (752 bp) is successfully synthesized with different primer concentrations ranging from 60 nM to 1 μM, as indicated by the sharp, narrow gel band of the desired size. (**a**) Fluorescence intensity versus cycle number plot for outer primers' concentrations of 60 nM (*open diamond*), 120 nM (*open square*), 200 nM (*open triangle*), 300 nM (*multiplication sign*), 400 nM (*plus sign*), and 1 μM (*open circle*). The inset shows the fluorescence signal for the first 20 cycles. (**b**) Corresponding agarose gel (1.5%) gel electrophoresis results.

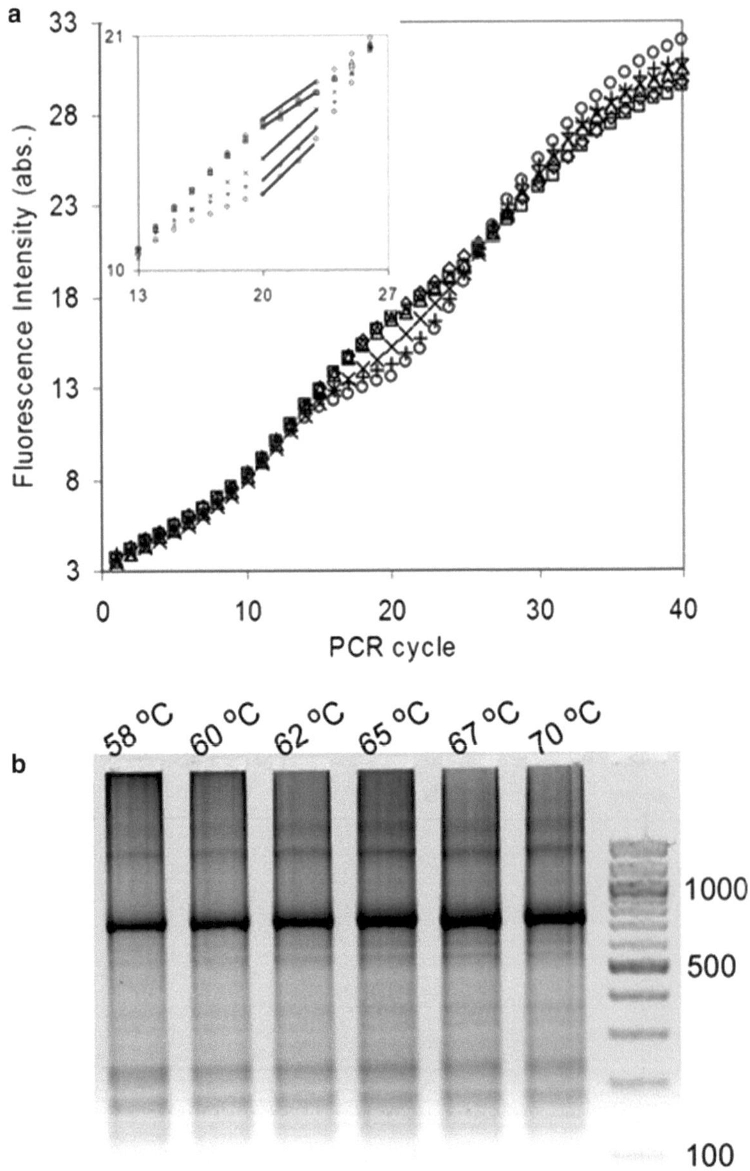


Fig. 5. Effect of annealing temperature on gene synthesis. S100A4 (752 bp) was synthesized using different assembly annealing temperatures ranging from 58 to 70°C for the first 20 cycles, followed by another 20 cycles at annealing temperature of 49°C. (**a**) Fluorescence intensity versus cycle number plot for different annealing temperatures: 58°C (*open diamond*), 60°C (*open square*), 62°C (*open triangle*), 65°C (*multiplication sign*), 67°C (*plus sign*), and 70°C (*open circle*). The inset shows the 15 midcycles (cycles 13–27). (**b**) Agarose gel (1.5%) electrophoresis results. A higher yield of gene synthesis was obtained with a stringent assembly annealing temperature (>67°C).

real-time gene synthesis. SYBR Green I binds preferentially to long DNA fragments (24) and can redistribute from short DNA fragments to large DNA fragments during thermal cycling, which makes it difficult to analyze the observed fluorescence signal, as the assembly mixture contains dsDNA of various sizes. The optimum concentration of LCGreen I is 1× for TopDown gene synthesis with 10–20 nM of oligos.

5. TmPrime gene design program is being optimized continuously. Hence, the average melting temperature and oligonucleotide sequences may be different from the provided data.

6. We recommend designing the oligos and outer primers with the following conditions: (1) design outer primers ($T_m$ = 50–55°C) and inner oligos ($T_m$ ~65°C) with distinct melting temperatures (i.e., $\Delta T_m \geq 8$°C); (2) oligos and outer primers concentration of 10 nm and 400 nM, respectively; and (3) 50 mM of $Na^+/K^+$, 4 mM of $Mg^{2+}$, and 4 mM of dNTPs.

7. Optimize the assembly cycle numbers: the number of PCR cycles influences the quality and quantity of PCR-based gene synthesis. The fluorescence curve (see 10-nM curve in Fig. 3) suggests that the assembly and amplification processes reach the plateau at around cycle 15 and cycle 35. Thus, the optimum number of PCR cycles is 30–15 cycles each for the assembly and the amplification reaction. The amplification efficiency of the PCR reaction decreases after it reaches the plateau. Additional PCR cycling will favor nonspecific annealing of the full-length product to either randomly assembled fragments or to itself. Additionally, we recommend conducting TopDown synthesis with an assembly annealing temperature that is 2–5°C higher than the average melting temperature of the constructed oligos. Such a stringent annealing condition usually leads to a better yield of the full-length DNA product (Fig. 5). The assembly efficiency of PCR gene synthesis depends on the gene length and sequence content. So far, the maximum gene length that we have successfully constructed using the TopDown approach is ~1.6 kb from a pool of 60 oligonucleotides.

## References


1. He M, Stoevesandt O, Palmer EA, Khan F, Ericsson O, and Taussig MJ (2008) Printing protein arrays from DNA arrays. *Nat. Methods* 5:175–177.
2. Cox JC, Lape J, Sayed MA and Hellinga HW (2007) Protein fabrication automation. *Protein Sc.* 16:379–390.
3. Sprinzak D, and Elowitz MB (2005) Reconstruction of genetic circuits. *Nature* 438: 443–448.
4. Basu S, Gerchman Y, Collins,CH. Arnold,FH and Weiss RA (2005) Synthetic multicellular system for programmed pattern formation. *Nature* 434:1130–1134.
5. Smith HO, Hutchison CA III, Pfannkoch C and Venter JC (2003) Generating a synthetic genome by whole genome assembly: ΦX174 bacteriophage from synthetic oligonucleotides. *Proc Natl Acad Sci USA* 100: 15440–15445.

6. Gibson DG, Benders GA, Andrews-Pfannkoch C, Denisova EA, Baden-Tillson H, Zaveri J, Stockwell TB, Brownley A, Thomas DW, Algire MA, Merryman C, Young L, Noskov VN, Glass JI, Venter JC, Hutchison CA.III and Smith HO (2008) Complete chemical synthesis, assembly, and cloning of a *Mycoplasma genitalium* genome. *Science* 319:1215–1220.
7. Au LC, Yang FY, Yang WJ, Lo SH and Kao,CF (1998) Gene synthesis by a LCR-based approach: High-level production of leptin-L54 using synthetic gene in *Escherichia coli. Biochem Biophys Res.Commun* 248:200–203.
8. Stemmer WP, Crameri A, Ha KD, Brennan TM and Heyneker HL (1995) Single-step assembly of a gene and entire plasmid from large numbers of oligodeoxyribonucleotides. *Gene* 164:49–53.
9. Gao X, Yo P, Keith A, Ragan TJ and Harris TK (2003) Thermodynamically balanced inside-out (TBIO) PCR-based gene synthesis: A novel method of primer design for high-fidelity assembly of longer gene sequences. *Nucleic Acids Res* 31:,e143.
10. Xiong A-S, Yao Q-H, Peng R-H, Li X, Fan H-Q, Cheng Z-M and Li Y (2004) A simple, rapid, high-fidelity and cost-effective PCR-based two-step DNA synthesis method for long gene sequences. *Nucleic Acids Res* 32:e98.
11. Sandhu GS, Aleff RA and Kline, BC (1992) Dual asymmetric PCR: One-step construction of synthetic genes. *Biotechniques* 12:14–16.
12. Toung L and Dong,Q.(2004) Two-step total gene synthesis method. *Nucleic Acids Res* 32:e59.
13. Prodromou C and Pearl L (1992) Recursive PCR: A novel technique for total gene synthesis. *Protein Eng* 5:827–829.
14. Xiong A-S, Yao Q-H, Peng R-H, Duan H, Li X, Fan H-Q, Cheng Z-M and Li Y (2006) PCR-based accurate synthesis of long DNA sequences. *Nat Protoc* 1:791–797.
15. Wu G, Wolf JB, Ibrahim AF, Vadasz S, Gunasinghe M and Freeland SJ (2006) Simplified gene synthesis: A one-step approach to PCR-based gene construction. *J. Biotechnol* 124: 496–503.
16. Ye H, Huang MC, Li M-H and Ying JY (2009) Experimental analysis of gene assembly with TopDown one-step real-time gene synthesis. *Nucleic Acids Res* 37:e51.
17. Hoover DM and Lubkowski J (2002) DNAWorks: An automated method for designing oligonucleotides for PCR-based gene synthesis. *Nucleic Acids Res* 30:e43.
18. Bode M, Khor,S, Ye H, Li M-H and Ying JY (2009) TmPrime: fast, flexible oligonucleotide design software for gene synthesis. *Nucleic Acids Res* 37:W214–W221.
19. Bang D and Church GM (2008) Gene synthesis by circular assembly amplification. *Nat Methods* 5:37–39.
20. Owczarzy R. Tataurov AV, Wu,Y, Manthey,JA, McQuisten KA, Almabrazi HG et al. (2008) IDT SciTools: a suite for analysis and design of nucleic acid oligomers. *Nucleic Acids Res* 36:W163–W169.
21. Ely JJ, Reeves-Daniel A, Campbell ML, Kohler S and Stone WH (1998) Influence of magnesium ion concentration and PCR amplification conditions on cross-species PCR. *Biotechniques* 25:38–40.
22. von Ahsen N, Wittwer CT and Schütz E. (2001) Oligonucleotide melting temperatures under PCR conditions: Nearest-neighbor corrections for $Mg^{2+}$, deoxynucleotide triphosphate, and dimethyl sulfoxide concentrations with comparison to alternative empirical formulas. *Clin Chem* 47:1956–1961.
23. Wittwer CT, Reed GH, Gundry,CN, Vandersteen JG and Pryor RJ (2003) High-resolution genotyping by amplicon melting analysis using LCGreen. *Clin Chem* 49:853–860.
24. Giglio S, Monis PT and Saint CP (2003) Demonstration of preferential binding of SYBR Green I to specific DNA fragments in real-time multiplex PCR. *Nucleic Acids Res.*31:e136.

# Chapter 4

# De Novo DNA Synthesis Using Single-Molecule PCR

Tuval Ben Yehezkel, Gregory Linshiz, and Ehud Shapiro

## Abstract

The throughput of DNA reading (i.e., sequencing) has dramatically increased recently owing to the incorporation of in vitro clonal amplification. The throughput of DNA writing (i.e., synthesis) is trailing behind, with cloning and sequencing constituting the main bottleneck. To overcome this bottleneck, an in vitro alternative for in vivo DNA cloning needs to be integrated into DNA synthesis methods. Here, we show how a new single-molecule PCR (smPCR)-based procedure can be employed as a general substitute for in vivo cloning, thereby allowing for the first time in vitro DNA synthesis. We integrated this rapid and high fidelity in vitro procedure into our previously described recursive DNA synthesis and error correction procedure and used it to efficiently construct and error-correct a 1.8-kb DNA molecule from synthetic unpurified oligonucleotides, entirely in vitro. Although we demonstrate incorporating smPCR in a particular method, the approach is general and can be used, in principle, in conjunction with other DNA synthesis methods as well.

**Key words:** DNA synthesis, In vitro cloning, In vivo cloning, DNA error correction, Single-molecule PCR, Synthetic biology

## 1. Introduction

The broad availability of synthetic DNA oligonucleotides enabled the development of many powerful applications in biotechnology. Longer synthetic DNA molecules and libraries (generated from assembly of these oligonucleotides) in the 0.5–5 kb range are now becoming increasingly available owing to newly developed synthesis and error correction methods (1–7). The wide availability of such molecules, in great need since the advent of synthetic biology and modern genetic engineering, is expected to enable the routine creation of new genetic material, as well as offer an alternative to obtaining DNA from natural sources.

Jean Peccoud (ed.), *Gene Synthesis: Methods and Protocols*, Methods in Molecular Biology, vol. 852,
DOI 10.1007/978-1-61779-564-0_4, 

Unfortunately, the synthetic DNA oligonucleotides (oligos) used as building blocks for the generation of the longer constructs are error-prone. Such errors accumulate linearly with the length of the constructed molecule and result in an exponential decrease in the fraction of error-free molecules. Hence, an exponentially increasing number of molecules have to be screened, i.e., cloned into a host organism and sequenced, in order to obtain ever longer error-free molecules. In order to mitigate this effect, a two-step assembly process (4, 7) is often used, in which fragments in the 500–1,000 bp range are first screened via cloning and sequencing before the error-free clones are synthesized.

In vivo cloning (1–7) is time consuming, labor intensive, and difficult to scale up and automate. These limitations combined with the sheer number of clones that needs to be screened to obtain long error-free synthetic DNA make the cloning phase a bottleneck in de novo DNA synthesis and prevent synthetic DNA from being routinely produced in a fast, cheap, and high-throughput manner. Reducing the number of clones required to obtain an error-free molecule is the subject of intensive ongoing research (1, 2, 4, 6), also recently addressed by us (5) with the development of a method that we believe relieves much of this burden.

In this chapter, we address the second major issue, namely, replacing the time consuming and labor intensive in vivo cloning procedure that is associated with synthetic DNA synthesis with a faster and less laborious in vitro cloning procedure.

Since its introduction, the polymerase chain reaction (PCR) (8) has been implemented in a myriad of variations, one of which is PCR on a single DNA template molecule (9), which essentially creates a PCR "clone." Single-molecule PCR (smPCR) is a faster, cheaper, scalable, and automatable alternative to traditional in vivo cloning. Its standard application in molecular biology has been nonsystematic, most commonly used for the amplification of single molecules for sequencing, genotyping, or downstream translation purposes (8–12). Recently, it has been systematically integrated into high-throughput DNA sequencing (13, 14). High-throughput DNA synthesis technologies can also benefit from smPCR, as demonstrated here for the use of smPCR in the context of our recently introduced DNA synthesis procedure (5), which combines recursive synthesis and error correction. In this chapter, we show that in vitro cloning based on smPCR can be used as a practical alternative to conventional in vivo cloning. In particular, we show the successful construction of a 1.8-kb-long DNA molecule from synthetic unpurified oligos using our recursive synthesis and error correction procedure combined with smPCR. As a control, we also constructed the same molecule using conventional in vivo cloning, and the results are compared below.

## 2. Materials

### 2.1. Core Recursive DNA Construction

1. T4 polynucleotide kinase (NEB, Ipswich, MA, USA).
2. Thermo-Start DNA polymerase (ABgene).
3. Lambda exonuclease (Epicenter).

### 2.2. smPCR

1. Hot-start Accusure (BioLINE, Taunton, MA, USA).
2. Taq polymerase (ABgene, UK).

### 2.3. Chemical Oligonucleotide Synthesis

Oligonucleotides for all experiments were ordered from commercial providers (Sigma Genosys & IDT) with standard desalting.

### 2.4. DNA Purification Kits

QIAGEN's QIAquick 96-well PCR purification kit and QIAGEN's MinElute PCR purification kit.

### 2.5. Cloning System

pGEM-T Easy Vector System and JM109 competent cells from PROMEGA.

## 3. Methods

### 3.1. Summary of the Procedure

1. Recursive (or other type of de novo) construction of the molecule from synthetic unpurified oligos (see Fig. 1), as specified below (see Subheading 3.4) and in ref. 5.
2. "Adaptor PCR" for the insertion of the CA primer sequence and the random bar-coding nucleotides (see Fig. 2) on templates from step 1. This is done using the PCR protocol and the primers specified in (15). Alternatively, these sequences can also be included as part of the original target sequence to avoid the additional PCR.
3. Early termination of the PCR of step 2 within the twofold exponential amplification phase (as shown in Fig. 3) to prevent heterodimer formation (see Figs. 4 and 5).
4. Optical density (OD) measurement and dilution of the PCR from step 3 according to the graphs depicted in Fig. 6. Alternatively, the real-time PCR-assisted calibration experiment (described in Subheading 3.7) can also be used to determine the required dilution instead of measuring the OD.
5. smPCRs using the CA primer and templates from the dilution prepared in step 4. The number of reactions prepared is determined according to the required number of clones as shown in Fig. 6, and the error rate as described in Figs. 7–9. The PCR

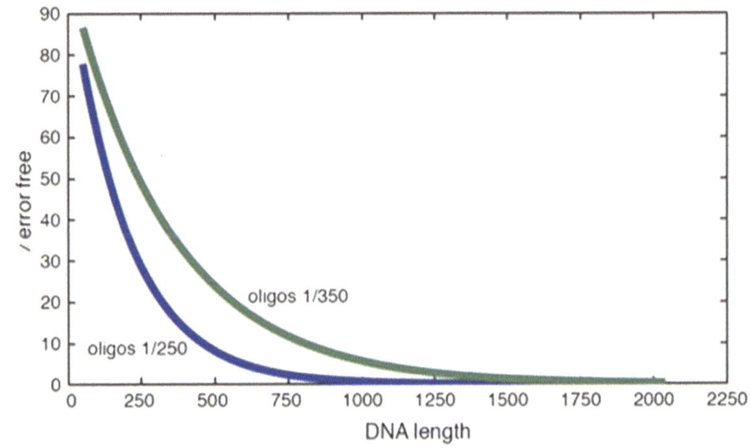


Fig. 1. Relationship between error rate and DNA length. Shown here is the percentage of molecules that are error-free as a function of construct length for two typical error rates of synthetic oligos (and hence of constructs); *left plot* with an error rate of 1/250 and the *right plot* with an error rate of 1/350. The high error rate results in a large drop in the fraction of error-free molecules, even in short fragments of 500–1,000 bp in length. Figure reproduced from ref. 15.

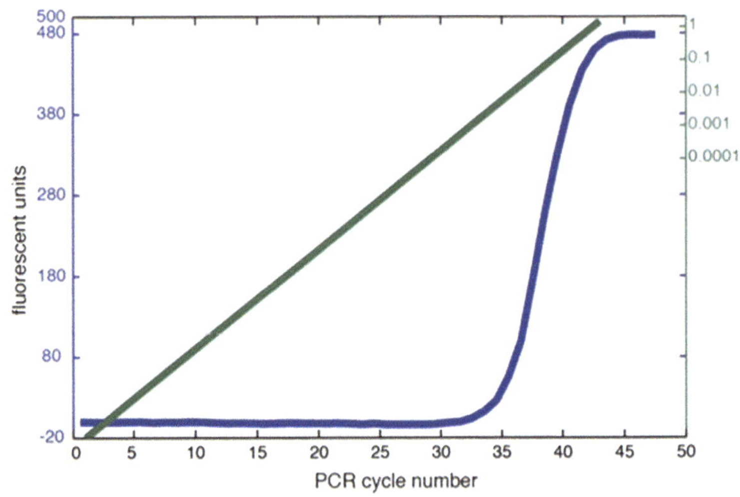


Fig. 2. Determination of required cycle number. The number of cycles required for single molecule amplification can be accurately anticipated from the initial and final amount of the DNA in a PCR with known amplification efficiency. The *straight line* (in *green*) gives the number of amplified DNA molecules in a PCR reaction that started from a single molecule as a function of the number of cycles assuming 100% amplification efficiency. The *curve* below (shown in *blue*) is an amplification curve obtained from a real-time smPCR. Figure reproduced from ref. 15.

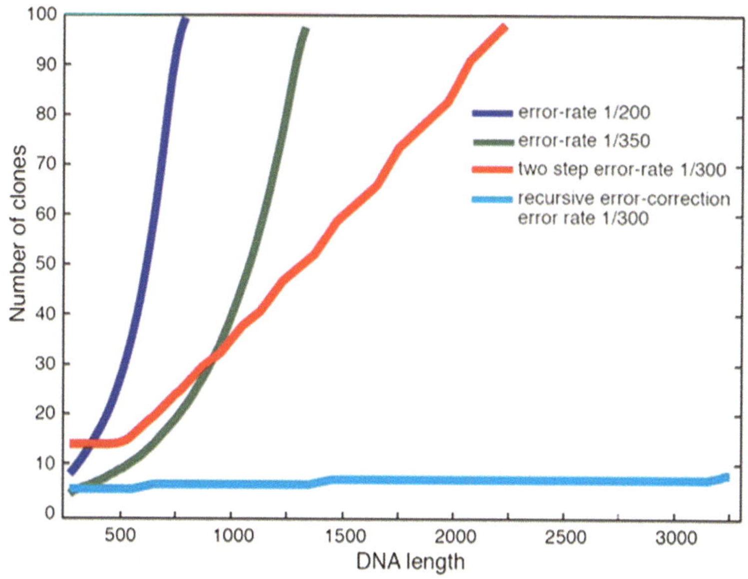


Fig. 3. Number of clones required to obtain error-free DNA molecules. Different DNA construction methods and error rates and their function of construct length are compared here in the plots shown from left to right: error rate of 1/200, error rate of 1/350, a two-step construction with error rate of 1/300 error rate, and recursive construction with an error rate of 1/300 and error correction. Figure reproduced from ref. 15.

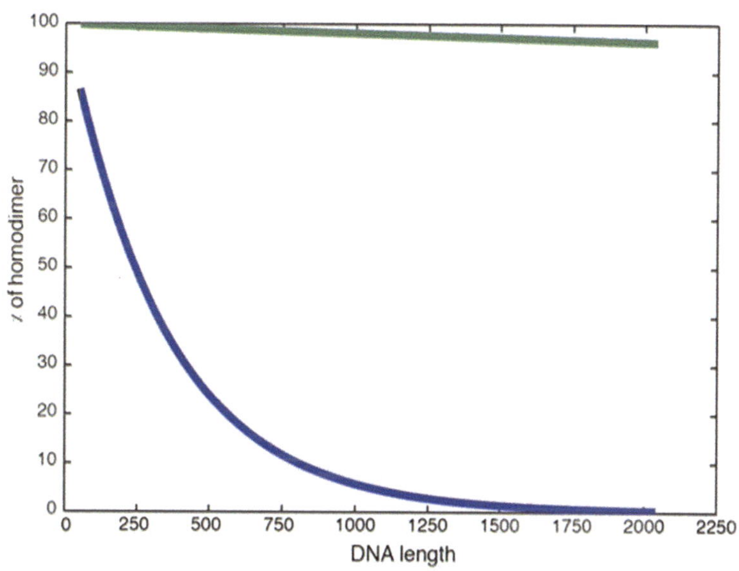


Fig. 4. Effect of overcycling on homoduplex formation. Shown here are two different PCR conditions. In the *lower plot*, a PCR is shown that has been allowed to cycle past the phase of 100% amplification efficiency, whereas the *upper plot* shows a PCR that has not been allowed to cycle past the phase of 100% amplification efficiency. Figure reproduced from ref. 15.

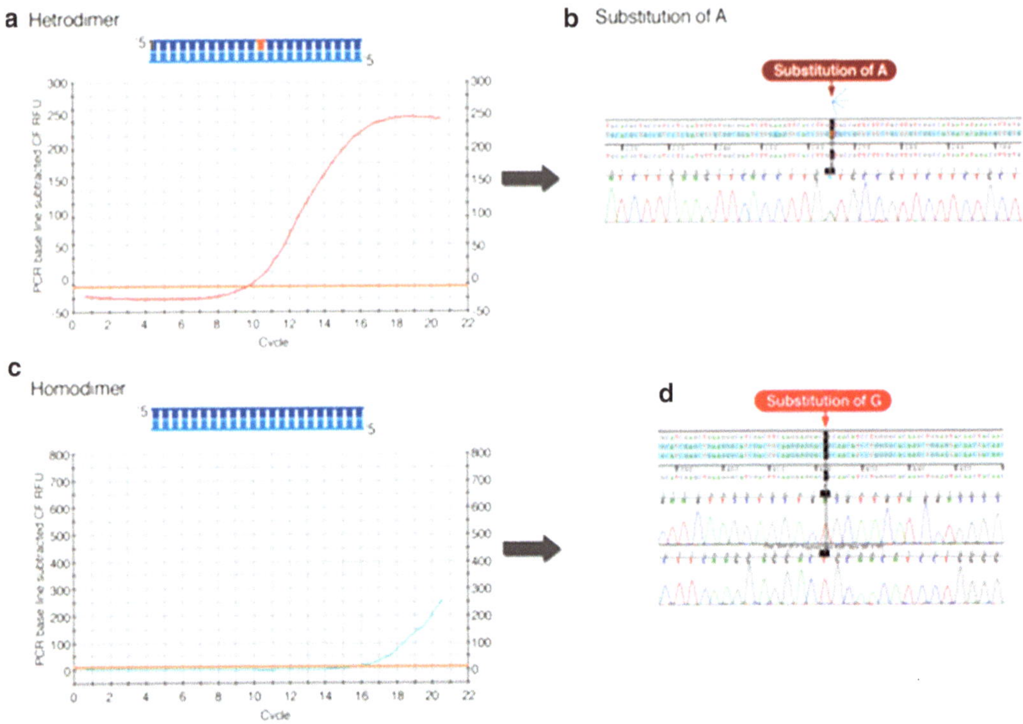


Fig. 5. Effects of PCR overcycling on sequence mutation. (**a**) Overcycling of the PCR past the phase of 100% amplification efficiency leads to the formation of heterodimers. (**b**) The sequencing chromatogram of a PCR-amplified substitution heterodimer shows two different base calls at the mutation, but these are not frame-shifted from the site of the mutation. (**c**) A PCR terminated before the end of 100% amplification efficiency generates homodimers and not heterodimers. (**d**) The sequencing chromatograms of homodimers are readable and not frame-shifted and always show a single base call at each base even if mutations (with respect to the target sequence). Figure reproduced from ref. 15.

protocol is executed as specified in Subheading 3.3 (preferably using real-time PCR).

6. Selection of positive amplification according to real-time PCR analysis or gel/capillary electrophoresis.
7. Sequencing of the true smPCR clones.
8. Computation of minimal cut from clones and reconstruction of the target molecule from the error-free segments, as shown in the manuscript and more generally in ref. 5.

### 3.2. Cloning

Fragments are cloned into the pGEM-T Easy Vector System 1. Vectors containing cloned fragments are transformed into JM109 competent cells and sequenced.

### 3.3. Single-Molecule PCR

smPCR is performed with hot-start Accusure for the longer mitochondrial DNA and with Taq polymerase for the GFP fragment. Template concentration is according to calculations described in (15) and dissolved in 5 μl $ddH_2O$; 10 pmol of the CA primer (see Subheading 3.12.2) dissolved in 10 μl $ddH_2O$. The PCR reaction

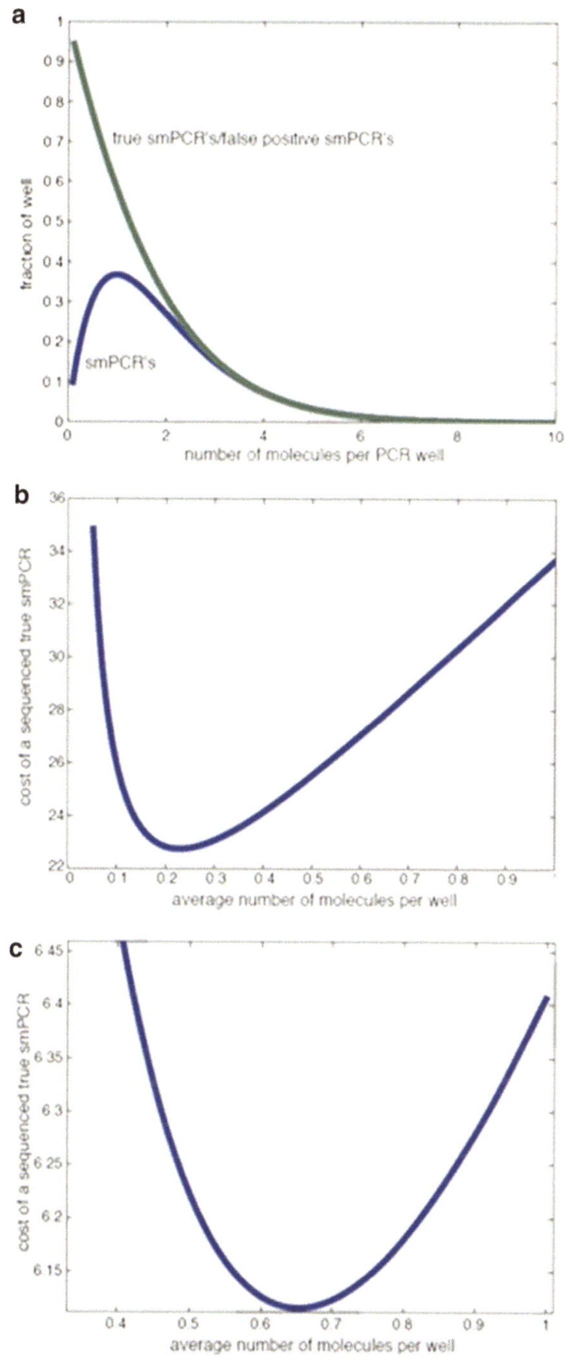


Fig. 6. Criteria for diluting of the PCR. (**a**) Shown here is the average number of molecules per PCR well versus the fraction of reactions. The *upper plot* shows those PCRs that have exactly one molecule out of all the PCRs performed. The *lower plot* shows the reactions that have exactly one molecule out of all the reactions that amplified (all PCRs excluding those yielding zero molecules, i.e., that did not amplify). *Panels* (**b**) and (**c**) show the costs associated with true smPCR sequencing. (**b**) Shown here is the cost of true smPCR sequencing, assuming it is 12 times higher than that of normal PCR. (**c**) Shown is the plot assuming that sequencing and PCR have equal cost. Higher ratios between sequencing and PCR costs (**b**) shift the minimum of the graph (i.e., minimal cost for obtaining a sequenced smPCR) to a lower number of molecules per well, and vice versa (see *lower plot* in **a**). Figure reproduced from ref. 15.

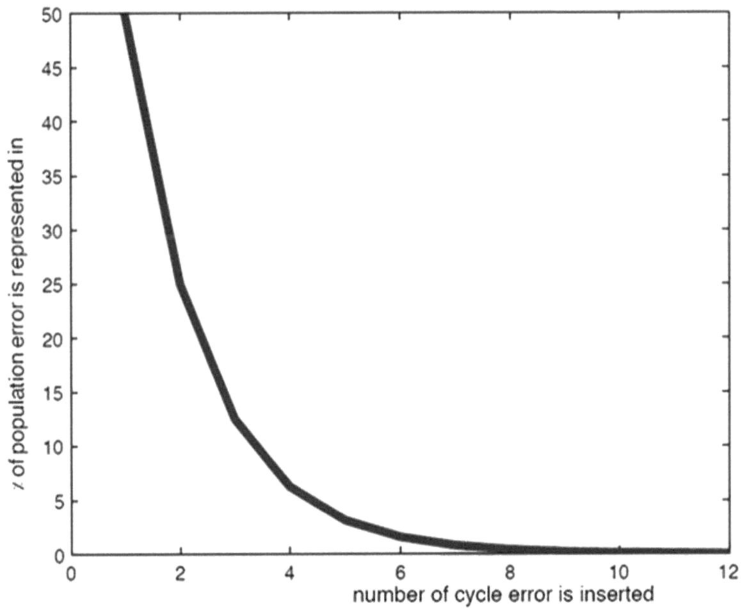


Fig. 7. Error incorporation. Shown here is a plot of the PCR cycle, in which an error was inserted versus the percentage of the population, in which this error is represented. Figure reproduced from ref. 15.

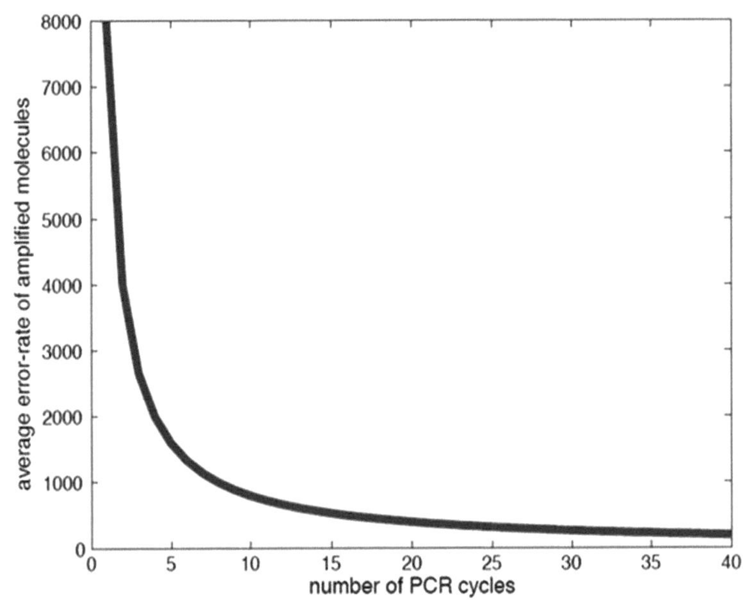


Fig. 8. Effect of PCR cycle numbers on error rates. The average error rate of DNA molecules that have been amplified from a single error-free molecule with PCR using Taq polymerase as a function of number of PCR cycles performed. Figure reproduced from ref. 15.

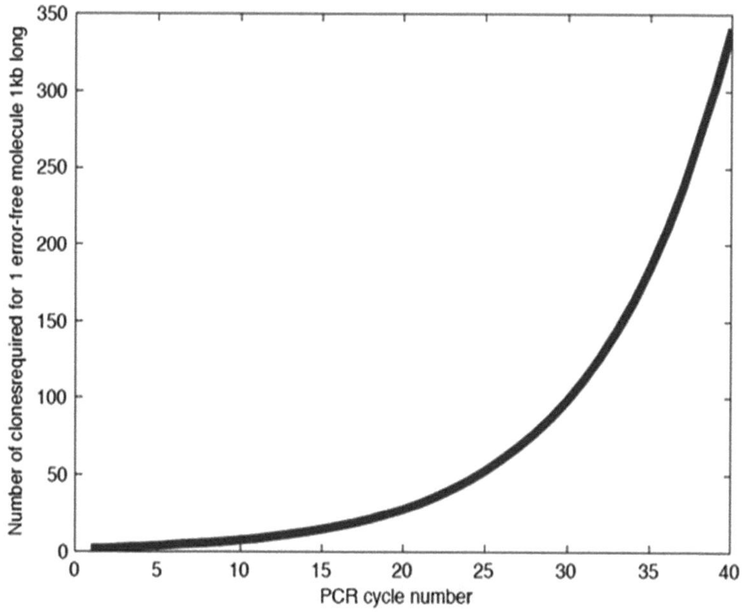


Fig. 9. Effect of the number of PCR cycles on the number of clones to be screened. Shown here is the minimum number of DNA molecules obtained from a smPCR that have to be screened to obtain at least one copy of the original single error-free molecule with a probability of 90% (using Taq polymerase) as a function of the number of amplification cycles. Figure reproduced from ref. 15.

contains 25 mM TAPS pH 9.3 at 25°C, 2 mM $MgCl_2$, 50 mM KCl, 1 mM β-mercaptoethanol, 200 μM each of dNTP, and 1.9 U AccuSure DNA Polymerase.

Real-time PCR (RT-PCR) thermal cycler program: enzyme activation at 95°C for 10 min, followed by 50 cycles of denaturation at 95°C for 30 s, annealing at $T_m$ of primers for 30 s, extension 72°C for 1.5 min/kb. It is important that the PCR reaction is prepared in a sterile environment using sterile equipment and uncontaminated reagents.

### 3.4. Pre-smPCR Recursive Construction and Error Correction

The core recursive construction and reconstruction (error correction) step requires four basic enzymatic reactions: phosphorylation, elongation, PCR, and Lambda exonucleation. These are described in the order of execution by the steps below.

#### 3.4.1. Phosphorylation

Phosphorylation of all PCR primers used by the recursive construction protocol is performed beforehand simultaneously, according to the following protocol: a total of 300 pmol of 5′ DNA termini in a 50-μl reaction containing 70 mM Tris–HCl, 10 mM $MgCl_2$, 7 mM dithiothreitol, pH 7.6 at 37°C, 1 mM ATP, and 10 U T4 polynucleotide kinase. Incubation is at 37°C for 30 min and inactivation at 65°C for 20 min.

*3.4.2. Overlap Extension Elongation Between Two ssDNA Fragments*

One to five picomoles of 5′ DNA termini of each progenitor in a reaction containing 25 mM TAPS pH 9.3 at 25°C, 2 mM $MgCl_2$, 50 mM KCl, 1 mM β-mercaptoethanol 200 μM each of dNTP, and 4 U Thermo-Start DNA polymerase. Thermal cycling program is as follows: enzyme activation at 95°C for 15 min, slow annealing 0.1°C/s from 95 to 62°C, and elongation at 72°C for 10 min.

*3.4.3. PCR Amplification of the Elongation Product with Two Primers, One of Which Is Phosphorylated*

A total of 1–0.1 fmol template, 10 pmol of each primer in a 25-μl reaction containing 25 mM TAPS pH 9.3 at 25°C, 2 mM $MgCl_2$, 50 mM KCl, 1 mM β-mercaptoethanol 200 μM each of dNTP, and 1.9 U AccuSure DNA Polymerase. Thermal cycling program is enzyme activation at 95°C for 10 min, followed by 20 cycles of denaturation 95°C, annealing at $T_m$ of primers, and extension at 72°C, each for 1.5 min/kb.

*3.4.4. Lambda Exonuclease Digestion of the PCR Product to Regenerate ssDNA*

One to five picomoles of 5′ phosphorylated DNA termini in a reaction containing 25 mM TAPS pH 9.3 at 25°C, 2 mM $MgCl_2$, 50 mM KCl, 1 mM β-mercaptoethanol, 5 mM 1,4-Dithiothreitol, and 5 U Lambda exonuclease. Thermal cycling program is enzyme activation at 37°C for 15 min, 42°C for 2 min and enzyme inactivation at 70°C 10 min.

### 3.5. Real-Time PCR

All PCRs are performed using the Bio-Rad's MyiQ Single-Color real-time PCR detection system with SyberGreen.

### 3.6. Fragment Analysis by Capillary Electrophoresis

Fragment analysis of PCR products is performed to single-base-pair resolution using an ABI analyzer and the LIZ500 (−250) size marker.

### 3.7. Calibration of smPCR Input DNA

The calibration experiment is performed to determine the dilution factor required to obtain an optimum DNA concentration in the smPCR. For this, RT-PCR amplification of the synthetic construct to be cloned is terminated within the exponential amplification phase. The terminated PCR is then diluted to different concentrations, and pools of 96 PCRs are generating using each of the dilutions as template. The ratio between amplified and nonamplified reactions is determined for each dilution pool. The dilution, which resulted in the desired amplification ratio (e.g., one positive well for every two negative wells is reasonable), is chosen for the required dilution in subsequent PCRs. A critical factor here is that the RT-PCR preceding the smPCR should always be terminated at the same stage at the beginning of the exponential amplification process, as determined by the RT-PCR curve. After this calibration, accurate dilutions for smPCR are made easy by terminating the PCR preceding the smPCR at the predetermined stage and preparing the predetermined dilution.

### 3.8. Chemical Oligonucleotide Synthesis

Oligonucleotides for all experiments were ordered from commercial providers with standard desalting.

### 3.9. DNA Purification

Manual DNA purification is performed with QIAGEN's MinElute PCR purification kit using standard procedures.

### 3.10. Recursive Construction Method

We apply "Divide and Conquer," the quintessential recursive problem-solving technique, to divide the target DNA sequence in silico into fragments short enough to be synthesized by conventional oligo synthesis, albeit with errors; these error-prone molecules are recursively combined in vitro, forming error-prone target DNA molecules; error-free parts of these molecules are identified, extracted, and used as new, typically longer and more accurate, inputs to another iteration of the recursive construction procedure. One execution of this procedure typically yields error-free molecules. Nevertheless, in principle, if errors remain, the entire process can be repeated until an error-free target molecule is formed.

### 3.11. Error Correction Method

In general, a composite object constructed from error-prone building blocks is expected to have a higher number of errors than each of its building blocks. However, if errors are randomly distributed among the building blocks and occur randomly during construction, and if several copies of an object are constructed, it is expected that few if not all of the error-prone copies would contain some error-free components with a certain minimal size. Moreover, based on the known rate and distribution of errors, we can predict a specific property of these error-free components, namely, the number of times they will occur in a given number of constructed objects. Furthermore, we can calculate the probability that a certain number of error-free components would collectively span the entire target object. Conversely (and more importantly), we can calculate the number of object copies (clones) required so that their error-free components span the entire target object with a desired probability. If such components could be identified and utilized from the faulty objects, they could be reused as building blocks for another recursive construction of the object.

Based on this observation, our recursive construction procedure can be reapplied to correct errors in synthetically constructed molecules as follows: error-free parts of the erroneous target DNA molecules are identified by cloning and sequencing and used as new, typically longer, inputs to the same recursive construction procedure. Since this construction starts from typically larger DNA building blocks that are error-free, the number of errors in the resulting reconstructed DNA is expected to decrease, possibly down to zero, eschewing additional screening of clones.

### 3.12. Minimal Cut

A cut in a tree is a set of nodes that includes a single node on any path from the root to a leaf. Let T be a recursive construction protocol tree and S a set of strings. We say that S covers T if there is a set of strings C such that every string in C is a substring of some string in S and C is a cut C of T. In such a case, we also say that S covers T with C.

Claim: if S covers T, then there is a unique minimal set C such that S covers T with C.

Proof: easy.

Error-free reconstruction algorithm: given an RC protocol T and a set of sequences (of molecular clones) S, find a minimal C such that S covers T with C.

Then, we lift C with PCR and do the recursive construction starting with C.

#### 3.12.1. Computing the Minimal Cut

We use a recursive approach for computing the minimal cut of a protocol tree. Each node in the tree represents a biochemical process with a product and two precursors. The algorithm starts with the root of the tree (target molecule) and for each node checks whether its product sequence exists with no errors in one of the clones. If such a clone exists, this product is marked as a new basic building block for reconstruction of the target molecule, and its primer pair and relevant clone (as template) are registered as its generating PCR reaction. If there is no clone which contains an error-free sequence of the node product, the reaction is registered as existing reaction in the new protocol, and the algorithm is recursively executed on the two precursors of the product. The output of such a protocol is a tree of reactions which comprises a minimal cut of the original tree. It contains leaves for which error-free products exist and that all its internal nodes are have no error-free clone that contain them. An automated program that utilizes these new error-free building blocks for recursive construction of the target molecule is generated for the robot.

#### 3.12.2. Sequences

CA Primer

CAACACACCACCCACCCAAC

>M1_Primer_1_smPCR_Adaptor

CAACACACCACCCACCCAACAGTAGATACAAGAGCAT ATTTTACTTC

>M1_Primer_2_smPCR_Adaptor

CAACACACCACCCACCCAACAAAACATAATTA TAACCTTACGGTCTG

>M1_Primer_3_smPCR_Adaptor

CAACACACCACCCACCCAACATAGATATTAAGAATAT CATTAATCCAGATATCCATGATAAAGGTAAAT

>M1_Primer_4_smPCR_Adaptor
CAACACACCACCCACCCAACACCTTTATCATGGATAT
CTGGATTAATGATATTCTTAATATCTATTGTTACAGC

## References


1. Bang D and Church GM (2008) Gene synthesis by circular assembly amplification. Nat Methods 5:37–39.
2. Carr PA, Park JS, Lee YJ, YuT, Zhang S and Jacobson JM (2004) Protein-mediated error correction for de novo DNA synthesis. Nucleic Acids Res, 32:e162.
3. Kodumal SJ, Patel KG, Reid R, Menzella HG, Welch M and Santi DV (2004) Total synthesis of long DNA sequences: synthesis of a contiguous 32 kb polyketide synthase gene cluster. Proc Natl Acad Sci USA 101:15573–15578.
4. Tian J, Gong H, Sheng N, Zhou X, Gulari E, Gao X and Church G (2004) Accurate multiplex gene synthesis from programmable DNA microchips. Nature 432:1050–1054.
5. Linshiz G. Yehezkel TB, Kaplan S, Gronau I, Ravid S, Adar R and Shapiro E (2008) Recursive construction of perfect DNA molecules from imperfect oligonucleotides. Mol Syst Biol 4:191.
6. Xiong AS, Yao QH, Peng RH, Duan H, Li X, Fan HQ, Cheng ZM and Li Y (2006) PCR-based accurate synthesis of long DNA sequences. Nat Protoc 1:791–797.
7. Xiong AS, Yao QH, Peng RH, Li X, Fan HQ, Cheng ZM and Li Y (2004) A simple, rapid, high-fidelity and cost-effective PCR-based two-step DNA synthesis method for long gene sequences. Nucleic Acids Res 32:e98.
8. Saiki RK, Gelfand DH, Stoffel S, Scharf SJ, Higuchi R, Horn GT, Mullis KB and Erlich HA (1988) Primer-directed enzymatic amplification of DNA with a thermostable DNA polymerase. Science 239:487–491.
9. Ohuchi S, Nakano H and Yamane T (1998) In vitro method for the generation of protein libraries using PCR amplification of a single DNA molecule and coupled transcription/translation. Nucleic Acids Res 26:4339–4346.
10. Nakano M. Komatsu J, Kurita H. Yasuda H, Katsura S and Mizuno A (2005) Adaptor polymerase chain reaction for single molecule amplification. J Biosci Bioeng 100:216–218.
11. Kraytsberg Y and Khrapko K (2005) Single-molecule PCR: an artifact-free PCR approach for the analysis of somatic mutations. Expert Rev Mol Diagn 5:809–815.
12. Lukyanov KA, Matz MV, Bogdanova EA, Gurskaya NG and Lukyanov SA (1996) Molecule by molecule PCR amplification of complex DNA mixtures for direct sequencing: an approach to in vitro cloning. Nucleic Acids Res 24:2194–2195.
13. Margulies M, Egholm M, Altman WE, Attiya S, Bader JS, Bemben LA, BerkaJ, Braverman MS, Chen YJ, Chen Z et al. (2005) Genome sequencing in microfabricated high-density picolitre reactors. Nature 437:376–380.
14. Shendure J, Porreca GJ, Reppas NB, Lin X, McCutcheon JP, Rosenbaum AM, Wang MD, Zhang K, Mitra RD and Churc,GM (2005) Accurate multiplex polony sequencing of an evolved bacterial genome. Science 309: 1728–1732.
15. Ben Yehezkel T, Linshiz G, Buaron H, Kaplan S, Shabi U and,Shapiro E (2008) De novo DNA synthesis using single molecule PCR, Nucleic Acids Res 36:e107.

# Part II

## Synthon Assembly

# Chapter 5

# SLIC: A Method for Sequence- and Ligation-Independent Cloning

**Mamie Z. Li and Stephen J. Elledge**

## Abstract

We describe here a method for sequence- and ligation-independent cloning (SLIC). SLIC uses an exonuclease, T4 DNA polymerase, to generate single-stranded DNA overhangs in insert and vector sequences. These fragments are then assembled in vitro and transformed into *Escherichia coli* to generate recombinant DNA of interest. SLIC inserts can also be generated by incomplete PCR (iPCR) or mixed PCR. As many as five inserts can be assembled in one reaction simultaneously with great efficiency using SLIC. SLIC circumvents sequence constraints for recombinant DNA using standard restriction enzyme-mediated cloning and previous ligation-independent cloning methods and provides a new approach for the efficient generation of recombinant DNA.

**Key words:** SLIC, Ligation-independent cloning, In vitro homologous recombination, iPCR, Sequence-independent cloning, Subcloning

## 1. Introduction

Since the invention of recombinant DNA technology (1–4), a wide variety of new cloning techniques such as the univector plasmid-fusion system (5), Gateway (6, 7), LIC (8, 9), and MAGIC (10) have evolved to allow various applications of synthetic biology. The univector plasmid-fusion system and MAGIC methods offer a seamless transfer of genes from one expression vector to another, but they lack a simple method for the initial transfer of the gene of interest into the original vector. Gateway can accommodate the initial transfer, but it requires expensive recombinases and specific sequences for recombination. Here we describe a sequence- and ligation-independent cloning (SLIC) technique (11). SLIC resembles LIC but has no sequence constraints in its homology region.

Jean Peccoud (ed.), *Gene Synthesis: Methods and Protocols*, Methods in Molecular Biology, vol. 852,
DOI 10.1007/978-1-61779-564-0_5, 

Thus, it eliminates any sequence restrictions that classical LIC requires and allows for insertion of any sequence into any other sequence with defined junctions.

Unlike some of the techniques described previously, SLIC relies on homologous recombination in vitro. Homologous recombination in vivo relies on a double-strand break, single-strand generation, homology searching, annealing, and gap repairing. There are two types of recombination in *E. coli* (12, 13), RecA-dependent and RecA-independent recombination (also called single-strand annealing), and both methods can be employed in vitro through SLIC. SLIC mimics homologous recombination observed in vivo by using an exonuclease (T4 DNA polymerase in the absence of dNTPs) to generate single-stranded DNA overhangs in insert and vector sequences in vitro. The length of the single-stranded DNA generated is controlled by the time of exonuclease treatment. These single-stranded overhangs are annealed in vitro with or without RecA and transformed into cells resulting in gap repair and generating recombinant DNA (Fig. 1). When a limited amount of DNA is used in the reaction (e.g., 3 ng), RecA is used to promote homology searching and annealing in vitro. However, RecA is unnecessary with sufficient amounts of DNA (e.g., ≥100 ng). Importantly, the homology need not be perfect for SLIC, and stretches of nonhomology on the ends of annealed sequences are trimmed back and removed in *E. coli* upon transformation. In addition, too much excision is also tolerated, providing a relatively wide margin of error in preparing the fragments unless they are very small fragments.

SLIC inserts can also be prepared by iPCR or mixed PCR by taking advantage of the fact that some of the PCR products in any reaction have a 5′ overhang due to incomplete DNA synthesis during later cycles of PCR. By denaturing and renaturing PCR products once after the cycles have been completed, we generate some inserts with proper single-stranded homologous ends suitable for SLIC with no additional enzymatic manipulations. SLIC inserts can also be prepared by mixing two PCR products, each having one of the two regions of homology needed for SLIC, and then denaturing and renaturing, which generate 25% correct overhangs (Fig. 2). A higher insert to vector ratio should be used if the insert is prepared by mixed PCR because only a quarter of these inserts have the correct overhangs. SLIC is so efficient that multiple inserts (up to 5) can easily be incorporated into a vector in one step and as many as ten has been accomplished in one step at a lower efficiency (Fig. 3).

Although SLIC works with a wide range of homology lengths, we typically use a 20-bp homology for routine subcloning. However, we recommend a 40-bp homology for subcloning with multiple inserts. SLIC is flexible with respect to the sequence

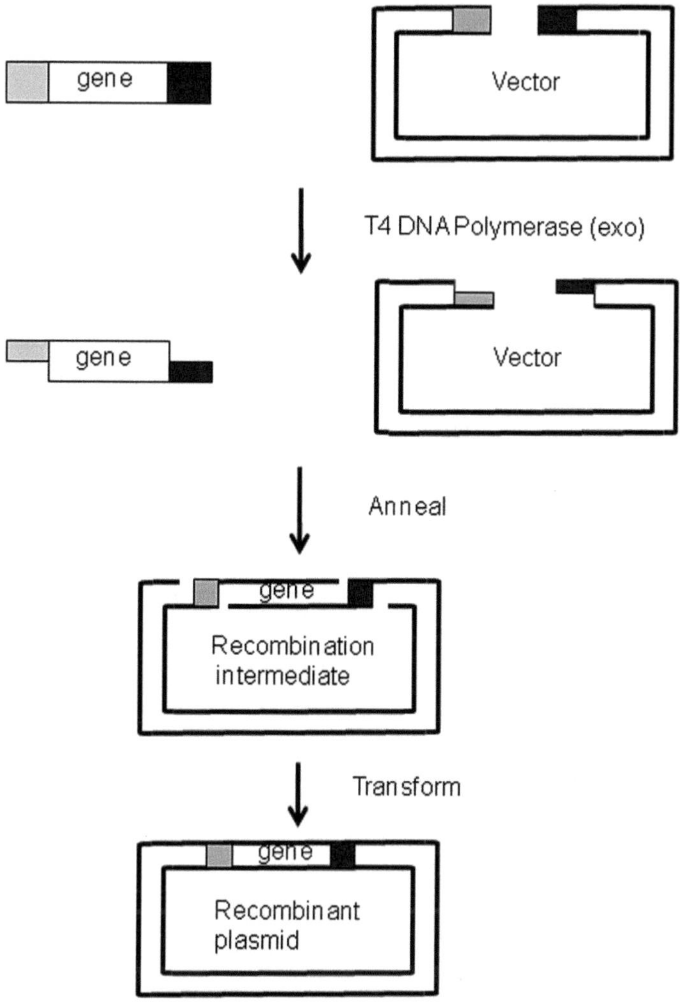


Fig. 1. In vitro homologous recombination. A schematic for production of recombinant DNA using SLIC.

junctions although incorporation of long inverted repeats in the homology region significantly reduces cloning efficiency, probably due to secondary structures in the homology region precluding annealing in trans. Nonhomology regions with up to 20 bp or longer at the ends can be assembled as long as the homology regions are single stranded.

In this chapter, we will discuss how to generate SLIC recombinant DNA, how to prepare SLIC inserts by iPCR and mixed PCR, and how to generate SLIC clones with RecA.

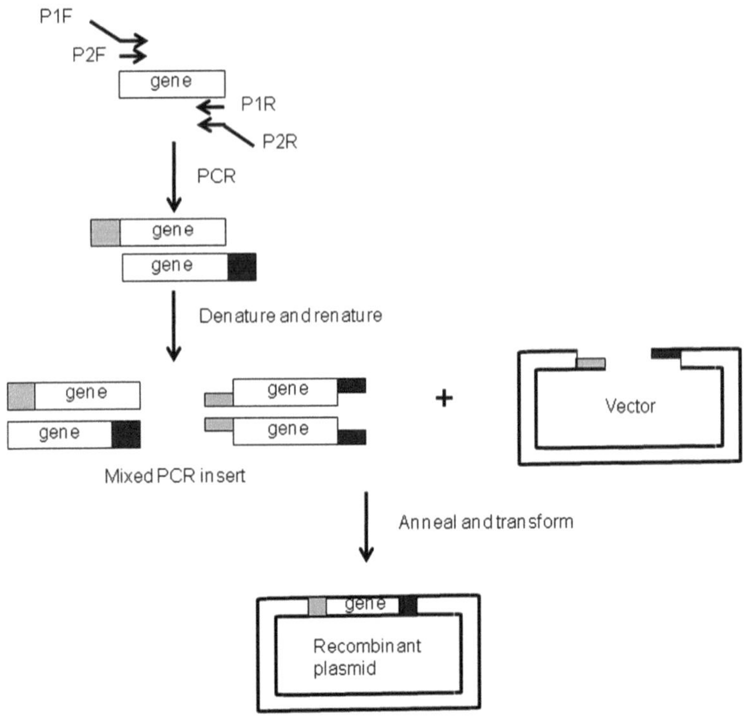


Fig. 2. Production of mixed PCR inserts. Two PCR products are independently prepared using the primer pair P1F-P1R and P2F-P2R. The two PCR products are mixed and denatured at 95°C for 5 min and slowly cooled to 22°C to renature. Since primers P1F and P2R are longer than primers P2F and P1R, the annealing of the fragments generates 5′ and 3′ overhangs, respectively. About 25% of the resulting mixtures have the correct overhangs.

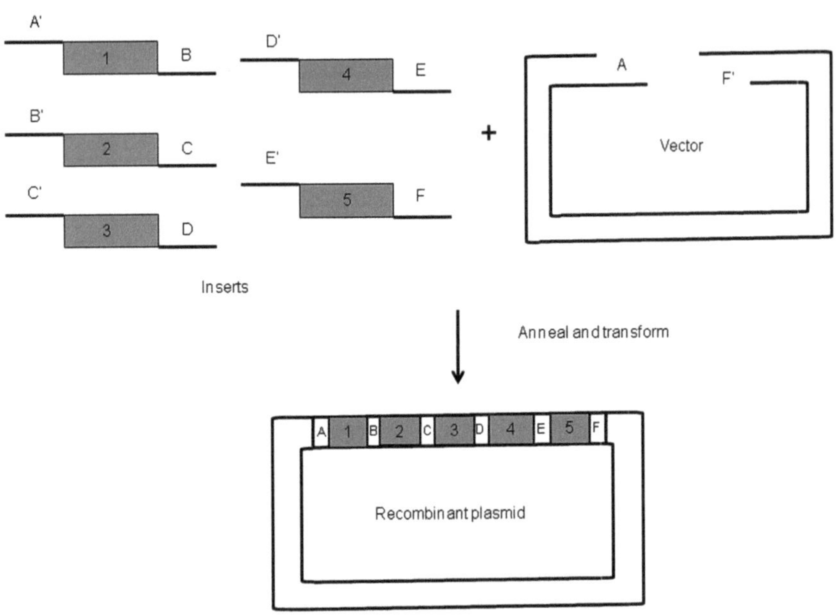


Fig. 3. SLIC with multiple inserts. A schematic illustrating the assembly of five inserts into a vector in a single step. The vector and inserts are prepared by T4 DNA polymerase proofreading excision to reveal the homologous overhangs, annealed in an equimolar ratio, and transformed into cells.

## 2. Materials

Restriction enzymes (New England Biolabs).

Agarose (Invitrogen).

QIAEX II gel extraction kit (Qiagen).

QIAquick PCR purification kit (Qiagen).

T4 DNA polymerase (New England Biolabs).

Taq DNA polymerase (Eppendorf).

RecA (Epicentre Biotechnologies).

dNTPs (Invitrogen).

DH5α competent cells.

DH10β competent cells.

BW23474 competent cells.

Antibiotics (Sigma) were used at these concentrations: ampicillin 100 μg/ml, kanamycin 50 μg/ml, chloramphenicol 30 μg/ml, carbenicillin 100 μg/ml.

SOC: 2% bacto tryptone, 0.5% yeast extract, 10 mM NaCl, 2.5 mM KCl, 10 mM $MgCl_2$, 10 mM $MgSO_4$, 20 mM glucose.

LB: 1% bacto tryptone, 0.5% yeast extract, 1% NaCl.

## 3. Methods

### 3.1. SLIC Subcloning Using T4 DNA Polymerase-Treated Inserts Without RecA

1. Digest 2 μg of vector DNA with restriction enzymes (see Note 1) and run on a 1% agarose gel. Excise the vector DNA from the agarose gel and extract the DNA using QIAEX II gel extraction kit. Determine the vector DNA concentration by measuring OD260 or running on a gel with abundance markers. If gel purification is not employed, linear vector can be prepared by phenol/chloroform extraction and ethanol precipitation and resuspended in TE (see Note 2).
2. Amplify the desired insert fragment of interest using Taq DNA polymerase or the thermostable polymerase of your choice. For Taq amplification, prepare a 100-μl PCR with 250 μM of each dNTP, 0.5 μM of each primer, and 2.5 U of Taq DNA polymerase and 10 ng of the template DNA of choice. Cycle as follows: 94°C for 45 s; 30 cycles of 94°C for 45 s, 54°C for 45 s, and 72°C for 1 min; and then a single 10-min incubation at 72°C after the 30 cycles. Add 20 units of DpnI to the 100 μl of PCR products after PCR and incubate at 37°C for 1 h in order to destroy any remaining template DNA (see Note 3). Purify PCR products by QIAquick PCR purification columns.

Determine the insert DNA concentration by measuring OD260 or running on a gel with abundance markers.

3. Treat 1 μg of vector and 1 μg of insert DNA with 0.5 U of T4 DNA polymerase (see Note 4) in NEBuffer 2 plus 100 μg/ml BSA in a 20 μl reaction at 22°C for 30 min (see Note 5). Stop the reaction by adding one-tenth volume of 10 mM dCTP (see Note 6) and place on ice or store in −20°C.
4. Set up a 10-μl annealing reaction using a 1:1 insert to vector molar ratio with 150 ng of a 3.1-kb vector (0.074 pmol) (see Note 7), 1× T4 DNA ligase buffer (NEB), an equimolar amount of insert, and water. Incubate the reaction at 37°C for 30 min and place on ice. Use the reaction immediately for the following steps.
5. Transform 5 μl of the annealed mixture into 100 μl of chemically competent *E. coli* cells, incubate on ice for 30 min, heat shock at 42°C for 45 s, return to ice for 2 min, add 0.9 ml of SOC, and recover at 37°C for 1 h.
6. Plate 100 μl of the transformation mix onto an LB plate containing the appropriate antibiotic and incubate at 37°C overnight.

### 3.2. SLIC Subcloning Using iPCR and Mixed PCR Inserts

1. Digest 2 μg of vector DNA with restriction enzymes (see Note 1) and run on a 1% agarose gel. Excise the vector DNA from the agarose gel and extract the DNA using QIAEX II gel extraction kit. Determine the vector DNA concentration by measuring OD260 or running on a gel with abundance markers. If gel purification is not employed, linear vector can be prepared by phenol/chloroform extraction and ethanol precipitation and resuspended in TE (see Note 2).
2. Amplify the desired insert fragment of interest using Taq DNA polymerase or the thermostable polymerase of your choice. For Taq amplification, prepare a 100-μl PCR with 250 μM of each dNTP, 0.5 μM of each primer, and 2.5 U of Taq DNA polymerase and 10 ng of the template DNA of choice. Cycle as follows: 94°C for 45 s; 30 cycles of 94°C for 45 s, 54°C for 45 s, and 72°C for 1 min; and then a single 10-min incubation at 72°C after the 30 cycles. Add 20 units of DpnI to the 100 μl of PCR products after PCR and incubate at 37°C for 1 h in order to destroy any remaining template DNA (see Note 3). Purify PCR products by QIAquick PCR purification columns. Determine the insert DNA concentration by measuring OD260 or running on a gel with abundance markers.
3. For preparation of the iPCR insert, heat the PCR product to 95°C for 5 min to denature, cool slowly to room temperature for 1 h to renature, dilute to 0.222 μM (~146 ng/μl for a 1-kb insert), and proceed to annealing reaction. For mixed PCR

inserts, mix the two PCR products in equal amounts and heat to 95°C for 5 min to denature, cool slowly to room temperature for 1 h to renature, dilute to 0.222 μM, and proceed to annealing reaction.

4. Treat 1 μg of vector with 0.5 U of T4 DNA polymerase (see Note 4) in NEBuffer 2 plus 100 μg/ml BSA in a 20 μl reaction at 22°C for 30 min (see Note 5). Stop the reaction by adding one-tenth volume of 10 mM dCTP (see Note 6) and place on ice or store in –20°C.
5. Set up a 10-μl annealing reaction using a 3:1 insert to vector molar ratio or higher with 150 ng of a 3.1-kb vector (0.074 pmol) (see Note 7), 1× T4 DNA ligase buffer (NEB), 0.222 pmol or more of insert, and water. Incubate the reaction at 37°C for 30 min and place on ice. Use the reaction immediately for the following steps.
6. Transform 5 μl of the annealed mixture into 100 μl of chemically competent *E. coli* cells, incubate on ice for 30 min, heat shock at 42°C for 45 s, return to ice for 2 min, add 0.9 ml of SOC, and recover at 37°C for 1 h.
7. Plate 100 μl of the transformation mix onto an LB plate containing the appropriate antibiotic and incubate at 37°C overnight.

### 3.3. SLIC Subcloning Using T4 DNA Polymerase-Treated Inserts with RecA

1. Digest 2 μg of vector DNA with restriction enzymes (see Note 1) and run on a 1% agarose gel. Excise the vector DNA from the agarose gel and extract the DNA using QIAEX II gel extraction kit. Determine the vector DNA concentration by measuring OD260 or running on a gel with abundance markers. If gel purification is not employed, linear vector can be prepared by phenol/chloroform extraction and ethanol precipitation and resuspended in TE (see Note 2).
2. Amplify the desired insert fragment of interest using Taq DNA polymerase or the thermostable polymerase of your choice. For Taq amplification, prepare a 100-μl PCR with 250 μM of each dNTP, 0.5 μM of each primer, and 2.5 U of Taq DNA polymerase and 10 ng of the template DNA of choice. Cycle as follows: 94°C for 45 s; 30 cycles of 94°C for 45 s, 54°C for 45 s, and 72°C for 1 min; and then a single 10-min incubation at 72°C after the 30 cycles. Add 20 units of DpnI to the 100 μl of PCR products after PCR and incubate at 37°C for 1 h in order to destroy any remaining template DNA (see Note 3). Purify PCR products by QIAquick PCR purification columns. Determine the insert DNA concentration by measuring OD260 or running on a gel with abundance markers.
3. Treat 1 μg of vector and 1 μg of insert DNA with 0.5 U of T4 DNA polymerase (see Note 4) in NEBuffer 2 plus 100 μg/ml

BSA in a 20-μl reaction at 22°C for 30 min (see Note 5). Stop the reaction by adding one-tenth volume of 10 mM dCTP (see Note 6) and place on ice or store in –20°C.

4. Set up a 10-μl annealing reaction using an equimolar 1:1 insert to vector molar ratio with 3 ng or less of a 3.1-kb vector (0.0015 pmol) (see Note 7), 1× T4 DNA ligase buffer with 1 mM ATP (NEB), the appropriate amount of insert, 20 ng of RecA protein (Epicentre Biotechnologies), and water. Incubate the reaction at 37°C for 30 min and place on ice. Use the reaction immediately for the following steps.
5. Transform 5 μl of the annealed mixture into 100 μl of chemically competent *E. coli* cells, incubate on ice for 30 min, heat shock at 42°C for 45 s, return to ice for 2 min, add 0.9 ml of SOC, and recover at 37°C for 1 h.
6. Plate 100 μl of the transformation mix onto an LB plate containing the appropriate antibiotic and incubate at 37°C overnight.

## 4. Notes

1. We typically use three- to fivefold more enzymes than the manufacturer suggests to achieve complete digestion.
2. Linear vector and inserts can be prepared by restriction enzyme cleavage or by PCR, but there is usually a cleanup step following to remove either the enzymes or dNTPs. Other than the QIAEX II gel extraction kit and QIAquick PCR purification column, one can use any other commercial and noncommercial method to purify the vector and insert DNA including phenol/chloroform extraction.
3. This step is to ensure that there is no carryover of the template DNA in the transformation. This is especially important if the antibiotic selection markers on the template and cloning vector are the same. It is necessary to clean up the PCR product with a commercial or a noncommercial method in order to avoid carrying over dNTPs.
4. The T4 DNA polymerase (NEB) should be diluted with 1× NEBuffer 2 buffer to 0.5 U/μl immediately before addition of the polymerase.
5. A reaction with 20 bp of homology should be incubated at least 30 min; a reaction with 40 bp of homology must be incubated at least 60 min under these conditions. One can try to use more units of T4 DNA polymerase per reaction for longer homologies such as over 75 bp or more. In that case, a T4

DNA polymerase treatment time course should be done to find the best condition for a particular homology length.

6. Any single dNTPs will work to stop excision by T4 DNA polymerase.
7. Depending on the size of vector, we have used between 150 and 300 ng of a vector for SLIC without RecA. If RecA is used, the amount of vector can be significantly reduced.

## References


1. Smith HO, Wilcox KW (1970) A restriction enzyme from Hemophilus influenzae. I. Purification and general properties. J Mol Biol 51:379–391
2. Danna K, Nathans D (1971) Specific cleavage of simian virus 40 DNA by restriction endonuclease of Hemophilus influenzae. Proc Natl Acad Sci USA 68:2913–2917
3. Cohen SN, Chang AC, Boyer HW, Helling RB (1973) Construction of biologically functional bacterial plasmids *in vitro*. Proc Natl Acad Sci USA 70:3240–3244
4. Backman K, Ptashne M (1978) Maximizing gene expression on a plasmid using recombination *in vitro*. Cell 13:65–71
5. Liu Q, Li MZ, Liebham D, Cortez D, Elledge SJ (1998) The univector plasmid fusion system, a method for rapid construction of recombinant DNA without restriction enzymes. Current Biology 8:1300–1309
6. Hartley JL, Temple GF, Brasch MA (2000) DNA cloning using *in vitro* site-specific recombination. Genome Res 10:1788–1795
7. Walhout AJ et al (2000) GATEWAY recombinational cloning: application to the cloning of large numbers of open reading frames or ORFeomes. Methods Enzymol 328: 575–592
8. Aslanidis C, de Jong PJ (1990) Ligation-independent cloning of PCR products (LIC-PCR). Nucleic Acids Res 18:6069–6074
9. Haun RS, Servanti IM, Moss J (1992) Rapid, reliable ligation-independent cloning of PCR products using modified plasmid vectors. Biotechniques 13:515–518
10. Li MZ, Elledge SJ (2005) MAGIC: An *in vivo* genetic method for the rapid construction of recombinant DNA molecules. Nat Gen 37:311–319
11. Li MZ, Elledge SJ (2007) Harnessing homologous recombination *in vitro* to generate recombinant DNA via SLIC. Nat Methods 4:251–256
12. Amundsen SK, Smith GR (2003) Interchangeable parts of the Escherichia coli recombination machinery. Cell 112:741–744
13. Kuzminov A (1999) Recombinational repair of DNA damage in *Escherichia coli* and bacteriophage lambda. Microbiol Mol Biol Rev 63: 751–813

# Chapter 6

# Assembly of Standardized DNA Parts Using BioBrick Ends in *E. coli*

**Olivia Ho-Shing, Kin H. Lau, William Vernon, Todd T. Eckdahl, and A. Malcolm Campbell**

## Abstract

Synthetic biologists have adopted the engineering principle of standardization of parts and assembly in the construction of a variety of genetic circuits that program living cells to perform useful tasks. In this chapter, we describe the BioBrick standard as a widely used method. We present methods by which new BioBrick parts can be designed and produced, starting with existing clones, naturally occurring DNA, or de novo. We detail the procedures by which BioBrick parts can be assembled into construction intermediates and into biological devices. These protocols are based on our experience in conducting synthetic biology research with undergraduate students in the context of the iGEM competition.

**Key words:** Synthetic biology, iGEM, BioBrick, Standardized parts, Undergraduate, Standardized assembly

## 1. Introduction

In 2003, Tom Knight and his colleagues developed the BioBricks method to standardize the assembly of DNA parts into devices and systems (1). The BioBricks method is convenient and cost-effective. More importantly, all BioBrick parts are compatible with each other. As a result, projects compliant with the BioBrick standard build on each other using interchangeable parts (2). The Registry of Standard Biological Parts and its associated online database (http://partsregistry.org) contain thousands of BioBrick parts built by undergraduates participating in the International Genetically Engineered Machines (iGEM) competition (3–6).

You can convert any DNA sequence into a BioBrick part by flanking the DNA with a BioBrick prefix and suffix (Fig. 1). The prefix contains the EcoRI, NotI, and XbaI restriction sites, while

Jean Peccoud (ed.), *Gene Synthesis: Methods and Protocols*, Methods in Molecular Biology, vol. 852,
DOI 10.1007/978-1-61779-564-0_6, © Springer Science+Business Media, LLC 2012

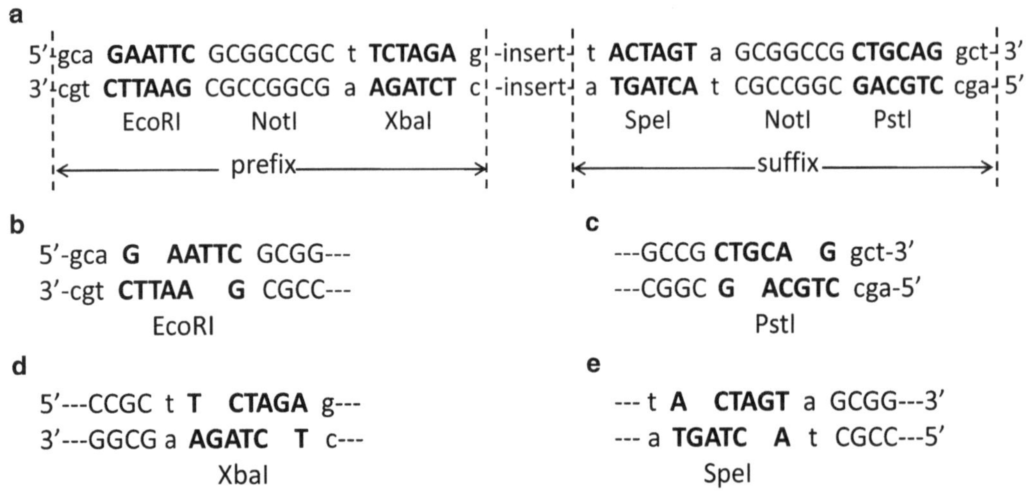


Fig. 1. Sequence of an uncut BioBrick part along with the sequences of cut restriction sites. (**a**) An uncut BioBrick part. (**b**)–(**e**) Cut restriction sites with overhangs for EcoRI, PstI, XbaI, and SpeI. Note that the XbaI and SpeI overhangs are complementary to each other.

the suffix contains the SpeI, NotI, and PstI restriction sites (1). It is important that you remove any sites for these enzymes found within the DNA part so that it will comply with the BioBrick standard. The EcoRI and PstI sites enable the transfer of BioBrick parts from one plasmid backbone to another. The restriction sites for XbaI and SpeI produce complementary sticky ends. Ligation of an XbaI sticky end with an SpeI sticky end produces a mixed site, or scar, that is not recognized by either XbaI or SpeI enzymes.

Synthetic biologists use several key terms for discussing the BioBrick system. A *part* is a basic unit with an indivisible biological function. Common examples are promoters, ribosome binding sites (RBS), coding sequences, and transcription terminators. A *construction intermediate* is formed when two or more parts are ligated together that do not constitute a functional device. A *device* is a combination of parts that carries out a biological function. Examples include reporters, inverters, and cell signal receivers. An *expression cassette* is a device that contains all the parts needed to express a gene. A common example includes a promoter, RBS, protein-coding gene, and a transcription terminator. A *composite part* is a general term for a device with more than one part in it.

Before designing and building a new device, you must answer a few questions. What is the purpose of the design? How many parts are needed? In what order should the parts get put together to make the assembly process most efficient? A simple example is related to the construction of the device shown in Fig. 2. If you followed method A, it would take you 6 days to rebuild the device with a new promoter. Method B would take you only 4 days since you would not have to repeat the ligation of coding + TT.

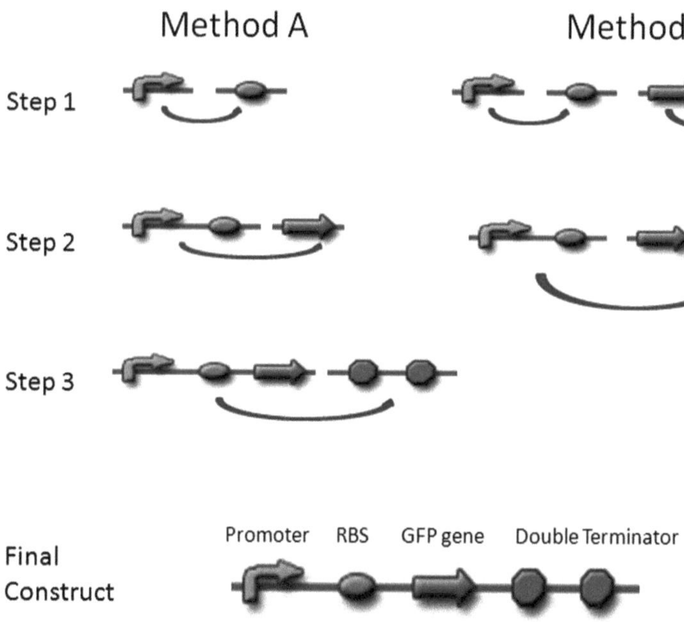


Fig. 2. This figure shows typical BioBrick representations of each of the parts with the *curved arrows* indicating ligation of the two parts together. A and B are two possible ways in which the final construct could be built. Method A shows a linear sequence of ligation in which it takes three sequential ligations to finish, but it produces intermediates that could be used in other devices. Method B shows how some ligations are performed simultaneously to reduce the number of days before the construction is completed.

Therefore, the general strategy illustrated by Method B reduces the time required to rebuild second-generation devices.

Assembly of parts using the BioBrick standard permits you to build from either the 5′ or 3′ end of a desired construct. For example, you can build the construct in Fig. 2 from promoter to terminator or terminator to promoter. The expression cassette shown is for green fluorescent protein (GFP) and consists of a promoter, RBS, GFP coding sequence, and transcriptional terminator (TT). The fastest way to build a GTP expression cassette is to ligate the promoter and RBS together while simultaneously ligating the GFP coding DNA with the terminator. Once the two halves are assembled, they can be ligated to build the final device. Alternatively, the BioBrick standard of assembly allows you to build one intermediate part (e.g., RBS + GFP + TT) and then add to the intermediate each of several different promoters to evaluate the strength of a set of promoters. Once two parts have been put together by the BioBrick assembly method, they cannot be taken back apart.

Because of the flexibility of the BioBrick system, standardized parts can be assembled de novo or from DNA isolated from nature. In order to standardize a new part, you must add BioBrick ends to the 5′ and 3′ ends of the new DNA sequence. If the part is being synthesized de novo, utilizing one of the growing number of companies that will manufacture genes (7), the requested sequence must be flanked by a BioBrick prefix and a suffix and devoid of internal restriction enzymes sites found in BioBrick ends. Sequences of interest within a genome, plasmid, or an existing part can be used as template for PCR with the BioBrick ends added to the primers.

## 2. Materials

### 2.1. Cleaning DNA

1. 3 M Sodium acetate: 40.8 g sodium acetate·3$H_2O$, 80 mL $H_2O$. Store at room temperature.
2. Ethanol: (a) 100% ethanol (b) 70% ethanol: 700 mL with 300 mL $H_2O$.

### 2.2. Minipreparation of Plasmid DNA

1. Wizard Plus SV Minipreps kit (Promega, Madison, WI).
2. LB media (low salt): 10 g tryptone, 5 g yeast extract, 5 g NaCl, 200 μL 5 M NaOH. Add water up to 1 L. Autoclave (see Note 1).

### 2.3. Enzyme Digestion of BioBrick Parts

1. Buffer H (Promega): 90 mM Tris–HCl pH 7.5, 10 mM $MgCl_2$, 50 mM NaCl.
2. Low buffer: 10 mM Tris–HCl pH 7.5, 10 mM $MgCl_2$, 0.1 mg/mL BSA, 50 mM NaCl.
3. Medium buffer: same as low buffer, but with 100 mM NaCl.
4. Restriction enzymes: EcoRI, XbaI, SpeI, and PstI (Promega, Madison, WI).

### 2.4. Gel Electrophoresis

1. Agarose, low EEO (Promega, Madison, WI).
2. TBE buffer: Prepare 5× stock solution with 54 g tris base, 27.5 g boric acid, 20 mL 0.5 M EDTA. Make up to 1 L with $H_2O$. Dilute 100 mL with 900 mL $H_2O$ for use. Use at 0.5× working concentration.
3. 1% Ethidium bromide (EtBr; Fisher, Pittsburgh, PA). EtBr is mutagenic so handle with care.
4. DNA loading dye (10×): 5 mL glycerol, a 0.2% w/v of bromophenol blue, 0.2% w/v xylene cyanol FF. Make up to 10 mL with $H_2O$. Store at 4°C. Dilute 100 μL with 900 μL $H_2O$ for use.
5. 1 kb DNA ladder (Invitrogen, Carlsbad, CA).

### 2.5. Gel Purification

1. NucleoSpin® Extract II gel extraction kit (Macherey-Nagel, Düren, Germany).

### 2.6. Ligation

1. 2× Rapid Ligation Buffer (Promega, Madison, WI).
2. T4 DNA ligase (Promega, Madison, WI).

#### 2.7. Transformation

1. Z-Competent™ *E. coli* cells (Zymo Research, Orange, CA), store at ≤70°C.
2. SOC medium: 20 g tryptone, 5 g yeast extract, 0.5 g NaCl, 200 μL 5 M NaOH, 10 mL 250 mM KCl (1.86 g KCl in 100 mL $dH_2O$). Add water up to 1 L. Autoclave for about 20 min. After cooling, add 5 mL sterile 2 M $MgCl_2$ (19 g $MgCl_2$ in 100 mL $dH_2O$).

#### 2.8. Colony PCR

1. GoTaq® 2× Green Master Mix (Promega, Madison, WI).

#### 2.9. Glycerol Stocks of E. coli Cells

1. CryoTube vials (Nunc, Roskilde, Denmark).
2. Sterile glycerol: Autoclave glycerol and store at room temperature.

#### 2.10. Single-Stranded Oligo Assembly

1. 10× Annealing buffer: 1 M NaCl, 100 mM Tris–HCl (pH 7.4).

## 3. Methods

#### 3.1. Minipreparation of Plasmid DNA

1. Grow 2–4 mL overnight culture for each miniprep. During incubation, culture tubes are shaken at 400 g and slanted to aerate the media.
2. Pour the contents of each culture tube into a microfuge tube.
3. Centrifuge the microfuge tubes for 2 min at 13,000 g.
4. Pour the liquid from the microfuge tubes, leaving the pellets. Gently shake the tubes to remove the remaining liquid. If there is more liquid culture, steps 2–4 can be repeated.
5. Mix each pellet with 250 μL of cell resuspension solution. Fully resuspend the cells by pipetting the solution up and down until the pellet is completely broken up and mixed with the solution.
6. Mix the contents of each microfuge tube with 250 μL cell lysis solution and invert several times.
7. To each microfuge tube, mix in 10 μL alkaline protease solution and invert several times. The tubes are incubated at room temperature for 3 min.
8. Centrifuge the tubes at 13,000 g for 10 min.
9. Insert the spin columns into the 2-mL collection tubes.
10. Use a micropipette to transfer the supernatant from each tube to the spin columns.
11. Centrifuge the spin columns at 13,000 g for 1 min. Discard the flow-through and reinsert the spin columns into the collection tubes.

12. Fill the spin columns with 750 μL wash solution (containing ethanol).
13. Centrifuge the spin columns at 13,000 g for 1 min. Discard the flow-through and reinsert the spin column into the collection tubes.
14. Fill the spin columns with 250 μL wash solution (with ethanol).
15. Centrifuge the spin columns at 13,000 g for 1 min. Discard the flow-through and insert the spin column into a clean 1.5-mL microfuge tube.
16. Fill the spin columns with 50–100 μL of nuclease-free water.
17. Centrifuge the spin columns at 13,000 g for 1 min.
18. Quantitate the plasmid DNA in the microfuge tube and store it at −20°C or use immediately (see Note 1).

### 3.2. Enzyme Digestion of BioBrick Parts

1. If you are digesting with EcoRI and PstI to verify the size of an insert, use 12 μL of miniprep DNA (see Subheading 3.1), or at least 400 ng. If digesting to make a vector or an insert, the amount of DNA to digest depends how much DNA you need for your ligation (see Notes 2 and 3).
2. Mix the DNA with 2 μL of the appropriate buffer (see Table 1) in a 500-μL microfuge tube. Add 1 μL of each enzyme (total enzyme volume cannot exceed 10% of reaction volume). Increase the volume to 20 μL with $dH_2O$.
3. For size verification, incubate the reaction at 37°C for at least half an hour. For inserts and vectors, incubate the reaction for at least 3 h. For maximum digestion, incubate the reaction overnight.

**Table 1**
**Salt conditions for double digestion. Useful enzyme combinations for digesting BioBrick parts with their optimal buffer and what the digestion produces. All five reactions are optimal at 37°C**

| Restriction enzymes | | Buffer | Product |
|---|---|---|---|
| EcoRI | XbaI | Low | Front vector |
| EcoRI | SpeI | Low | Front insert |
| SpeI | PstI | Medium | Back vector |
| XbaI | PstI | Low | Back insert |
| EcoRI | PstI | Buffer H | Whole insert |

### 3.3. Cleaning DNA

1. If the volume of the DNA is less than 200 μL, bring the volume up to 200 μL with sterile $dH_2O$ (see Subheading 3.2).
2. Add one-tenth volume of 3 M sodium acetate to the DNA solution and mix.
3. Add two volumes of 4°C 100% ethanol and vortex for 10 s. Put the tube in a −80°C freezer for 30 min (or overnight in a −20°C freezer).
4. Spin the tube in a microcentrifuge at 13,000 g for 10 min. Pour the ethanol out, keeping the pellet.
5. Wash the pellet with 500 μL of 4°C 70% ethanol. Gently roll the tube. Pour off the ethanol.
6. Dry the pellet in a centrifugal evaporator (SpeedVac).
7. Resuspend the DNA in 20 μL $dH_2O$ (adjust volume as necessary).

### 3.4. Gel Electrophoresis

1. Prepare a gel of appropriate agarose concentration (use the web tool at http://gcat.davidson.edu/iGEM08/gelwebsite/gelwebsite.html).
2. Run DNA samples (see Subheading 3.2) on the gel until there is adequate separation between the desired piece of DNA and the DNA that it was cut from (see Note 4).

### 3.5. Gel Purification

1. Place the gel (see Subheading 3.4) under UV light at an intensity just high enough to visualize the bands (see Note 5). Cut out the bands containing the insert and the vector to purify the DNA (see Note 6).
2. Place the gel slice in a 1.5-mL microfuge tube and weigh it. Add two volumes of Buffer NT to one volume of gel (100 mg = 200 μL). For gels >2% agarose, double the volume of Buffer NT.
3. Incubate the gel at 50°C for 5–10 min until the gel slice is completely dissolved. Vortex the tube every 2–3 min to speed up the dissolving process.
4. Place a spin column in one of the provided 2-mL collection tubes.
5. Place a NucleoSpin® column into a collection tube. Pipette the DNA solution onto the column. Centrifuge the DNA solution at 13,000 g for 1 min. The maximum volume the column can hold is 800 μL, so repeat this step using the same column if the volume is larger than that.
6. Discard flow-through from the previous step and place the column back in the collection tube.
7. Wash the DNA in the column by applying 600 μL of buffer NT3. Centrifuge the column for 1 min at 13,000 g.

8. Discard the flow-through and spin the column for 2 additional minutes to dry the column.
9. Place the spin column in a clean 1.5-mL microfuge tube.
10. To elute the DNA, add 10–30 μL of buffer EB to the center of the white matrix. Allow the column to sit for 1 min and then centrifuge it at 13,000 g for 1 min.
11. Quantify the DNA, which is eluted in the flow-through.

### 3.6. Ligation

1. Place 50 ng of digested vector (see Subheadings 3.3 or 3.5), 5 μL of ligation buffer, and 1 μL of T4 ligase into a 500-μL microfuge tube. The amount of insert to add is calculated from the following formula:

$$\text{ng of insert} = (2)(\text{bp insert})(50\,\text{ng linearized plasmid})/ (\text{size of plasmid in bp}).$$

Add water to increase the final volume to 10 μL (see Note 7).
2. Prepare both a positive ligation mixture that contains the digested vector and insert as well as a negative ligation mixture that contains only the digested vector. Add more water to the negative ligation mixture to prepare equal volumes.
3. Leave the ligation mixture at room temperature for 5 min, and then use it directly for transformation of *E. coli* competent cells, or store it by freezing until transformation.

### 3.7. Transformation

1. Prewarm culture plates to increase the drying rate of plated cells. The culture plates should contain the appropriate antibiotic for the transforming plasmid.
2. Store Z-Competent™ cells at −70°C or colder. Thaw a 100-μL tube of Z-Competent™ cells for 5 min on ice. At the same time, cool the tubes containing the ligation mixtures on ice.
3. Very gently add 25–50 μL of Z-Competent™ cells to each ligation mixture of 10 μL (see Subheading 3.6).
4. Let the mixtures incubate on ice for 5 min.
5. Add SOC media with no antibiotic to a final volume of 60–100 μL/tube. For plasmids using the ampicillin resistance marker, the cells will begin repairing their cell walls immediately and are ready to be plated. For plasmids with other antibiotic resistance markers, incubate without shaking for 20 min before plating.
6. Spread the cells on culture plates containing the appropriate antibiotic. Let the plates incubate overnight until colonies are visible and large enough to pick individually.
7. If transformation with Z-Competent™ cells is unsuccessful, traditional heat-shock transformation or electroporation with a different brand of competent cells may yield higher efficiencies.

### 3.8. Colony PCR

1. Compare your positive ligation plate to your negative ligation plate to estimate the number of background negative colonies (see Subheading 3.7). Plan to screen an appropriate amount so that it is highly probable at least one of the positive colonies will have the correct insert size given the amount of background negative colonies. Screen at least one negative colony also.
2. To conduct PCR, you will need a forward and reverse primer specific to the plasmid being screened. The primers should amplify the region where the insert was added (see Note 8).
3. For each colony to be picked, prepare a PCR tube with the following mixture: 12 μL 2× Green Master Mix, 10 μL $dH_2O$, 1 μL (20 pmol) forward primer, 1 μL (20 pmol) reverse primer.
4. Use a micropipette tip to pick a single colony off of the culture plate. Place the tip into a labeled PCR tube and mix by pipetting up and down.
5. Remove 1 μL of the PCR mixture and place it in a labeled test tube containing 200 μL culture media with antibiotic to reserve some of the cells from the colony to grow later. Incubate these cultures at 37°C.
6. Conduct the following PCR cycle: 95°C for 10 min, followed by 20 cycles of 95°C for 15 s, 46°C (or appropriate annealing temperature for your primers) for 15 s, 72°C for 60 s/kb of DNA of the expected size for a successful DNA ligation.
7. Run the reaction products on an agarose gel (see Subheading 3.4) appropriate for the size of the amplified product. You can use our gel optimization tool to choose the appropriate percent agarose (http://gcat.davidson.edu/iGEM08/gelwebsite/gelwebsite.html). Colonies containing unsuccessful ligations will have the same insert size as the negative control colony (see Notes 9 and 10). Successful ligations will be bigger than the negative control insert size (Fig. 3). For colonies that show the expected insert size, save the corresponding culture and discard all unsuccessful colonies.
8. The BioBrick part can be further verified by miniprepping (see Subheading 3.1) and digesting (see Subheading 3.2) and gel electrophoresis (see Subheading 3.4), or by sequencing (see Note 11). After successful ligation and transformation, the part can be used or manipulated more to construct more new parts.

### 3.9. Glycerol Stocks of E. coli Cells

1. Grow a 2-mL culture of a particular cell type (see Subheading 3.8) overnight for each frozen stock that is needed.
2. Microwave sterile glycerol for 30 s. Do not mix the glycerol so that the top portion remains as hot as possible.
3. Cut a 200-μL pipette tip with a clean razor blade to make a larger opening at the tip.

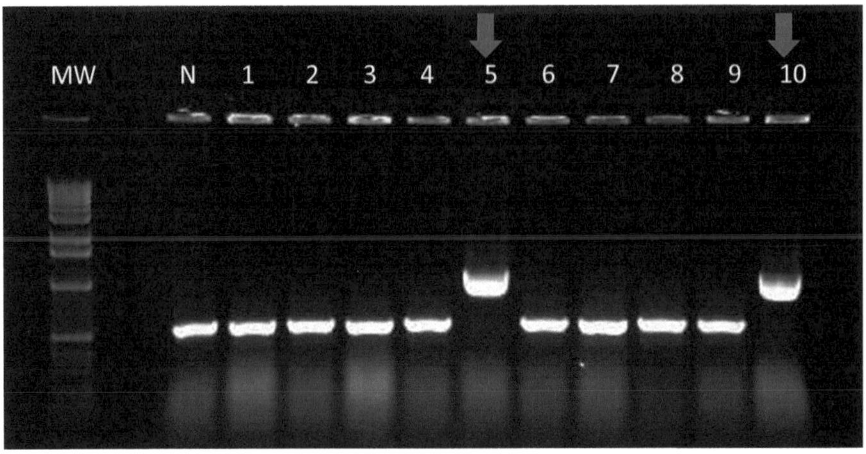


Fig. 3. After 20 cycles of amplification, colony PCR products for each colony were run on a 0.9% agarose gel. Selected colonies were numbered, and the negative colony labeled "N." Colonies #5 and #10 have an insert size bigger than the negative control and roughly the expected insert size for the DNA part (1,100 bp).

4. Label Nunc cryotubes as appropriate and add 150 μL of hot glycerol.
5. Allow the glycerol in the tubes to cool at room temperature for 1 min. Add 850 μL of bacterial culture to each tube, making 15% glycerol mixtures.
6. After putting the caps on, shake the tubes vigorously to ensure that the glycerol mixes evenly with the bacterial culture.
7. Immediately after step 6, put the tubes into a –80°C freezer.

### 3.10. Connecting Two BioBrick Parts

1. Obtain plasmid DNA for both parts (see Note 1).
2. Digest the plasmid DNA (see Subheading 3.2) for the two parts with the appropriate enzymes for the desired ligation (see Fig. 2, Table 1, and Note 12).
3. Run the digested DNA on a gel (see Subheading 3.4) until there is adequate separation between the desired piece of DNA and the DNA that it was cut from. You can use our online tool to optimize the percent agarose for your gel (http://gcat.davidson.edu/iGEM08/gelwebsite/gelwebsite.html).
4. Place the gel under UV light at an intensity just high enough to visualize the bands (see Note 5). Cut out the bands containing the insert and the vector with a razor and purify the DNA (see Subheading 3.5; see Note 6).
5. Ligate the purified insert and vector together (see Subheading 3.6).
6. Use the ligation mixture to transform into competent *E. coli* cells (see Subheading 3.7).

7. After overnight incubation of the transformed cells, pick colonies from the ligation and the control plates for colony PCR to quickly test if the ligation was successful in any of the colonies (see Subheading 3.8).
8. Grow clones with positive results from the colony PCR overnight in tubes of 2 mL LB media containing the appropriate antibiotic.
9. Miniprep the overnight cultures to obtain plasmid DNA. Digest the plasmid DNA with EcoRI and PstI as a conclusive test for whether the ligation was successful.
10. Store cells that have been confirmed to have successful ligations in glycerol stocks at −80°C (see Subheading 3.9).

### 3.11. Cloning a New Part into BioBrick Ends

There are three main approaches to the fabrication of a new BioBrick part. The easiest but most expensive method is to design the sequence of the part, add the BioBrick prefix and suffix sequences, and have the part synthesized and cloned (see Table 2). Companies that provided this de novo gene synthesis will generate the needed DNA oligonucleotides, assemble them, and provide the part cloned into a plasmid. The second approach is to add BioBrick ends to an existing DNA sequence using custom PCR primers (Fig. 4). The template used can be an existing clone or DNA from a natural source. The third approach is to assemble a part yourself using overlapping single-stranded oligonucleotides. Building parts from oligos work well with relatively small DNA parts, typically less than 300 base pairs long. In the synthesis of all new BioBrick parts, it is important to maintain the integrity of the DNA sequence and the BioBrick ends. Therefore, the DNA sequence of the part itself cannot contain restriction sites for NotI, EcoRI, XbaI, SpeI, or PstI. If a designed sequence contains one or more of these restriction sites, the DNA should be modified to remove the restriction sites. If the sequence is derived from an existing clone or a natural source, the offending restriction sites must be removed by mutagenesis. After the new part has been sequence verified, it can be manipulated through digestion, ligation, and transformation in order to assemble it with other BioBrick parts.

**Table 2**
**BioBrick prefix and suffix to flank de novo part**

| | |
|---|---|
| BioBrick prefix (if insert begins with ATG) | GAATTCGCGGCCGCTTCTAG |
| BioBrick prefix | GAATTCGCGGCCGCTTCTAGAG |
| BioBrick suffix | TACTAGTAGCGGCCGCTGCAG |

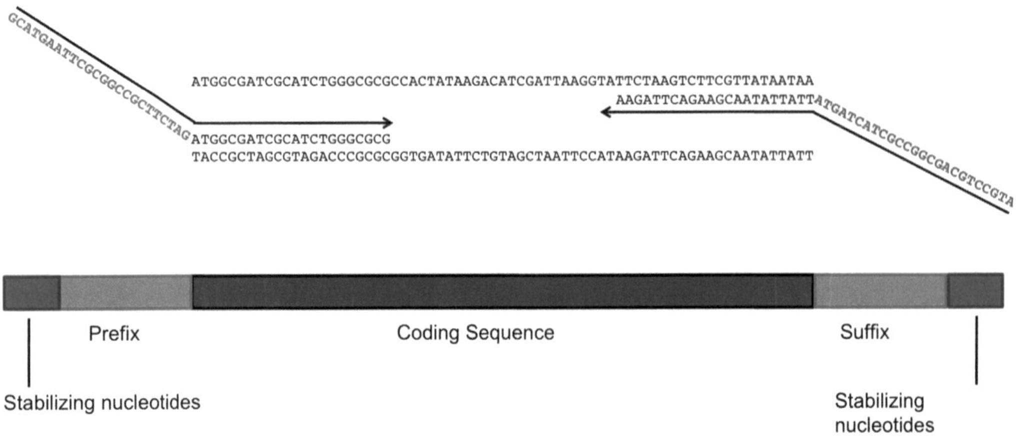


**Forward Primer:**
5′– GCATGAATTCGCGGCCGCTTCTAGATGGCGATCGCATCTGGGCGCG –3′

**Reverse Primer:**
5′– ATGCCTGCAGCGGCCGCTACTAGTATTATTATAACGAAGACTTAGAA –3′

Fig. 4. Designing primers for addition of BioBrick ends by PCR amplification. The given forward and reverse primers would be designed and assembled in order to add the BioBrick prefix and suffix to the given coding sequence. Four stabilizing nucleotides allow restriction enzymes needed to cut the BioBrick ends to bind securely to the target site. We like to use GCAT here, but any sequence that is not a restriction site will suffice.

If you make the part using PCR, the forward primer for the part must include the BioBrick prefix, the first 20–25 nucleotides of the part sequence, and four extra base pairs (see Fig. 4) on the 5′ end so that the EcoRI restriction enzyme can bind and cut the restriction site (see Notes 3 and 8). The reverse primer for the part must include the last 20–25 nucleotides of the template, the BioBrick suffix, and four extra base pairs. Reverse primers are the reverse complement of the template strand (see Fig. 4). Try to design the primers so that their melting temperatures in the PCR mix to be used are within 10°C of each other. You can calculate the melting temperatures using the following web tool: http://www.promega.com/biomath/calc11.htm. Compare the salt-adjusted melting temperatures. Lengthening (adding more base pairs that complement the template sequence) or shortening the primers will increase or decrease the melting temperatures, respectively. The desired DNA template can be part of a plasmid or chromosomal DNA. While cell extract containing the desired DNA can be used as template, you might want to purify the DNA so its concentration can be measured before beginning the PCR process. One to five nanograms of template DNA will be used in the PCR amplification. Perform PCR according to a standard amplification protocol. The final amplified product is ready for digestion with EcoRI and PstI to clone the new standardized part into a plasmid vector (see Subheading 3.2).

To produce a part of 300 bp of DNA or less, use oligo assembly:

1. Enter the desired 5′–3′ forward sequence into The Oligator A program at: http://gcat.davidson.edu/igem10/index.html. This tool generates single-stranded DNA oligonucleotide sequences that have very similar melting temperatures in their overlapping sequences. The overlap length for the oligos must be at least 20 base pairs. Limit the length of your oligos to 70 bases or shorter (see Note 13).
2. If you are cloning into EcoRI and PstI sticky ends of a BioBrick plasmid, check the box for adding prefix and suffix, choose EcoRI for the prefix and PstI for the suffix, and the web site will produce sticky ends when the oligos are fully assembled.
3. The output window will show the desired double-stranded DNA sequence, color-coded by the overlapping oligos. The top strand is the sequence submitted, beginning and ending with the BioBrick ends. Save the text from the output in a permanent document.
4. The resulting sequence and individual oligos have sticky ends equivalent to digestion with EcoRI and PstI. Have these individual oligos synthesized at a recommended 100 mM concentration. Store oligos at −20°C until ready to assemble.
5. To a 0.5-mL microcentrifuge tube, add 1 μL of a 100-μM solution of each oligo (the final concentration of each oligo will be 5 μM). Add a one-tenth volume of annealing buffer (2 μL in 20 μL) and water to a final volume of 20 μL.
6. Boil the oligos in 400–500 mL of water for 10 min.
7. Let the mixture cool slowly to room temperature overnight. The product is ready to use for ligation. To determine the concentration of the product, sum the amount of nanograms of each oligo added and divide by the total volume of the mixture. You may need to dilute an aliquot to avoid pipetting volumes below 0.5 μL (see Note 14).

## 4. Notes

1. When obtaining plasmid DNA for digestion and ligation, a yield of at least 15 ng/μL is desirable. Also, it is helpful to miniprep more than one sample of each type of plasmid DNA every time the miniprep procedure is done. Due to the overnight incubation step, minipreps are a bottleneck process, so minimize the number of times you perform minipreps. Another benefit of having a large volume of DNA is that you have the option of concentrating the DNA if needed (see Subheading 3.3).

2. We have found that digesting 50 times the calculated amount of DNA needed for a ligation generally yields appropriate concentrations. The digestion reaction volume can be increased above 20 μL for large DNA volumes as long as the volume of enzymes does not exceed 10% of the final reaction volume.
3. When preparing a receiving plasmid, you do not need to gel purify the vector. We have found that ethanol precipitation suffices (see Subheading 3.3). The small piece of DNA between the EcoRI and XbaI sites, or the SpeI and PstI sites, does not precipitate efficiently and is excluded from the DNA pellet.
4. We have run gels at the maximum of 120 V for the fastest run time, but that can cause a diffuse band especially with high DNA concentrations. We find a voltage of 80–100 V to be ideal. Secondly, we try to limit each DNA sample to 20 μL in one lane. However, if digestion volume is larger than 20 μL, two or more lanes may be used.
5. Excessive exposure to UV light is harmful. Wear gloves and protective eyewear when working with UV light. UV light also nicks DNA so minimize UV exposure to your DNA of interest.
6. Try to cut off most of the gel around the DNA bands. There is usually gel underneath the band that can be cut off as well. Decreasing the volume of the gel slice will improve yield for the gel purification step. The maximum weight of the gel slice is 400 mg. Higher weights will require more than one spin column during gel purification, and that is not recommended.
7. If the vector or insert DNA solutions is too dilute, the components of the ligation can be scaled up, or the DNA can be dried down to a smaller volume. Although you can perform ligations in as much as 20 μL, you can also vacuum-concentrate the DNA to dryness and resuspend in a small volume of water to keep the total reaction volume at 10 μL. The formula provided calculates the amount of insert for a 2:1 molar ratio of vector to insert, which is ideal. However, we have used ratios slightly lower than 2:1 if there was insufficient insert DNA. Alternatively, you can cut all values in half if purified DNA is limited. Ratios of 3:1 or 5:1 can be used if ligation fails the first time.
8. As a rule of thumb, PCR primer annealing temperatures should be about 5°C lower than the calculated melting temperature. Elongation times should be about 1 min/kb to reduce the number of new mutations.
9. If the insert and the plasmid are the same size, we digest the ampicillin resistance gene with ScaI to cut the plasmid into two pieces.
10. Some DNA sequences are so resistant to cloning as to be called "unclonable" (8). Often, the reason for this is not known, but suspected causes include unusual secondary structures serving

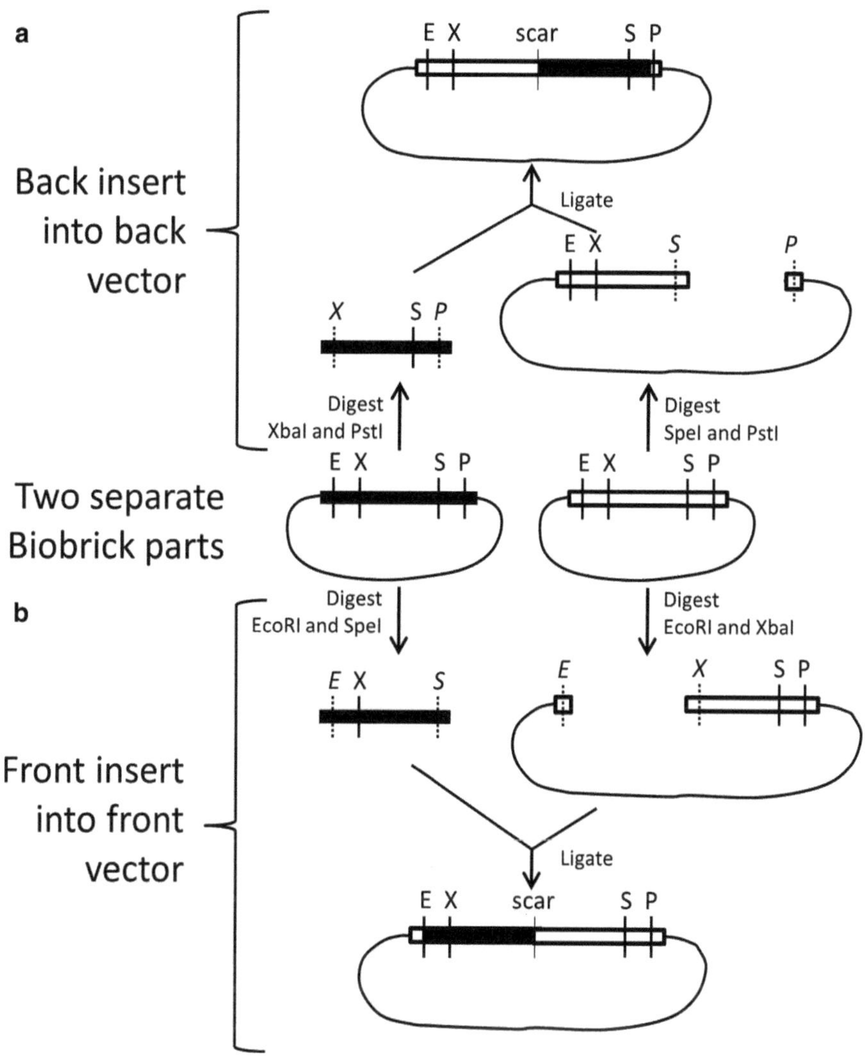


Fig. 5. Inserting a BioBrick part upstream or downstream of another BioBrick part. The letters E, X, S, and P stand for EcoRI, XbaI, SpeI, and PstI restriction sites, respectively. A *dotted line* indicates the restriction site has been digested and has an exposed overhang. (**a**) Inserting a part downstream of another part. (**b**) Inserting a part upstream of another part.

as substrates for recombination and generation of gene products that are toxic to the bacterial cells.

11. Sequence-verify all new constructs and those that contain more than one copy of a part. We have found that parts such as double transcription terminators cause unintended recombination.
12. Consider two hypothetical BioBrick parts, A and B (Fig. 5). Making part A the front insert and part B the front vector produces the same arrangement (part A:part B) as making part A the back vector and part B the back insert. However, the larger part is usually used as the insert for two reasons. First, it is

difficult to get adequate gel purification yields of small pieces of DNA. Second, a large insert will produce a more conspicuous band shift when the ligated product is run on a gel alongside the unligated vector to verify that the ligation occurred.

13. We order our DNA oligos using a small scale (10 or 50 nmol) and have them desalted. We have the oligos shipped in liquid form at 100 μM. This allows us to use 1 μL of the oligo directly in a 100 μL PCR amplification.
14. When ligating an insert, always perform a negative control with the same plasmid but no insert DNA. Ideally, you would see more colonies on the plate with the insert than on the negative control. However, we have found that when the negative control plate has more colonies, the smaller number of colonies on the experimental plate often contain the correct ligation product. If the plasmid recircularizes frequently, you can use a heat-sensitive alkaline phosphatase to remove the 5′ phosphates from your vector.

## References

1. Knight T, *et al.* (2003) Idempotent Vector Design for the Standard Assembly of BioBricks. http://people.csail.mit.edu/tk/sa3.pdf. (Accessed 13 March 2010).
2. Cai, Yizhi *et al.* (2007). A syntactic model to design and verify synthetic genetic constructs derived from standard biological parts. *Bioinformatics.* Vol. 23(20): 2760–2767.
3. Registry of Standard Biological Parts. http://partsregistry.org. (Accessed 11 April, 2010).
4. Haynes, Karmella A., *et al.* (2008). Engineering bacteria to solve the Burnt Pancake Problem. *Journal of Biological Engineering.* Vol. 2(8): 1–12.
5. Baumgardner, Jordan, *et al.* (2009). Solving a Hamiltonian Path Problem with a Bacterial Computer. *Journal of Biological Engineering.* Vol. 3:11.
6. Peccoud, Jean *et al.* (2008). Targeted development of registries of biological parts. *PLoS ONE.* Vol. 3(7): e2671.
7. Czar M.J., et al. (2009). Gene synthesis demystified. Trends Biotechnology. Vol. 27: 63–72.
8. Godiska R, Mead D, Dhodda V, Wu C, Hochstein R, Karsi A, Usdin K, Entezam A, Ravin N. (2010). Linear plasmid vector for cloning of repetitive or unstable sequences in *Escherichia coli. Nucleic Acids Research.* Vol. 38(6): e88.

# Chapter 7

# Assembling DNA Fragments by USER Fusion

**Narayana Annaluru, Héloïse Muller, Sivaprakash Ramalingam, Karthikeyan Kandavelou, Viktoriya London, Sarah M. Richardson, Jessica S. Dymond, Eric M. Cooper, Joel S. Bader, Jef D. Boeke, and Srinivasan Chandrasegaran**

## Abstract

Recent advances in DNA synthesis technology make it possible to design and synthesize DNA fragments of several kb in size. However, the process of assembling the smaller DNA fragments into a larger DNA segment is still a cumbersome process. In this chapter, we describe the use of the uracil specific excision reaction (USER)-mediated approach for rapid and efficient assembly of multiple DNA fragments both in vitro and in vivo (using *Escherichia coli*). For USER fusion in vitro assembly, each of the individual building blocks (BBs), 0.75 kb in size (that are to be assembled), was amplified using the appropriate forward and reverse primers containing a single uracil (U) and DNA polymerase. The overlaps between adjoining BBs were 8–13 base pairs. An equimolar of the amplified BBs were mixed together and treated by USER enzymes to generate complementary 3′ single-strand overhangs between adjoining BBs, which were then ligated and amplified simultaneously to generate the larger 3-kb segments. The assembled fragments were then cloned into plasmid vectors and sequenced to confirm their identity. For USER fusion in vivo assembly in *E. coli*, USER treatment of the BBs was performed in the presence of a synthetic plasmid, which had 8–13 base pair overlaps at the 5′-end of the 5′ BB and at the 3′-end of the 3′ BB in the mixture. The USER treated product was then transformed directly into *E. coli* to efficiently and correctly reconstitute the recombinant plasmid containing the desired target insert. The latter approach was also used to rapidly assemble three different target genes into a vector to form a new synthetic plasmid construct.

**Key words:** Synthetic yeast, Uracil excision, UDG, Endo VIII, USER enzymes, DNA assembly

## 1. Introduction

Development of efficient methods for rapid assembly of multiple DNA fragments into a larger DNA segment will not only have a significant impact on synthetic biology but also on basic biomedical research. The generation of large DNA segments from smaller DNA fragments by using restriction enzymes sites has limited utility because it requires unique sites for fusion of various fragments

Jean Peccoud (ed.), *Gene Synthesis: Methods and Protocols*, Methods in Molecular Biology, vol. 852,
DOI 10.1007/978-1-61779-564-0_7, © Springer Science+Business Media, LLC 2012

and the vector. Several methods that do not join DNA by ligation at restriction sites are being developed for seamless engineering of multidomain fusion protein constructs and multigene modular vector constructs. They include the In-Fusion assembly reaction (1), isothermal assembly reaction, and the uracil specific excision reaction (USER) fusion assembly reaction, discussed in detail below (2–5).

The In-Fusion assembly reaction joins any two pieces of DNA that have an overlap of 15 bp at their ends by using unique properties of the 3′–5′ exonuclease activity of poxvirus DNA polymerase (6), whereas the isothermal assembly reaction is a single-reaction method for assembling multiple DNA molecules that have a greater overlap of ~20–200 bp at their ends and is performed by the concerted action of a 5′ exonuclease (T5 exonuclease), a DNA polymerase (Phusion polymerase), and a DNA ligase (Taq DNA ligase) (2). As the 3′–5′ exonuclease activities of poxvirus DNA polymerase and the 5′ exonuclease activity of T5 exonuclease are not highly processive, the results of these assembly reactions can be quite variable, and these methods might thus not always be successful. Both of these approaches might require extensive and cumbersome optimization experiments before the desired product can be obtained.

Here, we describe rapid and efficient assembly of multiple DNA fragments using the USER-mediated approach (also known as USER fusion) for both in vitro (using Phusion polymerase and/or Taq ligase) and in vivo (using *E. coli*). The USER enzyme mix is a mixture of uracil DNA glycosidase (UDG) and DNA glycosylase-lyase endo VIII, and is commercially available. UDG selectively excises uracil bases while leaving the phosphodiester backbone intact. Endo VIII breaks the phosphodiester backbone at the 3′ and 5′ sites of an abasic site, releasing the base-free deoxyribose (7, 8). The basic assembly strategy employs USER to fuse about four building blocks (BBs) of ~0.75 kb in size at a time to yield larger DNA segments that are ~3 kb in size. Each of the BBs has been designed so that there is an overlap of ~8–13 bp between adjacent or neighboring BBs and has been assembled from oligonucleotides with its DNA sequence confirmed by sequencing as described elsewhere (9). The BBs are then amplified using BB-specific start and stop primers, each of which contains a single uracil residue to generate unique single-strand overhangs after treatment with USER. These unique single-strand ends of each BB can be ligated to adjoining BBs, since they have compatible ends. We have used the USER-mediated approach to assemble several sets of four BBs at a time to yield larger overlapping DNA fragments of ~3 kb in size, which are then assembled into even larger DNA segments by direct transformation into yeast.

For in vitro USER fusion assembly reactions, each of the individual BBs is amplified using the appropriate forward and reverse primers that contain a single deoxyuracil nucleotide and Pfu polymerase (Fig. 1). Here, an equimolar amount of amplified BBs in each set are mixed and treated with USER enzyme mix to generate 3′ single-strand overhangs between adjoining BBs, which are then simultaneously ligated and amplified using Taq DNA ligase and Phusion polymerase to generate the larger 3-kb fragments (Fig. 2). For the in vivo USER fusion assembly reaction, the 3-kb fragments were generated in situ by direct transformation of *E. coli* with the corresponding BBs and a synthetic plasmid (pJHU1) after USER treatment (Fig. 3). The 5′-end of the synthetic plasmid has a region that is homologous with one of the BBs, while the 3′-end of the plasmid is homologous to another BB in the reaction mixture. Correct recombination of all the fragments in *E. coli* reconstituted the plasmid with the desired 3-kb insert, which then could be

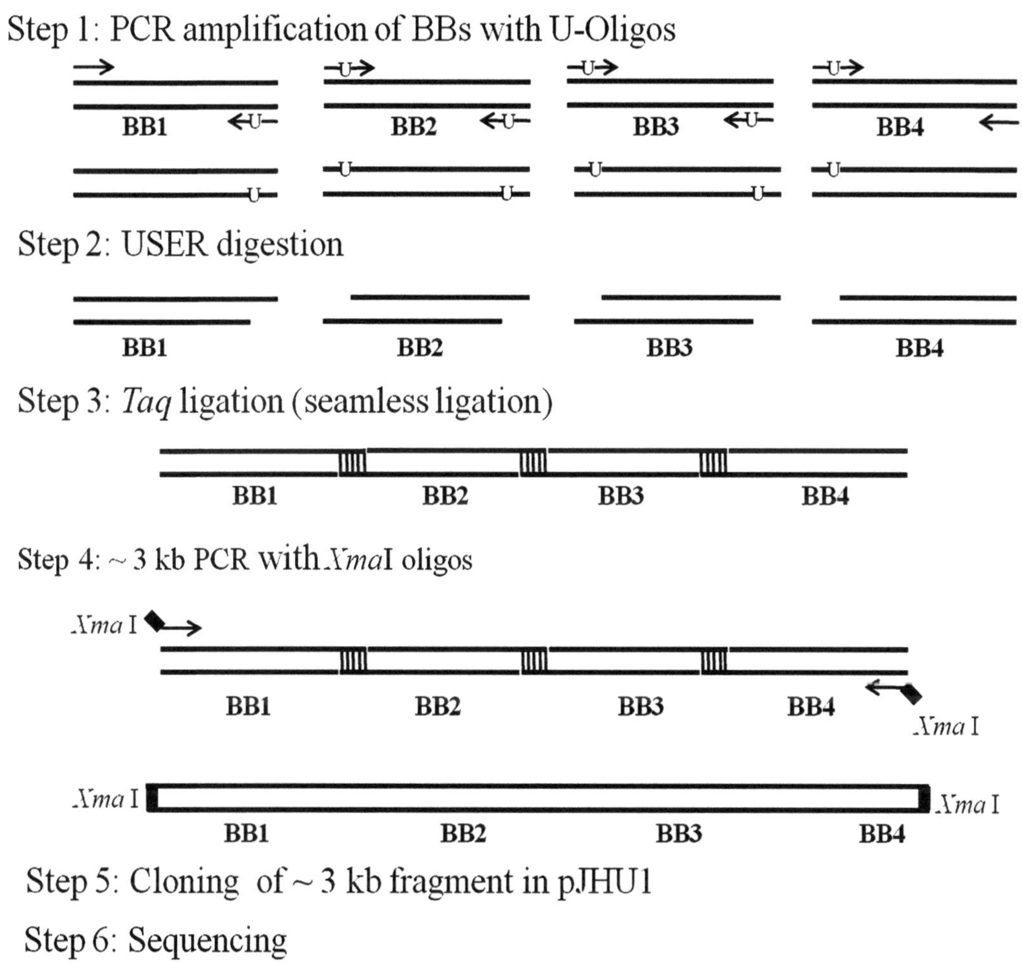


Fig. 1. In vitro USER fusion schematic diagram showing the various steps.

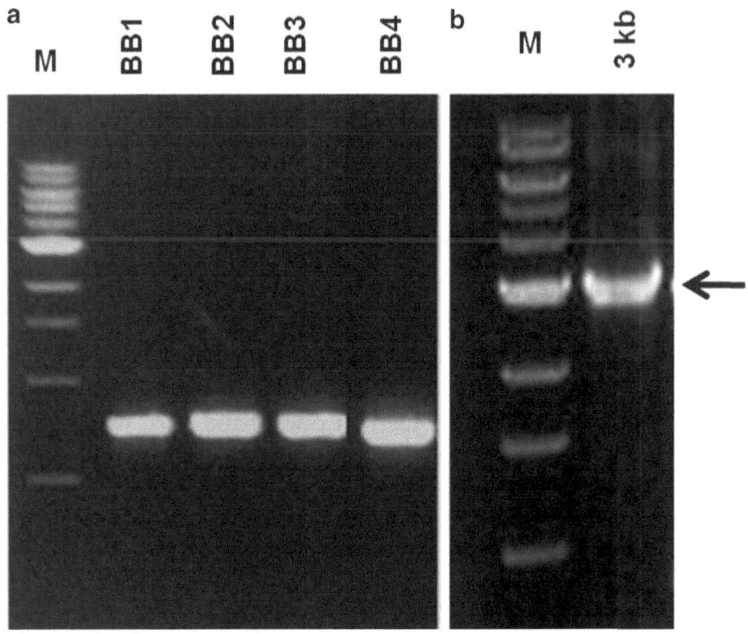


Fig. 2. Example of product analysis after in vitro USER fusion assembly reaction. Agarose gel electrophoresis of the products of USER fusion in vitro assembly reaction that are obtained in the various steps. (**a**) This *panel* shows the fours BBs that have been amplified by PCR using forward and reverse primers containing a single uracil. (**b**) Shown here is the correct amplified PCR product of 3 kb in size from step 4 using *Xma*I forward and reverse primers. *Arrows* indicate the expected products.

released by using an appropriate restriction enzyme (*Xma*I in this case) digestion (Fig. 4). This approach can also be used to rapidly assemble three target genes into a vector to form a new plasmid construct (Figs. 5 and 6).

## 2. Materials

*Reagents*

1. Luria–Bertani (LB) medium: sodium chloride (NaCl, 10 g/L), Bacto tryptone (10 g/L), and Bacto yeast extract (5 g/L). Add 15 g/L Bacto agar to LB medium to prepare LB agar for plating. Dissolve in $ddH_2O$ and autoclave.
2. SOC medium: complement 1 L of LB medium with 10 mL of 1 M $MgSO_4$ (10 mM final), 10 mL of 1 M $MgCl_2$ (10 mM final), and 18 mL of 20% dextrose (0.36% w/vol).
3. Plasmid vector—pJHU1.
4. Polymerases: *PfuTurbo*® *Cx* Hotstart DNA polymerase (Agilent Technologies), *Phusion*® Hotstart II High-Fidelity DNA polymerase (New England Biolabs), and *PfuUltra*® II Fusion Hotstart DNA polymerase (Agilent Technologies).

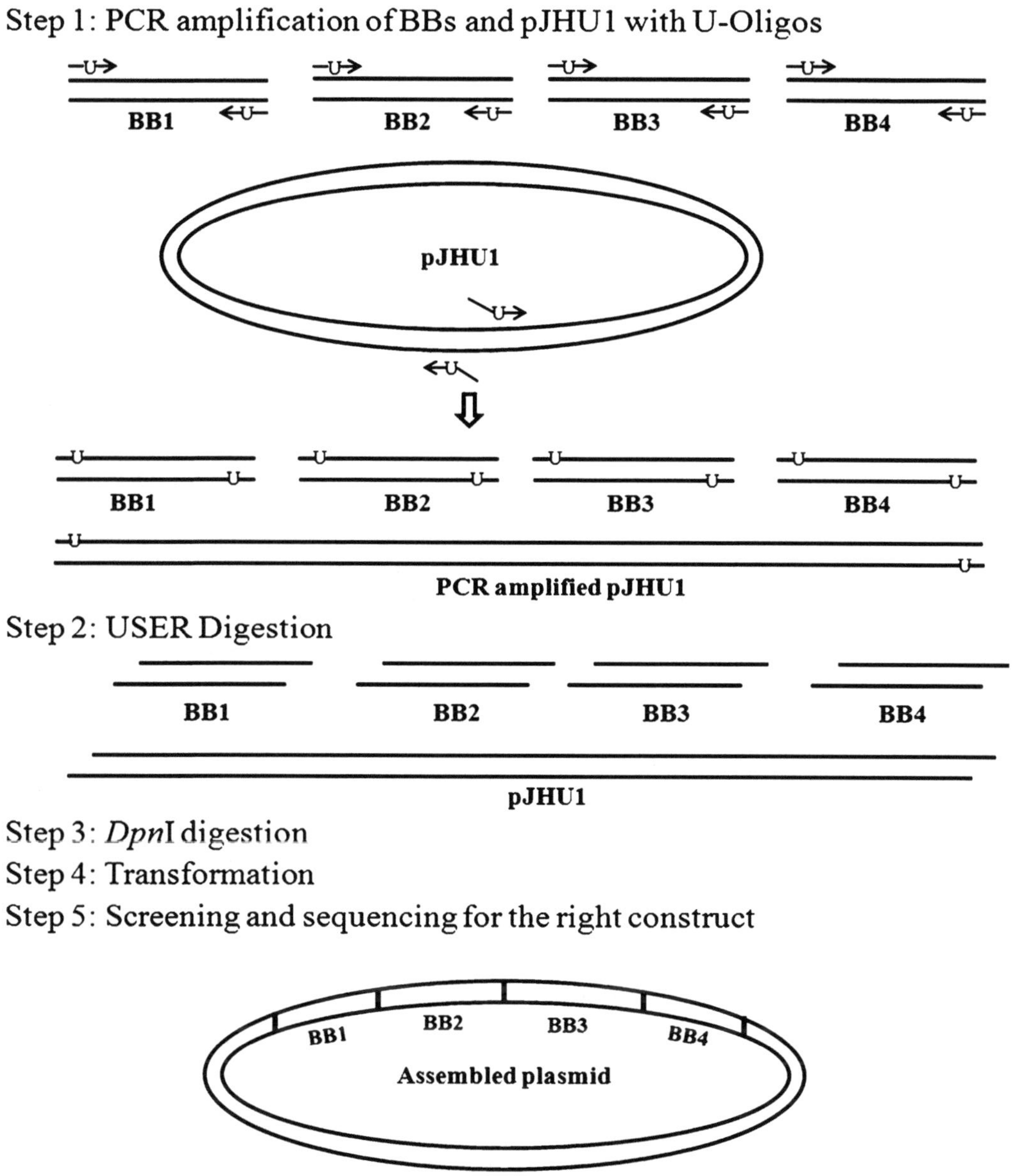


Fig. 3. In vivo USER fusion Schematic diagram showing the various steps.

5. Purification of PCR products: QIAquick PCR Purification Kit (Qiagen) or similar.
6. PCR Grade 10 mM dNTP Mix (Invitrogen). Prepare a 2.5-mM working stock solution of dNTPs in sterile milli-Q (or $ddH_2O$) water for regular use. Store all PCR reagents at −20°C.
7. Agarose and buffer used in gel electrophoresis: agarose, 50× Tris-acetate-EDTA (TAE) buffer [242× g Tris-HCl base,

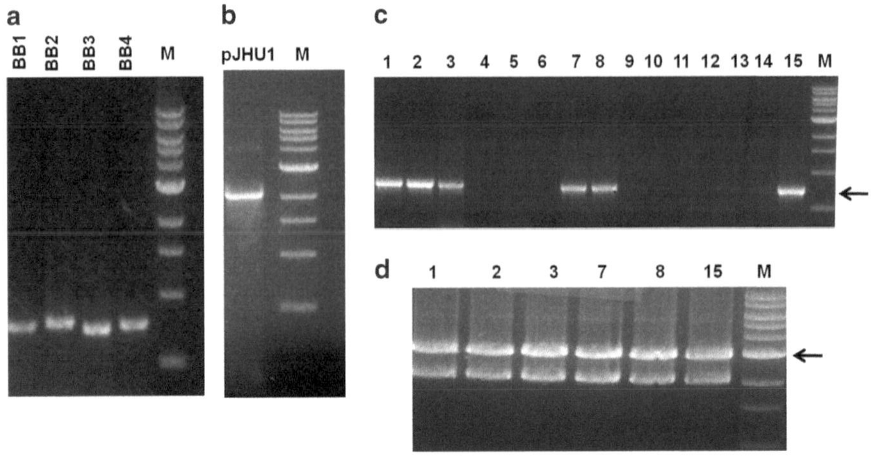


Fig. 4. Example of product analysis after in vivo USER fusion assembly reaction. Analyses of the products from USER fusion in vivo assembly reaction by agarose gel electrophoresis. (**a**) Shown here are the four BBs amplified by PCR using forward and reverse primers containing a single uracil. (**b**) This *panel* shows the plasmid pJHU1 amplified by PCR using forward and reverse primers containing a single uracil. (**c**) Shown here is the result of a colony-screening PCR (CSPCR) of the clones obtained in step 5 of Subheading 3.2 that assayed for the presence of one of the BBs within the insert using the corresponding forward and reverse primers. The *arrow* indicates the size of the expected product. (**d**) Confirmation of the correct insert size. Here, recombinant plasmids of the positive clones identified by CSPCR as shown in *panel* (**c**) were isolated and digested with *Xma*I to release the 3-kb insert that has been assembled from the four BBs. The *arrow* indicates the expected product.

100 mL of 0.5 M EDTA (pH 8.0), and 57.1 mL glacial acetic acid per 1 l solution]. Autoclave and store at room temperature. Prepare a 1× TAE in milli-Q water for regular use. BioRad gel apparatus system or similar is used for gel electrophoresis.

8. Gel purification of PCR products: PureLink™ Quick Gel Extraction Kit (Invitrogen).
9. The purified DNA was quantified using the Nanodrop ND-1000 Spectrophotometer.
10. 1 kb DNA ladder.
11. Restriction enzymes and buffers: *Dpn*I (New England Biolabs), *Xma*I (New England Biolabs), and *Bst*XI (New England Biolabs).
12. Taq DNA ligase (New England Biolabs).
13. USER™ enzyme (New England Biolabs).
14. 10× PCR buffer (GeneAmp).
15. Dithiothreitol.
16. 3 M Sodium acetate buffer solution, pH 5.5.
17. Nicotinamide adenine dinucleotide (NAD).
18. Phenol-chloroform, chloroform (see Note 1).
19. Ethidium bromide (EtBr) (see Note 2).

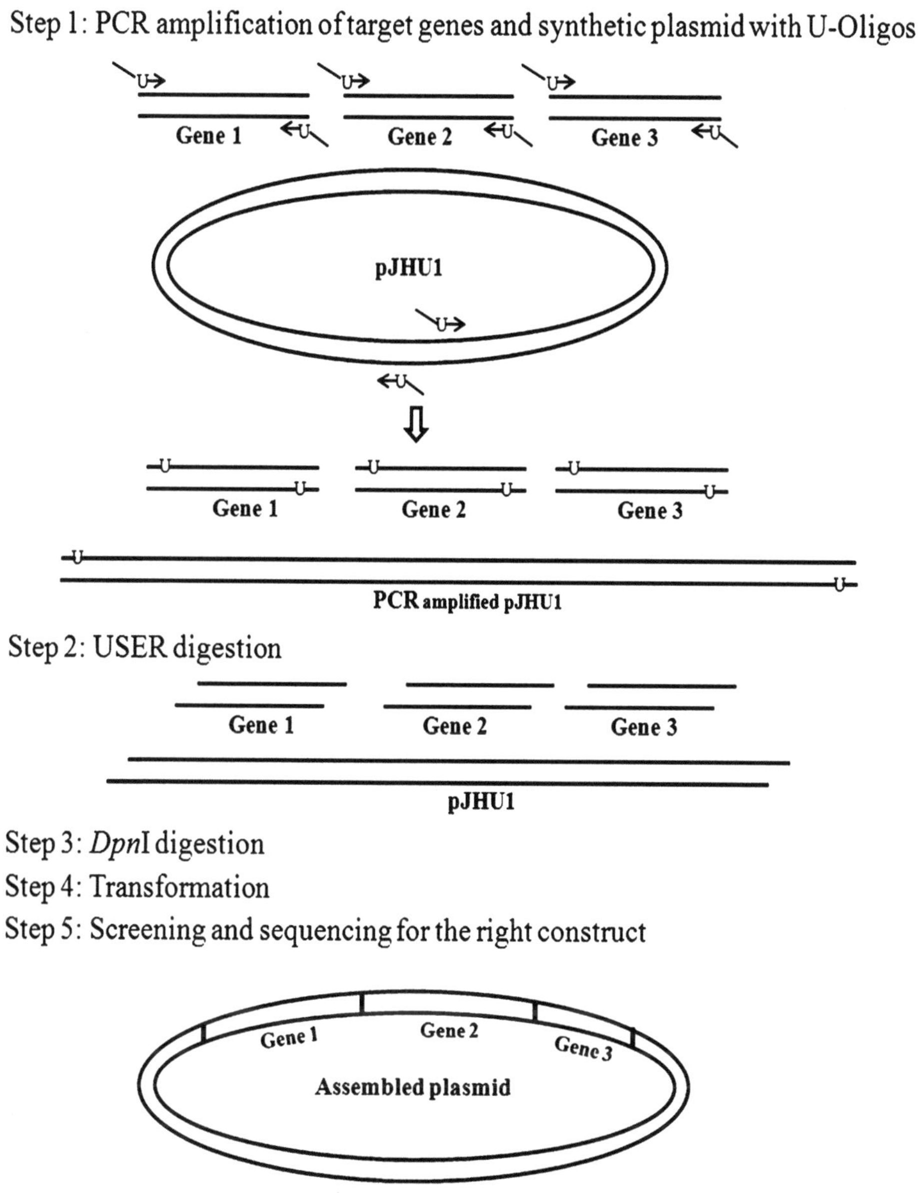


Fig. 5. USER fusion for the assembly of new plasmid construct with multiple target genes.

20. *E. coli* strain for transformation: Fusion-Blue™ competent cells (from Clontech), Genotype: *end*A1, *hsd*R17 ($r_{k12}^-$, $m_{k12}^+$), *sup*E44, *thi*-1, *rec*A1, *gyr*A96, *rel*A1, *lac* F′[*pro*A⁺B⁺, *lacI*$^q$ZΔ *M15*::Tn*10*(tet$^R$)]; JM109 competent cells (from Promega), Genotype: *end*A1, *rec*A1, *gyr*A96, *thi*, *hsd*R17 ($r_k^-$, $m_k^+$), *rel*A1, *sup*E44, Δ(*lac*-*pro*AB), [F′ *tra*D36, *pro*AB, *laq*I$^q$ZΔM15].
21. Plasmid isolation from *E. coli* cells: QIAprep Spin Miniprep Kit (Qiagen).

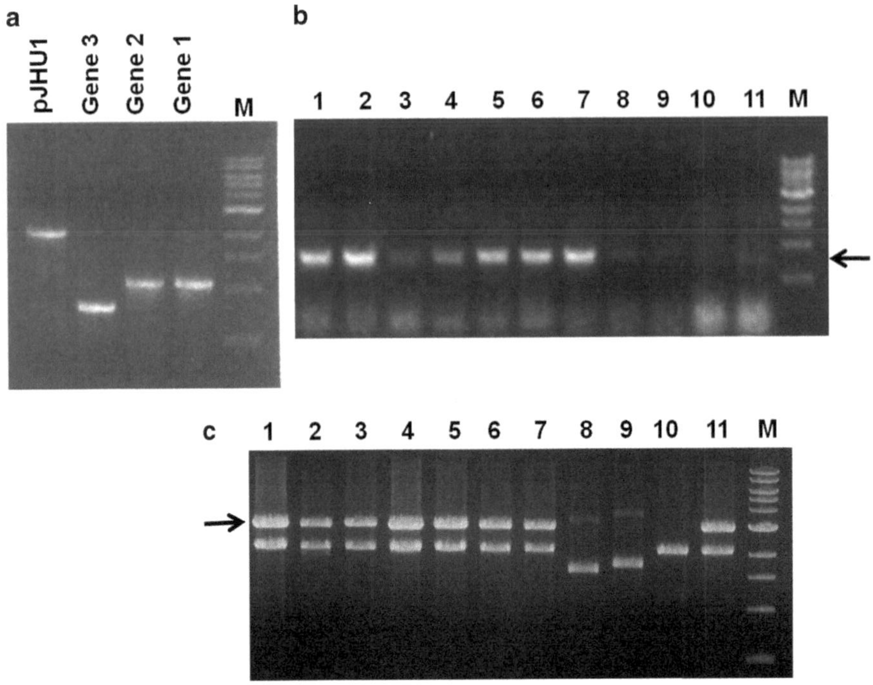


Fig. 6. Example of product analysis after USER-mediated plasmid assembly. (**a**) pJHU1 and three different target genes were amplified by PCR using appropriate forward and reverse primers containing a single uracil. (**b**) Shown here is the result of a colony-screening PCR of the clones obtained from step 5 of Subheading 3.3, which confirmed the presence of one of the target genes (here Gene3) in the insert using the corresponding forward and reverse primers. (**c**) The recombinant plasmids from 11 clones obtained from *panel* (**b**) were isolated and digested with *Bst*XI to release the insert of ~2.9 kb in size that has been assembled from the three target genes. *Arrows* indicate the expected products.

22. All synthetic oligonucleotides and uracil-containing oligonucleotides used were purchased from Integrated DNA Technologies. Prepare 10 μM working stocks.
23. Sequencing primers: M13F-*5′ GCCAGGGTTTTCCCAGTC ACGA 3′*, M13R-*5′ GAGCGGATAACAATTTCACACAGG 3′*.
24. 10× Loading dye: Dissolve 0.24× g of Bromophenol Blue and 0.42× g of Xylene Cyanol FF in $ddH_2O$, add 30 mL of 30% glycerol, and adjust the volume to 100 mL with $ddH_2O$.

## 3. Methods

### 3.1. USER Fusion In Vitro Assembly

1. Perform polymerase chain reactions (PCRs) with *PfuTurbo® Cx* Hotstart DNA polymerase of the four BBs using the corresponding start and stop primers that contain a single deoxyuracil residue (dU). For this, the following components should be mixed in a PCR tube.

| | |
|---|---|
| Template (plasmid DNA with BB) | 1 μL (2–10 ng) |
| *PfuTurbo*® *Cx* Hotstart DNA polymerase buffer 10× | 10 μL |
| dNTP mix (2.5 mM each) | 10 μL |
| Primer-1 (10 μM) | 5 μL |
| Primer-2 (10 μM) | 5 μL |
| *PfuTurbo*® *Cx* Hotstart DNA polymerase | 2 μL |
| $ddH_2O$ | 67 μL |
| Total | 100 μL |

PCR program

| | |
|---|---|
| 95°C | 4 min (initial denaturation step) |
| ↓ | |
| 30 Cycles | |
| 95°C | 30 s |
| 55°C | 30 s |
| 72°C | 1 min |
| ↓ | |
| 72°C | 10 min (final elongation step) |
| ↓ | |
| 4°C | Forever |

2. Prepare 1% agarose gel in 1× TAE buffer and pour the gel in a gel tray. The gel should harden in 30 min. Once the gel has set, carefully remove the comb, then add running buffer (1× TAE) to the unit.
3. Add 1 μL of loading dye (5×) to 5 μL of each PCR product and load the samples, as well as 0.5 μg of 1 kb DNA ladder into wells. Run the gel at 80–100 V until the DNA bands are well separated. Analyze the agarose gel under UV light to verify the size and yield of the amplified fragments. If other, unexpected bands or product smear is observed, optimize the PCR to maximize the yield of the desired product by changing the annealing temperature. Alternatively, if the correct product band is present in sufficient yield, extract it from the gel by using the PureLink™ Quick Gel Extraction Kit or a similar product following manufacturer's instructions.
4. Purify the PCR products with QIAquick PCR Purification Kit or similar. Follow the manufacturer's instructions and elute the product in the final step using 30 μL of $ddH_2O$.

5. Measure the DNA concentrations of the PCR products using a Nanodrop.
6. Set up the USER reactions in a 1.5-mL eppendorf tube.

| | |
|---|---|
| Equimolar mix of PCR products | 13.0 μL (see Note 3) |
| 10× PCR buffer | 2.0 μL |
| DTT (100 mM) | 1.0 μL |
| USER™ enzyme | 4.0 μL |
| Total | 20.0 μL |

7. Incubate at 37°C overnight.
8. Denature the reaction mix at 70°C for 10 min.
9. Add 80 μL of ddH$_2$O as it may be difficult to perform the subsequent extraction with volumes smaller than 100 μL. The DNA can be concentrated again after ethanol precipitation.
10. Add 100 μL of phenol-chloroform to the tube, vortex vigorously to mix the phases (see Note 1).
11. Spin in a microcentrifuge at 16,000×g for 4 min to separate the phases.
12. Transfer the aqueous phase to a new tube. Add 100 μL of chloroform to the tube and vortex vigorously to mix the phases.
13. Spin in a microcentrifuge at top speed for 1 min to separate the phases.
14. Repeat steps 12 and 13 to remove any trace phenol.
15. Transfer the aqueous phase to another 1.5-mL eppendorf tube and then add 1/10 of the volume of 3 M sodium acetate (pH 5.5). Mix the solution by inverting the tube.
16. Add 2.5 volumes of ethanol. Mix by inverting the tube. Store it at −80°C for 1–2 h.
17. Spin the precipitate down using a microcentrifuge at maximum speed at 4°C for 30 min.
18. Decant (or carefully pipette off) the supernatant.
19. Add 500 μL of 70% ethanol (ice cold, stored at −20°C) to wash the precipitate.
20. Spin in a microcentrifuge at maximum speed for 10 min at 4°C (see Note 4).
21. Decant supernatant and air-dry the pellet by leaving the caps open for ~15 min or dry using a speedvac.
22. Add 13 μL of ddH$_2$O. Vortex to dissolve the precipitate and spin down the solution.

23. Set up the following Taq ligase reaction in PCR tubes to be used in a thermal cycler.

| | |
|---|---|
| USER-digested product (from step 22) | 13.0 μL |
| 10× Taq ligase buffer | 3.0 μL |
| DTT (100 mM) | 1.0 μL |
| NAD (40 mM) | 1.0 μL |
| Taq DNA Ligase | 2.0 μL |
| $ddH_2O$ | 10.0 μL |
| Total | 30.0 μL |

PCR program

| | |
|---|---|
| 72°C | 4 min |
| ↓ | |
| 30 Cycles | |
| 72°C | 1 min |
| 22°C | 5 min |
| 45°C | 20 min |
| ↓ | |
| 45°C | Forever |

24. Set up a PCR with *PfuUltra*® II Fusion Hotstart DNA polymerase or *Phusion*® Hotstart II High-Fidelity DNA polymerase using the product from the Taq ligation step as the template to amplify the ~3-kb product.

PCR mix

| | |
|---|---|
| Taq ligation product (from step 23) | 2 μL |
| *PfuUltra*® II Fusion Hotstart DNA polymerase buffer 10× | 5 μL |
| dNTP mix (2.5 mM each) | 5 μL |
| Primer-1 (10 μM) | 2.5 μL |
| Primer-2 (10 μM) | 2.5 μL |
| *PfuUltra*® II Fusion Hotstart DNA polymerase | 1 μL |
| $ddH_2O$ | 32 μL |
| Total | 50 μL |

PCR program

| | |
|---|---|
| 98°C | 2 min (initial denaturation step) |
| ↓ | |
| 30 Cycles | |
| 94°C | 1 min |
| 55°C | 2 min |
| 70°C | 7 min |
| ↓ | |
| 72°C | 35 min (final elongation step) |
| ↓ | |
| 4°C | Forever |

25. Run a 1% agarose gel using 5 μL of the PCR sample to determine whether the obtained products are of the correct size and if there are any secondary bands. If other bands or product smear is observed, optimize PCR to maximize the yield of the desired product by changing the annealing temperature. Alternatively, if the right product is present in sufficient yield, extract the product from the gel using PureLink™ Quick Gel Extraction Kit following manufacturer's instructions.
26. Purify the PCR products with QIAquick PCR Purification Kit following manufacturer's instructions.

### 3.2. USER Fusion In Vivo Assembly

1. Set up a PCR each for the four building blocks (BBs) and the synthetic plasmid (pJHU1) using *PfuTurbo*® *Cx* Hotstart DNA polymerase and start/stop primers containing a single deoxyuracil residue (dU). For use as template DNA, dilute the purified plasmid DNA containing the building blocks or the purified pJHU1 DNA (~1:500).

| | |
|---|---|
| Template (plasmid DNA with BB) | 1 μL (2–5 ng) |
| *PfuTurbo*® *Cx* Hotstart DNA polymerase buffer 10× | 10 μL |
| dNTP mix (2.5 mM each) | 10 μL |
| Primer-1 (10 μM) | 5 μL |
| Primer-2 (10 μM) | 5 μL |
| *PfuTurbo*® *Cx* Hotstart DNA polymerase | 2 μL |
| $ddH_2O$ | 67 μL |
| Total | 100 μL |

PCR program

| | |
|---|---|
| 95°C | 4 min (initial denaturation step) |
| ↓ | |
| 30 Cycles | |
| 95°C | 30 s |
| 55°C | 30 s |
| 72°C | 1 min (1 min/kb) |
| ↓ | |
| 72°C | 10 min (final elongation step) |
| ↓ | |
| 4°C | Forever |

2. Analyze 5 μL of each of the PCRs by agarose gel electrophoresis to verify the size and yield of the amplified fragments. If there are secondary bands, optimize the annealing temperature or extract the desired bands by using PureLink™ Quick Gel Extraction Kit following manufacturer's instructions as described above.
3. Purify the PCR products with QIAquick PCR Purification Kit and elute DNA with 30 μL of $ddH_2O$.
4. Measure the DNA concentrations of the PCR products using a Nanodrop.
5. Set up the USER reaction as follows:

| | |
|---|---|
| Equimolar mix of PCR products | 13.0 μL (see Note 5) |
| 10× PCR buffer (GeneAmp) | 2.0 μL |
| DTT (100 mM) | 1.0 μL |
| USER™ enzyme | 4.0 μL |
| Total | 20.0 μL |

6. Incubate at 37°C overnight.
7. Add 80 μL of $ddH_2O$. It may be difficult to perform extraction with volumes smaller than 100 μL. The DNA can be concentrated later after ethanol precipitation.
8. Heat the sample at 70°C for 15 min.
9. Add 100 μL of phenol-chloroform to the tube, then vortex vigorously to mix the two phases (see Note 1).
10. Centrifuge the precipitate down in a microcentrifuge by spinning at 16,000 × g for 4 min to separate the two phases.

11. Transfer the aqueous layer to a new tube and add 100 µL of chloroform to the tube, then vortex vigorously to mix the phases.
12. Spin at 16,000×g for 1 min to separate the two phases.
13. Repeat step 9 and 10 to remove any trace of phenol.
14. Remove the aqueous layer to another 1.5-mL eppendorf tube and then add 10 µL volume of 3 M sodium acetate (pH 5.5). Mix by inverting the tube.
15. Add 2.5 volumes of ethanol. Mix by inverting the tube. Store at −80°C for 1–3 h.
16. Spin the precipitate down in a microcentrifuge by spinning at 16,000×g at 4°C for 30 min.
17. Decant (or carefully pipette off) the supernatant.
18. Add 500 µL 70% ethanol (ice cold, stored at −20°C).
19. Spin the precipitate in the microcentrifuge at 16,000×g for 10 min at 4°C.
20. Decant supernatant and air-dry the pellet by leaving the cap open for ~15 min or dry in a speedvac (see Note 4).
21. Add 10 µL of $ddH_2O$. Vortex to dissolve the DNA and then spin down the solution.
22. Set up the *Dpn*I digestion reaction at 37°C for ~2 h to degrade the template plasmid DNA to reduce background from transformation experiments of *E. coli* (see Note 6).

| | |
|---|---|
| USER-digested product (from step 21) | 10 µL |
| 10× Buffer 4 | 5 µL |
| *Dpn*I | 2 µL |
| $ddH_2O$ | 33 µL |
| Total | 50 µL |

23. Add 50 µL $ddH_2O$ to the reaction. Extract the reaction mix once with an equal volume of phenol-chloroform/chloroform, followed by two extractions with an equal volume of chloroform and then precipitate the DNA with 2.5 volumes of ethanol as described above. Repeat steps 7–19 above. Resuspend the DNA in 10 µL of $ddH_2O$ in the final step.
24. Keep the USER-digested products at 55°C for 30 min and then transfer to room temperature for an hour.
25. For *E. coli* transformation experiments, add 5 µL of USER-digested product to a 50-µL aliquot of Fusion-Blue™ chemically competent cells, pipette gently to mix and incubate on ice for 30 min.
26. Heat shock for 45 s in a 42°C water bath and keep on ice for 2 min. Add 1 mL of SOC medium to the tube and incubate at 37°C with continuous shaking.

27. After 1 h, pellet the cells in a microcentrifuge at 2,200 × g for 5 min.
28. Decant 700 μL of SOC and mix gently.
29. Resuspend the cells and plate on LB plates containing 100 μg/mL carbenicillin.
30. Place the plates in a 37°C incubator overnight.
31. Perform colony screening PCR on at least 24 colonies using the start and stop primers for one of the BBs.
32. Isolate plasmid from any positive clones found in step 31 and analyze the recombinant plasmid by restriction enzyme digestion for the presence of correct-sized insert.
33. Sequence the positive clones that have the correct-sized insert to confirm their sequence identity.

### 3.3. Generating New Plasmid Constructs with USER Fusion

1. Set up a PCR for three target genes and pJHU1 using oligonucleotide primers containing single deoxyuracil residue (dU) and *PfuTurbo*® Cx Hotstart DNA polymerase. Dilute the purified building blocks containing plasmid DNA or the purified pJHU1 DNA (~1:500) to use as template for PCR.

| | |
|---|---|
| Template (plasmid DNA with target gene) | 1 μL (2–10 ng) |
| *PfuTurbo*® *Cx* Hotstart DNA polymerase buffer 10× | 10 μL |
| dNTP mix (2.5 mM each) | 10 μL |
| Primer-1 (10 μM) | 5 μL |
| Primer-2 (10 μM) | 5 μL |
| *PfuTurbo*® *Cx* Hotstart DNA polymerase | 2 μL |
| $ddH_2O$ | 67 μL |
| Total | 100 μL |

PCR program

| | |
|---|---|
| 95°C | 4 min (initial denaturation step) |
| ↓ | |
| 30 Cycles | |
| 95°C | 30 s |
| 55°C | 30 s |
| 72°C | 1 min (1 min/kb) |
| ↓ | |
| 72°C | 10 min (final elongation step) |
| ↓ | |
| 4°C | Forever |

2. Analyze 5 μL of each of the PCRs by agarose gel electrophoresis to verify the size and yield of the amplified fragments. If there are secondary bands, optimize the annealing temperature by gradient PCR to maximize the desired product. Otherwise, extract the desired bands by using PureLink™ Quick Gel Extraction Kit following manufacturer's instructions as described above.
3. Purify the PCR products with QIAquick PCR Purification Kit and elute with 30 μL of $ddH_2O$.
4. Measure the DNA concentrations of the PCR products using a Nanodrop.
5. Set up the USER reaction as follows:

| | |
|---|---|
| Equimolar mix of PCR products | 13.0 μL (see Note 5) |
| 10× PCR buffer (GeneAmp) | 2.0 μL |
| DTT (100 mM) | 1.0 μL |
| USER™ enzyme | 4.0 μL |
| Total | 20.0 μL |

6. Incubate at 37°C overnight.
7. Add 80 μL of $ddH_2O$. It may be difficult to perform extraction with volumes smaller than 100 μL. The DNA can be concentrated later after ethanol precipitation.
8. Heat the sample at 70°C for 10 min.
9. Add 100 μL of phenol-chloroform to the tube, vortex vigorously to mix the two phases (see Note 1).
10. Spin in a microcentrifuge at 16,000 × g for 4 min to separate the phases.
11. Transfer the aqueous layer to another 1.5-mL eppendorf tube and then add 100 μL of chloroform to the tube, then vortex vigorously to mix the phases.
12. Spin in a microcentrifuge 16,000 × g for 1 min to separate the phases.
13. Repeat steps 11 and 12 to remove any trace of phenol.
14. Transfer the aqueous layer to another 1.5-mL eppendorf tube and then add 10 μL of 3 M sodium acetate (pH 5.5). Mix by inverting the tube.
15. Add 2.5 volumes of ethanol. Mix by inverting then tube. Store it at −80°C for 1–3 h.
16. Spin the precipitate down in a microcentrifuge at 16,000 × g at 4°C for 30 min.
17. Decant (or carefully pipette off) the supernatant.
18. Add 500 μL of 70% ethanol (stored at −20°C).

19. Spin in microcentrifuge at 16,000 × g for 15 min at 4°C.
20. Decant supernatant and air-dry the pellet by leaving the cap open for ~15 min or dry in a speedvac (see Note 4).
21. Add 10 μL of $ddH_2O$. Vortex to dissolve the DNA and then spin down the liquid.
22. Set up the *Dpn*I digestion reaction at 37°C for ~2 h to degrade the template plasmid to reduce background from *E. coli* transformation experiments (see Note 6).

| | |
|---|---|
| USER-digested product (from step 21) | 10 μL |
| 10× Buffer 4 | 5 μL |
| *Dpn*I | 2 μL |
| $ddH_2O$ | 33 μL |
| Total | 50 μL |

23. Add 50 μL $ddH_2O$ to the reaction. Extract the reaction mix once with an equal volume of phenol-chloroform/chloroform, twice with chloroform and then precipitate the DNA with 2.5 volumes of ethanol as described above. Repeat steps 7–19. Resuspend the DNA in 10 μL of $ddH_2O$ in the final step.
24. Keep the USER-digested products at 55°C for 30 min and then transfer to room temperature for an hour.
25. For *E. coli* transformation experiments, add 5 μL of USER-digested product to a 50 μL aliquot of Fusion-Blue™ chemically competent cells, pipette gently to mix and incubate on ice for 30 min.
26. Heat shock for 45 s in a 42°C water bath and then keep on ice for 2 min. Add 1 mL of SOC medium to the tube and incubate at 37°C with continuous shaking.
27. After 1 h, pellet the cells in a microcentrifuge at 2,200 × g for 5 min.
28. Decant 700 μL of SOC and resuspend the cells gently in the remaining medium.
29. Spread them on LB plate containing 100 μg/mL carbenicillin.
30. Store the plates in a 37°C incubator overnight.
31. Perform colony screening PCR on at least 24 colonies using the start and stop primers for one of the target genes.
32. Isolate plasmid from the positive clones of colony screening PCR and analyze the recombinant plasmid by restriction enzyme digestion for the presence of correct-sized insert.
33. Sequence the positive clones with the correct-sized insert to confirm their sequence identity.

## 4. Notes

1. All phenol- and chloroform-containing solutions are highly toxic and corrosive. Phenol can induce chemical burns. Always wear gloves and goggles when handling phenol/chloroform solutions and use a well-ventilated area, such as a fume hood to conduct the experiments.
2. Ethidium bromide is a mutagen. Always wear gloves when handling gels and any solution containing EtBr.
3. Mix 200–300 ng of each BB to prepare an equimolar mix. If the resulting volume is greater than 13 μL, lyophilize the solution and then resuspend the DNA in 13 μL of $ddH_2O$.
4. Remove the supernatant carefully immediately after centrifugation to avoid dislodging the pellet. Keep in mind that the pellet may not be visible.
5. Mix 200–300 ng of each of the BBs, or the target genes, with 400–600 ng of pJHU1. If the volume is greater than 13 μL, lyophilize the mix and then resuspend the DNA in 13 μL of $ddH_2O$.
6. *Dpn*I restriction enzyme cleaves only Dam-methylated DNA (template); it does not cleave the nonmethylated PCR-amplified products.

## Acknowledgments

This work was supported by grants from National Science Foundation (MCB0718846) to J.D.B., J.S.B., and S.C.; from Microsoft to J.S.B. and J.D.B.; and from National Institutes of Health (GM077291) to S.C.

### References

1. Zhu, B, Cai, G, Hall, EO, and Freeman, GJ (2007) In-fusion assembly: seamless engineering of multidomain fusion proteins, modular vectors, and mutations. *Biotechniques* 43:354–359.
2. Gibson, DG, Young, L, Chuang, RY, Venter, JC, Hutchison, CA 3rd., and Smith, HO (2009) Enzymatic assembly of DNA molecules up to several hundred kilobases. *Nat Methods* 6:343–345.
3. Nour-Eldin, HH, Hansen, BG, Nørholm, MH, Jensen, and JK, Halkier, BA (2006) Advancing uracil-excision based cloning towards an ideal technique for cloning PCR fragments. *Nucleic Acids Res* 34:e122.
4. Geu-Flores, F, Nour-Eldin, HH, Nielsen, MT, and Halkier, BA (2007) USER fusion: a rapid and efficient method for simultaneous fusion and cloning of multiple PCR products. *Nucleic Acids Res* 35:e55.

5. Villiers, BR, Stein, V, and Hollfelder, F (2010) USER friendly DNA recombination (USERec): a simple and flexible near homology-independent method for gene library construction. *Protein Eng Des Sel* 23:1–8.
6. Hamilton, MD, Nuara, AA, Gammon, DB, Buller, RM, and Evans, DH (2007) Duplex strand joining reactions catalyzed by vaccinia virus DNA polymerase. *Nucleic Acids Research* 35:143–151.
7. Jiang, D, Hatahet, Z, Melamede, RJ, Kow, YW, and Wallace, SS (1997) Characterization of *Escherichia coli* Endonuclease VIII. *J Biol Chem* 272:32230–32239.
8. Melamede, RJ, Hatahet, Z, Kow, YW, Ide, H, and Wallace, SS (1994) Isolation and characterization of endonuclease VIII from *Escherichia coli*. *Biochemistry* 33:1255–1264.
9. Dymond, JS, Scheifele, LZ, Richardson, S, Lee, P, Chandrasegaran. S, Bader, JS, and Boeke, JD (2009) Teaching synthetic biology, bioinformatics and engineering to undergraduates: the interdisciplinary Build-a-Genome course. *Genetics* 181:13–21.

# Chapter 8

## Fusion PCR via Novel Overlap Sequences

**Kamonchai Cha-aim, Hisashi Hoshida, Tomoaki Fukunaga, and Rinji Akada**

### Abstract

Overlap extension or fusion PCR is thought to be a simple and easy method to produce fusion DNA fragments without the need for restriction enzyme digestion and DNA ligation. However, this method has not been used frequently, probably as it is not always reliable. When natural sequences are used for overlap sequences, sometimes either no fusion DNA is produced or only faint DNA bands are detected owing to low annealing between the overlap sequences selected. Here, we introduce several artificial overlap sequences, most of which are GC-rich, that can be used for reliable fusion PCR. We describe how these overlap sequences can be used for fusion DNA construction, in-frame gene fusion, and cloning in yeast.

**Key words:** Overlap extension PCR, GC-rich annealing sequences, Recombinant DNA construction, Gene splicing, Homologous recombination, Yeast

## 1. Introduction

Genetic engineering is based on recombinant DNA technology, which is the generation of fused DNA from two or more DNA fragments. Conventional recombinant DNA techniques commonly require *Escherichia coli* plasmid vectors, restriction enzyme digestions, and DNA ligation processes (1). To simplify these processes and to reduce failed results by the traditional method, many alternative methods were developed. The "Gateway" system utilizes recombination sequences and a recombination reaction for DNA linkage (2), "TA" cloning uses an additional dA overhang at the terminal ends of DNA after PCR to ligate with a plasmid vector that contains a dT overhang (3), "In-Fusion" cloning utilizes in vitro recombination reaction between homologous sequences of DNA to be inserted with plasmid vector (4), and uracil DNA glycosylase (UDG) cloning uses UDG to form efficient annealing between fragment and plasmid (5).

Jean Peccoud (ed.), *Gene Synthesis: Methods and Protocols*, Methods in Molecular Biology, vol. 852,
DOI 10.1007/978-1-61779-564-0_8, © Springer Science+Business Media, LLC 2012

These systems nevertheless still require *E. coli* cloning, which involves the preparation of *E. coli* competent cells, their transformation, and the selection of correct clones by time-consuming plasmid preparations. If *E. coli* plasmid is not necessary for the procedure, these time-consuming steps can be eliminated, facilitating most of the molecular genetics work. In addition, the genomics era requires DNA manipulations of thousands of genes, preferably using researcher-friendly, cost-effective, and automated pipelines for which easy DNA fabrication methods, such as a versatile and reliable DNA fusion without the need for *E. coli* plasmid amplification, will be required.

To construct fused DNA without steps involving *E. coli*, overlap extension or fusion PCR has been introduced (6–10). Identical overlap sequences belonging to the two DNA fragments, of which the 3′ and 5′ ends are to be fused, have enabled the construction of fused DNA fragments by PCR (Fig. 1).

Although the concept of fusion PCR is simple and appears to be easy, fusion PCR has not been used frequently thus far. Fusion PCR often proved difficult in constructing large DNA fragments of greater than 2 kb (11). Even if PCR conditions and length of the overlap sequences are modified, some of the selected authentic overlap sequences have resulted in failed fusion or only yielded low-fusion DNA products (12). These unsatisfactory results and the consequently limited adoption of this method have suggested that the approach per se is afflicted with problems. We expected that this

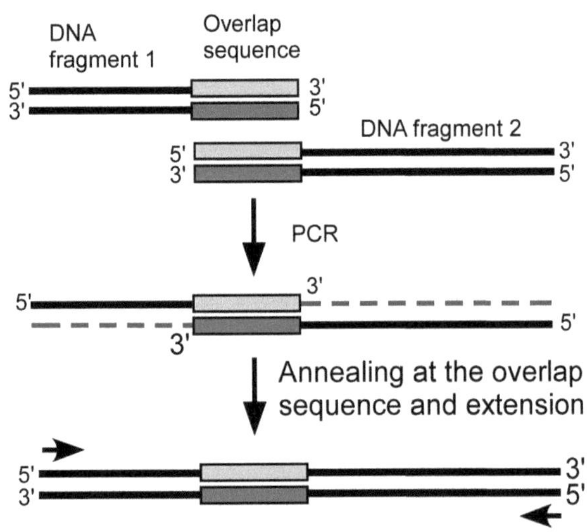


Fig. 1. Fusion PCR by overlap extension. DNA fragment 1 and 2 contain an overlap sequence at the 3′ and 5′ end, respectively. PCR allows annealing of the sequence that overlaps between fragment 1 and 2, followed by DNA polymerase extension from the 3′ end of the overlap sequence. Primers (*arrows*) at the both ends of the fused fragment produce fusion DNA after the extension.

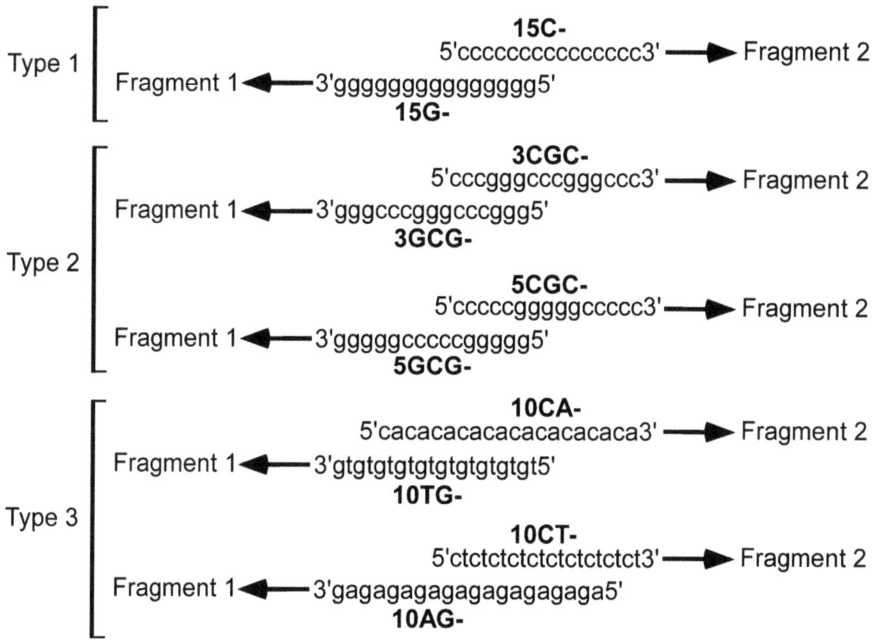


Fig. 2. Sets of overlap sequences for fusion PCR. Each set of primers, such as 15C and 15G, should be used together to amplify the two fragments to be fused, one at the 5′ end and another at the 3′ end. Type 1 sequence is used for fusions that allow mutations in 15C/15G region (see Note 2). Type 2 sequence is useful for in-frame fusion owing to its low mutation frequency. Type 3 sequence is used for GC-rich templates that sometimes prevent amplification with GC-rich primers.

difficulty might be caused by weak annealing between overlap sequences and have solved this issue with the establishment of reliable overlap sequences (12).

In this protocol, we introduce reliable fusion PCR mediated by novel overlap sequences, many of these are short GC-rich sequences (Fig. 2). This approach cannot only be applied for conventional DNA fusion but also for in-frame gene fusion or gene splicing to produce chimeric proteins. Linear DNA obtained by fusion PCR can be used for plasmid constructions, but it is more advantageous if they are used directly without circularization. As an example of linear DNA manipulations, gene construction by targeting through homologous recombination in the yeast *Saccharomyces cerevisiae* was introduced. The resulting transformants are the clones of fused DNA fragments, which can be used for subsequent gene manipulations after their amplification from chromosomes. Other examples of linear DNA manipulations are in the yeast *Kluyveromyces marxianus*, in which a linear DNA without any homologous sequences is randomly and efficiently integrated into chromosomes (13), and in popular mammalian cell lines, in which linear DNA constructs can be directly transfected and expressed (unpublished results). Therefore, we believe that the construction of linear genes formed by fusion PCR has the potential to change numerous molecular genetics studies as it does not require the use of circular plasmids.

## 2. Materials

### 2.1. Oligonucleotide Primers

1. Primers used in this protocol are listed in Table 1. The naming of the oligonucleotide primers is in accordance with naming conventions and directly related to the oligonucleotide features (see Note 1).
2. Oligonucleotide primers were dissolved in sterile water to give 10 μM and stored at −20°C.
3. Oligonucleotides were ordered from commercial oligo DNA providers. Oligonucleotides of low quality were purified by a simple spin column before use but no special quality was required. It should be noted that oligonucleotides obtained from commercial suppliers sometimes have mutations.

### 2.2. PCR and Agarose Gel Electrophoresis

1. KOD Plus DNA polymerase (1 U/μL, Toyobo, Osaka, Japan). Store at −20°C.
2. KOD Plus buffer (10×), 25 mM $MgSO_4$, 2 mM dNTPs. Store at −20°C.
3. DNA molecular weight markers: 1 kb DNA ladder. Store at −20°C.

**Table 1. Oligonucleotide primers**

| Primer name[a] | Sequence (5′–3′)[b] |
|---|---|
| URA3-423 | gatgacaatacagacgatgataaca |
| URA3+1160c | cagtctgtgaaacatctttctacca |
| 15G-yEGFP+717c | GGGGGGGGGGGGGGGttatttgtacaattcatcc |
| 15C-URA3-223 | CCCCCCCCCCCCCCCaagcttttcaattcatctttttttttttttg |
| CDC28+147 | gggtgttcccagtacagccatcaga |
| 5GCG-CDC28+894c | GGGGGCCCCCGGGGGtgattcttggaagtaggggtggatg |
| 5CGC-yEGFP+1 | CCCCCGGGGGCCCCCatgtctaaaggtgaagaatt |
| 15C-PpHIS3-81 | CCCCCCCCCCCCCCCactctggctccagaaggagaaa |
| CDC28+937c-PpHIS3+875c | agtagcatttgtaatataatagcgaaatagattataatgccagtaggtcatctcagcccc |
| TDH3-40(40)-NcSU9+1 | caagaacttagtttcgaataaacacacataaacaaacaaaatggcctccactcgtgtcct |
| 5GCG-NcSu9+87c | GGGGGCCCCCGGGGGagcaacgcggacagcagggc |
| 5GCG-NcSu9+81c | GGGGGCCCCCGGGGGgcggacagcagggcgggcaa |
| 5GCG-NcSu9+60c | GGGGGCCCCCGGGGGcttggcggaagcagccatct |
| 5CGC-yEGFP+4 | CCCCCGGGGGCCCCCgtctctaagggtgaagaattg |
| URA3+1179c | tgttgtgaagtcattgacacag |

[a] See Note 1 for naming

[b] Overlap sequences are indicated by capital letters

4. TAE buffer (50×): Mix 242 g Tris base, 57.1 mL of acetic acid, and 100 mL of 0.5 M EDTA, pH 8.0, and adjust to 1 L with sterile water. Store at room temperature. Dilute to 1× TAE for electrophoresis and agarose gel preparation.
5. Agarose gel: 0.7–1.0% (w/v). Certified molecular biology grade agarose was dissolved by heating in 1× TAE buffer with 0.05 μg/mL ethidium bromide.
6. Sterile filter tips, 100 μL.
7. DNA concentration assay: Qubit fluorometer and Quant-iT dsDNA Assay Kit, broad range (Invitrogen).

### 2.3. Yeast Gene Manipulations

1. Yeast strains: For template preparation, the following *S. cerevisiae* strains were used: RAK3940 (*MAT***a** *his3Δ200 leu2Δ0 met15Δ0 trp1Δ63 ura3Δ0::GAL10*p-yEGFP-$_{15C}$Pp*HIS3*) and BY4704 (*MAT***a** *ade2Δ::hisG his3Δ200 leu2Δ0 lys2Δ0 met15Δ0 trp1Δ63*) (14). For mitochondrial signal fusion, *S. cerevisiae* RAK3655 (*MAT***a** *his3Δ1 leu2Δ0 met15Δ0 ura3Δ0::TDH3*p-*PGK1ter*-Pp*HIS3*) was used as transformation host. Chromosomal DNA from RAK5915 (*MAT***a** *his3Δ1 leu2Δ0 met15Δ0 ura3Δ0::TDH3*p-NcSu9$_{5CGC}$yEGFP2-$_{15C}$*URA3*) and pVT100U-mtGFP plasmid (15) were used for template preparation.
2. Yeast extract-peptone-dextrose (YPD) medium: 1% yeast extract, 2% polypeptone, 2% glucose, and 2% agar (if necessary).
3. Transformation selection medium: synthetic dextrose (SD) dropout media (0.17% yeast nitrogen base without amino acids and without ammonium sulfate, 0.5% ammonium sulfate, 2% glucose, and required nutrients) (1).
4. Polyethylene glycol 3350 (PEG3350): PEG3350 was dissolved in hot sterile water (60–80°C), sterilized by filtration, and adjusted to a final concentration of 60% (w/v).
5. Carrier DNA: 10 mg/mL salmon testes DNA was dissolved in TE buffer (10 mM Tris–HCl, pH 7.5, 1 mM EDTA, pH 8.0), boiled for 10 min, and then cooled immediately on ice for 2–3 min. Store in aliquots at −20°C.
6. Lithium acetate solution: 4 M lithium acetate was dissolved in water and autoclaved. Store at room temperature.

## 3. Methods

Reliable fusion PCR can be achieved with the appropriate design of overlap annealing sequences. Three types of overlap sequences are used for different purposes (Fig. 2). Type 1 is the 15C/15G sequence. This sequence produces reliable and reproducible results even if it is shorter than 15 nucleotides (12), but then might lead to mutations and a low resolution in the sequencing (see Note 2). This sequence type is effective for joining two DNA fragments that have independent functions. Type 2 is 5CGC/5GCG and 3CGC/3GCG. These sequences produce reliable results with low mutation frequency. Therefore, these sequences can be applied to in-frame fusion. No problematic sequencing results were observed in these sequences (see Note 2). Type 3 is 10CA/10TG and 10CT/10AG. Annealing of these sequences is weaker than that in the Type 1 and 2 sequences but effective for amplification from GC-rich templates (see Note 3).

All these overlap sequences have some specificity and therefore can be used simultaneously for multiple fusions, or for fusion to a fragment containing different overlap sequences. Conversely, overlap sequences cannot be used with a fragment that already contains the same sequences (see Note 4).

Basically, conventional PCR DNA polymerases and reaction conditions can be used for fusion PCR without the need for specific adaptation.

### *3.1. Preparation of DNA Fragments to Be Fused*

1. To obtain the PCR products that are to be fused by PCR, DNA fragments are amplified with primers containing overlap sequences (Fig. 2), one with the overlap at the 3′ end and the other at the 5′ end.
2. For templates, plasmid and chromosomal DNA can be used equally well. To obtain chromosomal DNA from yeast, the conventional Zymolyase method (16) or the colony PCR method (17) can be used.
3. PCR amplifications are carried out using KOD Plus DNA polymerase.
4. The PCR mixture contains the following solutions: 6 μL of sterile water, 1 μL of 10× KOD Plus buffer, 1 μL of 2 mM dNTPs, 0.4 μL of 25 mM $MgSO_4$, 0.2 μL of 10 μM forward primer, 0.2 μL of 10 μM reverse primer, 1 μL of template DNA (1–5 ng/μL), and 0.2 μL of KOD Plus polymerase. Total volume: 10 μL.
5. The PCR cycles are comprised of an initial denaturation step at 94°C for 1 min, followed by 30 cycles of 94°C for 20 s, 47–68°C for 30 s, and 68°C for extension time calculated by 1 min for 1 kb.
6. Prepare 0.7% (w/v) agarose gel.
7. The PCR products are loaded and separated through the agarose gel at 100 V for 30 min. In the example shown in Fig. 3, fragment 1 was amplified using the primers URA3-423 and 15G-yEGFP+717c and chromosomal DNA template from RAK3940 strain, and fragment 2 was amplified using

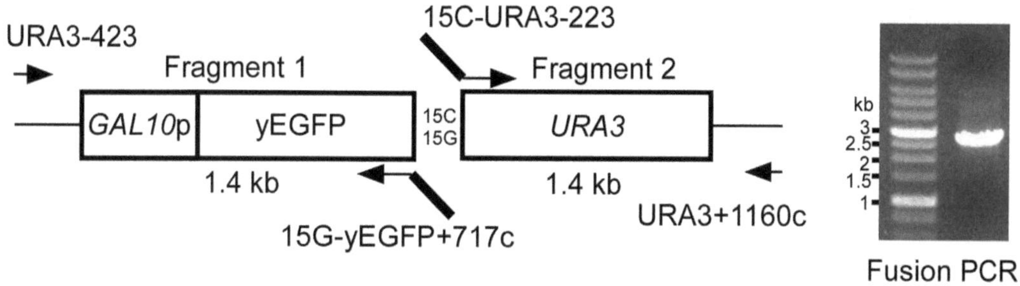


Fig. 3. Fusion PCR between *GAL10*p-yEGFP and *URA3* transformation marker gene using the 15C/15G overlap sequence. *Thick bars with arrows* indicate the 15C and 15G sequences in the primers. PCR of fragment 1 and 2: annealing at 47°C and 55°C, respectively, with 2-min extension. Fusion PCR: 60°C annealing with 3-min extension. One microliter of ten-microliter reaction mixture was subjected to agarose gel electrophoresis.

15C-URA3-223 and URA3+1160c primers and chromosomal DNA template from BY4704 strain.

8. To extract DNA products from agarose gel, expose the agarose gel to ultraviolet light (see Note 5), excise the products by spatula, and transfer the DNA-gel piece onto the top of a sterile tip filter, which is placed in a 1.5-mL microtube by cutting the tip end to fit to a microtube (18).
9. Centrifuge the filter tip in a microtube to obtain DNA solution in a microfuge at 8,000 rpm (4,000 g) for 3 min. Remove filter tip.
10. The DNA concentration of extracted DNA solutions is quantified by fluorometry (Quant-iT dsDNA Assay Kit as recommended by the supplier, Invitrogen, Carlsbad, CA). DNA concentration is adjusted to 0.5 ng/μL (see Note 6).

### 3.2. Fusion PCR

1. Mix 1 μL of (10×) KOD Plus buffer, 1 μL of 2 mM dNTPs, 0.4 μL of 25 mM $MgSO_4$, 0.2 μL of 10 μM forward primer, 0.2 μL of 10 μM reverse primer, 0.4 μL of DNA fragment 1, 0.4 μL of DNA fragment 2, 0.2 μL KOD Plus DNA polymerase, and sterile water to give a final volume of 10 μL. In the example shown in Fig. 3, URA3-423 and URA3+1160c primers were used.
2. The PCR cycles are similar to the conditions in Subheading 3.1. PCR cycles are comprised of an initial denaturation step at 94°C for 1 min, followed by 30 cycles of 94°C for 20 s, 60°C for 30 s, and 68°C for 3 min. For fusion PCR, the annealing temperature was usually increased (see Note 7). Extension time was calculated according to the resulting DNA size.
3. After fusion PCR, 1 μL of sample is mixed with gel loading buffer and subjected to agarose gel electrophoresis (Fig. 3).

### 3.3. Fusion PCR of Three Fragments

Fusion PCR with GC-rich overlap sequences can be applied to the fusion of multiple fragments because each overlap sequence is specific to its counterpart. We obtained successful fusions of three fragments with the simultaneous use of overlaps between 15C/15G and 5CGC/5GCG, between 15C/15G and 3CGC/3GCG, and between 5CGC/5GCG and 3CGC/3GCG (12). However, three-fragment fusions are more difficult to achieve than two-fragment fusions. One complicating factor is the required adjustment for DNA concentration (Fig. 4). We also succeeded at four-fragment fusions, but the method was not reliable (data not shown).

1. Two different overlap sequences are designed in three-fragment fusions. In Fig. 4, DNA fragment 1 (*CDC28*) was amplified using 5GCG-containing primer (5GCG-CDC28+894c) at the 3′ end. Fragment 3 (Pp*HIS3*) was amplified using 15C-containing primer (15C-PpHIS3-81) at the 5′ end. Fragment 2 (yEGFP) was amplified using 5CGC-containing primer (5CGC-yEGFP+1) at the 5′ end and 15G-containing primer (15G-yEGFP+717c) at the 3′ end. *CDC28* and yEGFP were fused by 5CGC/5GCG, and yEGFP and Pp*HIS3* were fused by 15C/15G.

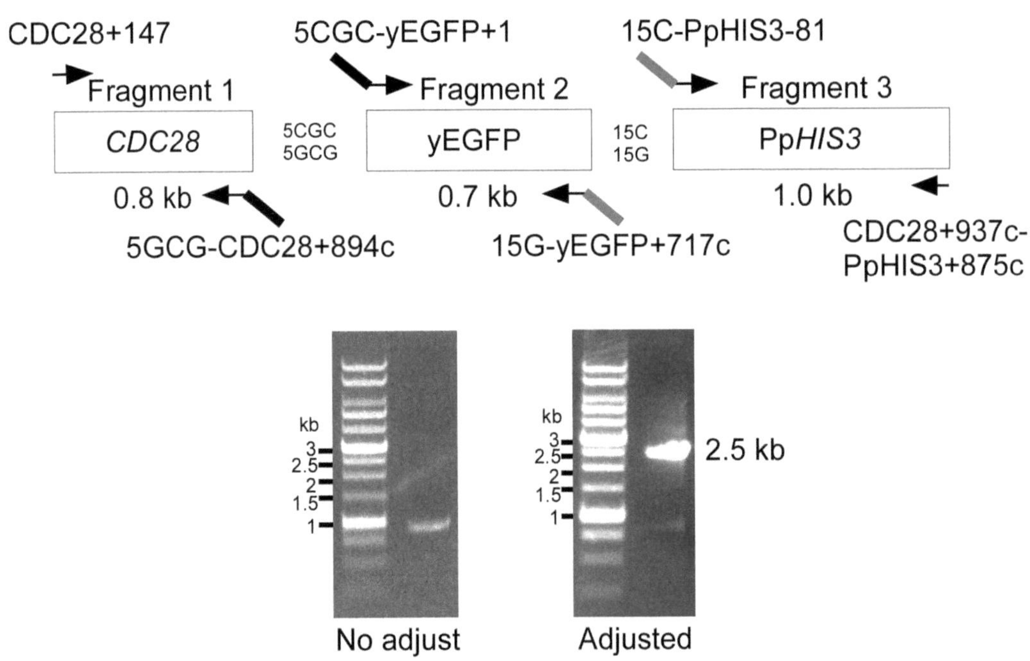


Fig. 4. Fusion PCR of three DNA fragments. Three DNA fragments were amplified in the first round PCR with primers containing overlap sequences. *CDC28*, yEGFP, and Pp*HIS3* DNA concentrations were measured: 2.38 ng/μL, 1.72 ng/μL, and 0.5 ng/μL, respectively. For DNA fusion, either 0.4 μL each of nonadjusted DNA solutions or 0.4 μL each of adjusted DNA (0.5 ng/μL each) was mixed in a reaction tube to give a final volume of 10 μL and amplified using CDC28+147 and CDC28+937c-PpHIS3+875c primers. An incorrect DNA fusion band appeared in nonadjusted DNA mixture.

2. The DNA fragments to be fused are amplified using the same method described above. All DNA solutions are adjusted to 0.5 ng/μL, and the fusion PCRs are performed using similar conditions, with the annealing temperature adjusted according to primers used. Extension time was 3 min. Higher annealing temperatures between 65°C and 68°C are usually used for multiple fusions (see Note 7).
3. When the DNA concentration of the fragments was not adjusted, the fusion PCR failed (see "No adjust" lane in Fig. 4), but when it was adjusted, the correct band of 2.5 kb could be detected, indicating that DNA concentration of the fragments needs to be adjusted for multiple-fragment fusion.

#### 3.4. Construction of a Chimeric Gene by In-Frame Fusion PCR in Yeast

DNA fusion with GC-rich overlap sequences can also be applied to in-frame gene fusion. This is useful for gene splicing (deletion), chimeric gene construction, and in-frame marker or tag addition. For in-frame fusion, continuous C/G, CA/TG, and CT/AG sequences are not appropriate because these showed lower efficiency of correct in-frame fusion owing to mutations at the overlap sequences (12). Annealing between these sequences causes errors by slipped annealing. In contrast, the sequences 5CGC and 3CGC were successfully used for in-frame fusion. These sequences produced the N-ProProGlyAlaPro-C peptide in the case of 5CGC (Fig. 5a) and N-ProGlyProGlyPro-C peptide for 3CGC, which

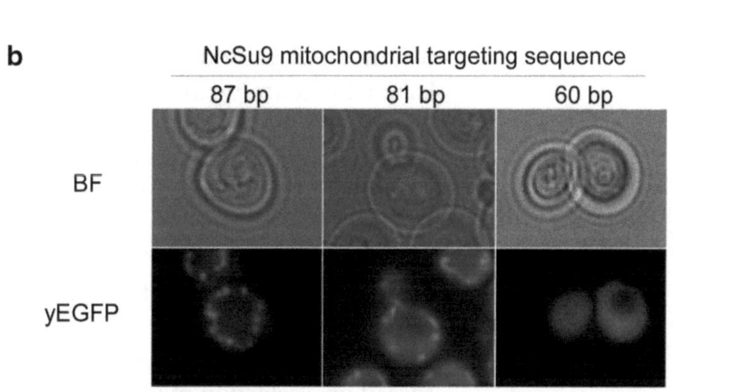


Fig. 5. In-frame gene fusion with a 5CGC/5GCG overlap sequence. Mitochondrial-targeting sequence (NcSu9) from *Neurospora crassa* (15) was deleted, fused with yEGFP, and expressed in yeast. The transformants were subjected to fluorescence and bright field (BF) microscopy.

were expected to have negative effects on protein function (see Note 8):

1. In-frame primers containing 5CGC/5GCG sequences were designed (Fig. 5a). NcSu9 is a mitochondrial-targeting signal derived from *Neurospora crassa* (15). To fuse NcSu9 with yEGFP (yeast codon-optimized EGFP) and clone it in yeast, yEGFP was amplified using 5CGC-yEGFP+4 containing the 5CGC sequence at its 5′ end. The NcSu9 sequence was amplified using 5GCG-NcSu9+nnc; nn denotes sequence numbers for deletion (Table 1).
2. The DNA fragments to be fused were amplified from chromosomal DNA of RAK5915 (yEGFP2-*URA3*: 5CGC-yEGFP+4 and URA3+1179c primers) and from pVT100U-mtGFP plasmid (NcSu9: TDH3-40(40)-NcSu9+1 and one of 5GCG-NcSu9+nnc primers).
3. The two DNA fragments were extracted from agarose gel, and their concentrations were adjusted. The fragments were then mixed with TDH3-40(40)-NcSu9+1 and URA3+1179c primers in a PCR mixture. Fusion PCR was carried out in similar conditions as described above, at 60°C annealing temperature and with a 2-min extension time.
4. The PCR product was analyzed by agarose gel electrophoresis to confirm a correct fusion size. This product contained 40 bp of *TDH3* promoter sequence and ~300 bp of *URA3* downstream sequences at the ends for homologous recombination in yeast.
5. The fusion product then was transformed into yeast strain RAK3655 that contained constitutive strong promoter of *TDH3* gene in the *ura3* locus (*ura3Δ0::TDH3*p-*PGK1ter*-Pp*HIS3*). For transformation, yeast cells were initially grown in YPD overnight, after which 1 mL of the culture was taken and mixed with 9 mL of fresh YPD and incubated for a further 5 h at 30°C without shaking. The yeast cells were then collected by centrifugation, washed once with 5 mL of sterile water, and the supernatant was removed by decanting. The yeast cells were vortexed for suspension in the residual water. The final yeast volume was approximately 150 μL. To the transformation solution (a mixture of 120 μL of 60% PEG 3350, 10 μL of 10 mg/mL carrier DNA, and 5 μL of 4 M lithium acetate), 61 μL of the yeast suspension and 4 μL of the in-frame fusion product (directly from PCR mixture) were added. The mixture was incubated at 42°C for 1 h and then spread on a selection medium plate. The plate was incubated at 30°C for 3 days.
6. The transformant colonies were picked on a YPD plate. The yeast cells were cultured in 1 mL YPD for 1 day at 30°C, before

being centrifuged and washed with sterile water. The yeast cells were then observed by fluorescence microscopy (Carl Zeiss, Axio Imager) using a GFP filter (38HE 489038-0000) (Fig. 5).

7. NcSu9 fragments containing the nucleotides 1–87 and 1–81 showed mitochondrial localization with a frequency of 1/5 and 3/5 clones, respectively, whereas the 1–60 showed cytoplasmic localization in 6/6 clones. The NcSu9 mitochondrial localization signal was initially thought to be 207 bp long, but this fusion result indicated that the first 1–81 nucleotides encoding for 27 amino acids are sufficient to direct mitochondrial localization.

## 4. Notes

1. Oligonucleotide primers are designated by rules to make it easy to understand their features:

   *Rule 1.* Oligonucleotide names are given according to the sequence from the 5′ to the 3′ end. For example, "9C" indicates 5′-ccccccccc-3′ but not 3′-ccccccccc-5′ in the oligonucleotide sequence.

   *Rule 2.* Many of the oligonucleotides are sequences related to their appearance in genes. These oligonucleotides can be identified by their distance from the ATG start codon of the respective gene. When position A of ATG start codon is counted as +1, the 5′ ends of many of the primers can be indicated by the distance from this +1 position. For example, URA3-423 indicates the location that is 423-bp upstream from position A of the ATG of the *URA3* gene, or CDC28+147 indicates the location that is 147-bp downstream from ATG of *CDC28* gene.

   *Rule 3.* Oligonucleotides that relate to genes are also designed to be complementary to these genes. This can be marked by including "c" at the end of their name. For example, URA3+1179c indicates that the 5′ end is located at position 1,179 downstream from the A of ATG of *URA3* gene and that the sequence is complementary to the *URA3* coding sequence.

   *Rule 4.* Names can be combined according to the 5′–3′ sequence and joined with a hyphen. For example, 15C-URA3-223 indicates that this oligonucleotide contains 15C at the 5′ end, followed by the URA3-223 sequence. CDC28+937c-PpHIS3+875c indicates that CDC28+937c is located at the 5′ region and PpHIS3+875c is at the 3′ region. Primer length or annealing sequence length is indicated by parenthesis, if necessary.

For example, TDH3-40(40)-NcSU9+1 indicates that TDH3-40 contains 40 nucleotides from the –40 site.

*Comment 1.* If URA3-423 and URA3+1160c were used for normal amplification, this will produce a 1,583-bp fragment fragment (i.e., 423 + 1,160 = 1,583). Length of PCR products can therefore often be estimated from primer names.

*Comment 2.* Researchers can find desired primers from the list of primers by names or design similar primers as speculated from their names.

*Comment 3.* Accuracy of the sequence position may not be important as there are often many small mutations, such as deletions and additions, in template sequences. Even if named primer sites are not completely correct, speculation based on their names might still be useful.

2. Sequencing of 15C/15G fusion region, but not of 5CGC/5GCG and 3CGC/3GCG, was problematic to resolve (Fig. 6), and the long stretches of single nucleotide might inhibit normal DNA amplification. Low resolution in the region downstream from the 15G stretch was also observed.
3. Low PCR amplification, or none at all, from certain GC-rich templates, such as mammalian and virus genes, was experienced with primers containing Type 1 and 2 sequences. This suggests that primer with a high GC content anneal to several places on the templates, resulting in weak amplification. For such templates, primers with Type 3 sequences are more effective for the conventional amplification and fusion PCR.
4. Indication of the presence of the overlap sequences in template fragments or clones is important. For this reason, we indicated the presence of these sequences in the genotype of yeast strains, i.e., *ura3Δ0::GAL10*p-yEGFP-$_{15C}$Pp*HIS3* of RAK3940 (see Subheading 2.3), by subscript. In this case, 15C sequence is located between yEGFP and Pp*HIS3* genes.

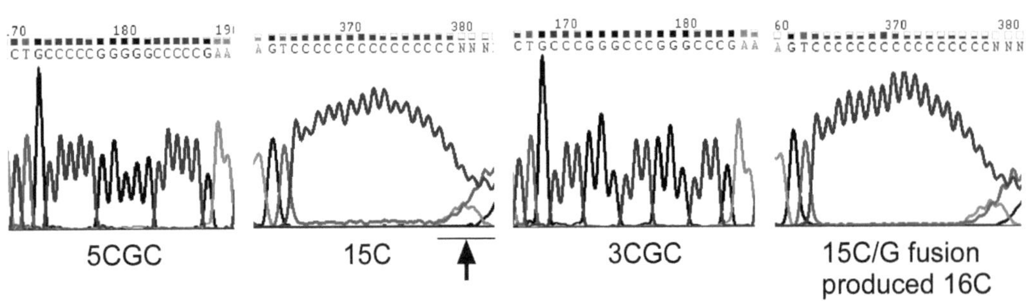


Fig. 6. Sequencing of fusion regions. DNA sequencing of 15C sequence showed low resolution in the downstream region (*arrow*). This was also observed in 15G (*data not shown*). 15C/15G overlap regions often produced incorrect sequences, such as 14C, 16C, and 17C. DNA sequencing was performed with BigDye terminator V3.1 cycle sequencing kit (Applied Biosystems, Foster City, CA).

5. To reduce DNA damage by UV light, agarose gel containing DNA should not be exposed to a strong UV light source for any extended period of time. Using a UV light of 365 nm reduces DNA damage.
6. DNA concentrations in conventional PCR are approximately 1–5 ng/μL. In this concentration range, DNA can be used directly for fusion PCR of two fragments without further adjustment.
7. The annealing temperatures usually range from 60°C to 68°C. This reduces unspecific annealing. This condition requires longer primers for the fusion PCR. We usually design primers longer than 22 nucleotides, preferably containing more than 50% GC content. Without high enough GC content, longer primers of 25–30 nucleotides are designed.
8. The GC-rich overlap sequences can be located in the protein-coding region as described in Subheading 3.4 and at the boundary between an open reading frame (ORF) and a terminator without having any negative effect, whereas gene expression was often affected when the overhang is located at the boundary between the promoter and an ORF (unpublished results).

## Acknowledgments

The authors thank Sachiko Watanabe, Moemi Yoshiura, Ryota Sakai, and Yukie Misumi for their technical assistance. This work was supported in part by the New Energy and Industrial Technology Development Organization (NEDO) and by the Program for Promotion of Basic Research Activities for Innovative Biosciences (PROBRAIN), Japan.

## References

1. Ausubel FM, Brent R, Kingston RE et al (1999) Short protocols in molecular biology. John Wiley & Sons, Inc, New York
2. Hartley JL, Temple GF, Brasch MA (2000) DNA cloning using *in vitro* site-specific recombination. Genome Res 10:1788–1795
3. Marchuk D, Drumm M, Saulino A, Collins FS (1990) Construction of T-vectors, a rapid and general system for direct cloning of unmodified PCR products. Nucleic Acids Res 19:1154
4. Benoit RM, Wilhelm RN, Scherer-Becker D, Ostermeier C (2006) An improved method for fast, robust, and seamless integration of DNA fragments into multiple plasmids. Prot Express Purific 45:66–71
5. Rashtchian A (1995) Novel methods for cloning and engineering genes using the polymerase chain reaction. Curr Opin Biotech 6:30–36
6. Horton RM, Hunt HD, Ho SN, Pullen JK, Pease LR (1989) Engineering hybrid genes without the use of restriction enzymes: gene splicing by overlap extension. Gene 77:61–68
7. Horton RM, Ho SN, Pullen JK, Hunt HD, Cai Z, Pease LR (1993) Gene splicing by overlap extension. Method Enzymol 217:270–279

8. Warrens AN, Jones MD, Lechler RI (1997) Splicing by overlap extension by PCR using asymmetric amplification: an improved technique for the generation of hybrid proteins of immunological interest. Gene 186:29–35
9. Lu Q (2005) Seamless cloning and gene fusion. Trends Biotechnol 23:199–207
10. Heckman KL, Pease LR (2007) Gene splicing and mutagenesis by PCR-driven overlap extension. Nat Protocol 2:924–932
11. Vallejo AN, Pogulis RJ, Pease LR (1994) *In vitro* synthesis of novel genes: mutagenesis and recombination by PCR. Cold Spring Harbor Laboratory Press; PCR Methods and Applications, s123–s130
12. Cha-aim K, Fukunaga T, Hoshida H, Akada R (2009) Reliable fusion PCR mediated by GC-rich overlap sequences. Gene 434:43–49
13. Abdel-Banat BMA, Nonklang S, Hoshida H, Akada R. (2010) Random and targeted gene integrations through the control of non-homologous end joining in the yeast *Kluyveromyces marxianus*. Yeast 27:29–39
14. Brachmann CB, Davies A, Cost GJ, Caputo E, Li J, Hieter P, Boeke JD (1998) Designer deletion strains derived from *Saccharomyces cerevisiae* S288C: a useful set of strains and plasmids for PCR-mediated gene disruption and other applications. Yeast 14:115–132
15. Benedikt W, Walter N (2000) Mitochondria-targeted green fluorescent proteins: convenient tools for the study of organelle biogenesis in *Saccharomyces cerevisiae*. Yeast 16:1421–1427
16. Amberg DC, Burke D, Strathern JN (2005) Method in yeast genetics: A Cold Spring Harbor Laboratory Course Manual, New York
17. Akada R, Murakane T, Nishizawa Y (2000) DNA extraction method for screening yeast clones by PCR. Biotechniques 28:668–670
18. Dean AD, Greenwald JE (1995) Use of filtrated pipet tips to elute DNA from agarose gels. Biotechniques 18:980

# Chapter 9

# Using Recombineering to Generate Point Mutations: The Oligonucleotide-Based "Hit and Fix" Method


## Suhwan Chang, Stacey Stauffer, and Shyam K. Sharan



## Abstract

Ability to manipulate the genome or design genes with desired mutation is critical for functional studies. Recombineering has made genetic manipulation of large genomic fragments very feasible and efficient. In the bacteriophage lambda-based recombineering system, three prophage genes, *exo*, *bet*, and *gam*, under the control of a temperature-sensitive lambda cI-repressor, provide the recombination function. The high efficiency of recombineering by oligonucleotides allows generation of subtle alterations in the bacterial chromosomal DNA as well as episomal DNA. We describe here a two-step "Hit and Fix" method, in which a short heterologous sequence is inserted to the target site first (Hit) and this sequence is replaced with the desired mutation in the second step (Fix). Insertion and replacement of the heterologous sequence allows screening of the recombinant clones by PCR or colony hybridization.



**Key words:** Recombineering, "Hit and Fix" method, Oligonucleotide, Bacterial artificial chromosome, Point mutation


## 1. Introduction

Recombineering is rapidly becoming a standard technology for genetic engineering (1–4). It can be used as an efficient method to generate site-specific mutations in bacterial artificial chromosomes for functional studies. Many different recombineering systems have been described (2, 3, 5). The method described here utilizes the bacteriophage lambda Red recombination system (2). The bacteriophage lambda genes needed for recombineering are *exo*, *bet*, and *gam*. The *exo* gene product has 5′–3′ exonuclease activity, and the *bet* gene product is a single-strand DNA binding protein that promotes annealing. The *gam* gene product inhibits the recBCD nuclease, thus preventing the degradation of linear DNA fragments.


Jean Peccoud (ed.), *Gene Synthesis: Methods and Protocols*, Methods in Molecular Biology, vol. 852, DOI 10.1007/978-1-61779-564-0_9, © Springer Science+Business Media, LLC 2012

The minimum homology required for recombination is about 35 bases (2). An *Escherichia coli* strain, DY380, along with its derivative SW102, harboring a defective lambda prophage, has been developed such that it promotes much higher recombination efficiencies, as described in the previous chapter (2, 6). In these bacterial strains, the prophage provides the recombination genes *exo*, *bet*, and *gam* that are under the control of a temperature-sensitive lambda cI-repressor. Therefore, these genes are switched on by inactivation of the repressor by transiently shifting the culture from 32°C to 42°C.

Dy380 and its derivative SW102 have been successfully used to modify exogenous DNA cloned into a bacterial artificial chromosome (BAC) vector. However, occasionally, it can be difficult to transfer some BACs into these cells. To overcome this hurdle, a more versatile phage system was sought that would generate high yields of recombinants but at the same time could be easily moved into strains that contain the BAC clones. A few such mobile recombineering systems have been generated, such as the mini-lambda (mini-λ), the pSIM plasmid systems, and the replication-defective λ phage, λTetR (7–9). In this chapter, we describe the use of the mini-λ circular DNA for BAC engineering.

In the mini-λ, all replication and lysis genes have been deleted, creating a replication-defective, smaller version of the lambda prophage (12.5 kb). The mini-λ provides an inducible Red recombination system that can be introduced by electroporation into nearly any *E. coli* strain, including the *recA* mutant DH10B and its derivatives carrying BACs (7). The mini-λ also contains an antibiotic resistance gene (e.g., tetracycline, kanamycin) as a selection marker. Unlike the lambda prophage in DY380 or SW102 cells, the attachment sites *attL* and *attR* are present in the mini-λ. Consequently, inactivation of the repressor at restrictive temperature activates *int* and *xis* gene expression, causing site-specific excision of the prophage DNA circle. Thus, the cells can be cured of the mini-λ DNA and do not remain temperature sensitive. The excision of the mini-λ DNA from the chromosomal DNA also allows purification of the DNA circles from bacterial cells using standard plasmid purification protocols. The mini-λ therefore provides a tractable system to transiently express the phage recombination genes in bacterial cells.

In this chapter, we describe the generation of point mutations without the use of any selection maker (e.g., *galK*, as described in the previous chapter). This approach is particularly helpful when multicopy episomal DNA is being manipulated where selection-counterselection method is not feasible. To generate point mutations without the use of any selectable marker by recombineering, we describe the use of 180-mer oligonucleotides in a two-step "Hit and Fix" method (10). This method is based on the high efficiency of recombineering mediated by single-stranded oligonucleotides (11, 12). The two-step "Hit and Fix" method was developed to facilitate the screening of recombinant clones (Fig. 1a). This is achieved

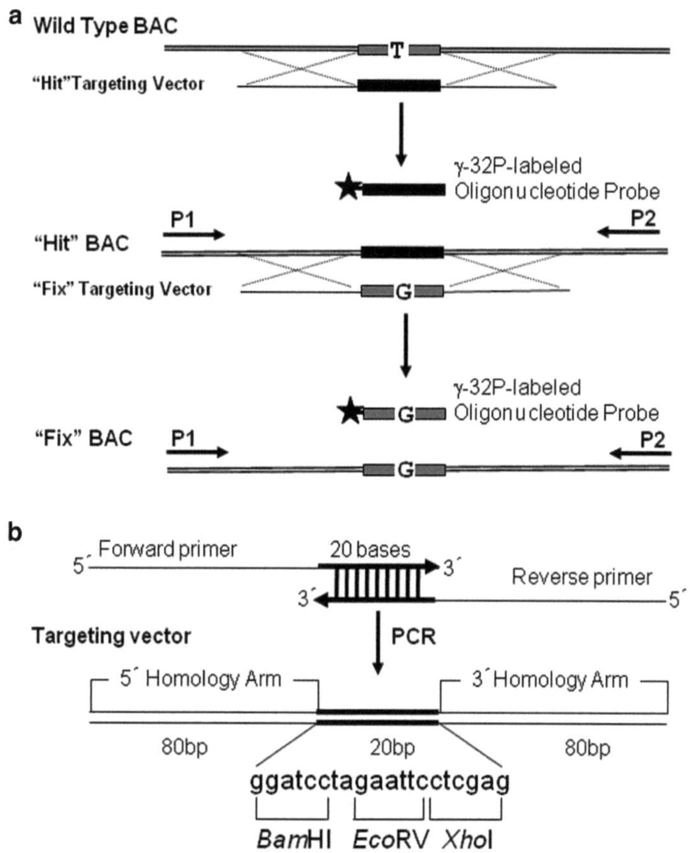


Fig. 1. Schematic representation of the "Hit and Fix" method to generate subtle mutations using single-stranded oligonucleotides as targeting vector. (**a**) Scheme to generate subtle alterations (e.g., T to G) without the use of a selectable marker using 180-mer oligonucleotides by the two-step "Hit and Fix" method. In step 1, a 180-mer single-strand oligonucleotide is used to replace 20 nucleotides (*gray box*) around the target site with 20 bases of a heterologous sequence (*black box*). Recombinants can be identified by colony hybridization using an end-labeled 20-mer oligonucleotide. Generation of a correct recombinant clone can be confirmed by digesting the PCR product (~300–500 bp) of primers P1 and P2 with *Bam*HI, *Eco*RV, or *Xho*I (recognition sequence present in the heterologous sequence). In step 2, the 20 nucleotides are restored to the original sequence, except for the desired mutation. Such clones can be identified by colony hybridization using a 20-mer oligonucleotide as probe. The generation of correct recombinant clones can be tested by the loss of the restriction sites by digesting the PCR product of primers P1 and P2, and confirmed by sequencing. (**b**) The single-stranded oligonucleotides containing 80 bases of 5′ and 3′ homologies and 20 bases of a heterologous sequence (contains restriction sites *Bam*HI, *Eco*RV, *Xho*I) can be generated by using two 100-mer oligonucleotides as forward and reverse primers in a PCR. The two 100-mer oligonucleotides have 20 complementary bases at their 3′ ends. To obtain single-stranded oligonucleotides, the 180-bp PCR product can be denatured at 94°C for 5 min and chilled on ice.

by the introduction of 20 unique nucleotides (unique heterologous sequence) to the site where a mutation has to be generated in the first "Hit" step. In the second "Fix" step, the original sequence is restored with the exception of the point mutation. In both steps, 180-bases-long single-stranded DNA fragments generated by PCR are used as targeting vectors (Fig. 1b). The 20-nucleotide sequence can be used for initial screening by PCR or colony hybridization. Here, we describe the use of the hybridization approach because it allows screening of more than 5,000 individual colonies from a single 15-cm agar plate (13). It also allows simultaneous screening of multiple mutations (in "Hit" step) by using the same hybridization probe. In addition, the heterologous sequence is designed to contain several restriction sites that facilitate rapid confirmation of the correct targeting; 180-mer single-stranded oligonucleotides that are used as targeting vectors are generated by PCR using two 100-mer oligonucleotides with 20 base complementary sequences at their 3′ ends, followed by denaturation of the PCR product.

## 2. Materials

### 2.1. Bacterial Strain

1. *E. coli* DH10B strain containing the bacteriophage mini-λ DNA. This strain is tetracycline resistant. It is temperature sensitive and must be grown at 32°C.
2. *E. coli* DH10B containing the BAC clone of interest.

### 2.2. Equipment

Standard laboratory equipments are used, some of which are described below:

1. An incubator set at 32°C.
2. A shaking incubator set at 32°C and a shaking water bath (200 rpm) set at 42°C.
3. Electroporator and cuvettes with 0.1 cm gap.
4. High-speed centrifuge and a refrigerated microcentrifuge.
5. Thermal cycler and accessories for PCR.
6. Spectrophotometer.
7. Hybridization oven.
8. UV crosslinker 1800.
9. 1.5-ml microfuge tubes.
10. 0.2-ml flat-cap PCR tubes.
11. Insulated ice buckets.
12. Sterile glass culture tubes (16 × 150 mm) for overnight growth of bacterial cultures.
13. Stainless steel closures for culture tubes.
14. Pipetters of various volumes with aerosol-resistant sterile tips.
15. Petri plates, 100 × 15 mm.

### 2.3. Other Reagents

1. Plasmid DNA purification reagents (Qiagen).
2. Gel extraction kit or reagents to purify DNA from agarose gel (Qiagen).
3. Expand High Fidelity (HiFi) PCR System (Roche).
4. Restriction enzymes (New England Biolabs).
5. Standard Taq polymerase (Invitrogen).
6. dNTP mixture, 10 mM each, PCR grade (Invitrogen).
7. Agarose (SeaKem LE, ISC Bioexpress).
8. Primers for PCR amplification of recombineering substrates, 25 pmol/μl in $H_2O$.
9. Tetracycline (Sigma).
10. Chloramphenicol (Sigma).
11. T4 polynucleotide kinase and buffer (New England Biolabs).
12. Quick Spin Sephadex G-25 columns (Roche).
13. *LB (Luria broth) agar plate*: for 1 L broth, 10 g Bacto-tryptone; 5 g yeast extract; 5 g NaCl, pH 7.2; and 15 g agar.
14. *"Superbroth special"*: for 1 L broth, 35 g Bacto-tryptone, 20 g yeast extract, and 5 g NaCl.
15. *SOC medium*: for 1 L media, 20 g Bacto-tryptone, 5 g yeast extract, 2 ml of 5 M NaCl, 2.5 ml of 1 M KCl, 10 ml of 1 M $MgCl_2$, 10 ml of 1 M $MgSO_4$, and 20 ml of 1 M glucose.
16. [γ-32P] ATP (5,000 Ci/mmole, 10 mCi/ml, Amersham).
17. X-ray film.

## 3. Methods

### 3.1. Preparation of Mini-λ DNA

1. Pick an isolated colony of *E. coli* DH10B containing mini-λ DNA and inoculate 125 ml of "Superbroth special" media containing tetracycline (12.5 μg/ml) in a 1-L flask. Culture overnight in a shaking incubator at 32°C.
2. Next morning, induce excision of mini-λ DNA circles by culturing at 42°C for 15 min in a shaking water bath.
3. Chill the culture on ice/water slurry for 15 min, with occasional shaking.
4. Extract the mini-λ DNA from the cells using a plasmid purification kit.
5. Resuspend DNA pellet in 50 μl of 1× TE buffer (see Note 1).

### 3.2. Preparation of Electrocompetent DH10B Cells Containing Desired BAC

1. Pick an isolated colony of DH10B cells containing the desired BAC clone and grow overnight in 3 ml of "Superbroth special" media at 32°C.
2. Next morning, transfer 1 ml of the overnight culture to 50 ml of "Superbroth special" medium in a 250-ml flask and grow at 32°C to an $OD_{600}$ of 0.55–0.60. Transfer 10 ml of the culture to a 50-ml Oak Ridge centrifuge tube and centrifuge at 6,000 × *g* in a prechilled rotor for 10 min at 1°C.
3. Gently discard the supernatant and wash the cells twice with 25 ml of ice-cold, sterilized water followed by centrifugation at 6,000 × *g* in a prechilled rotor for 10 min at 1°C. Carefully remove the supernatant using a pipette and resuspend the pellet in 1 ml of water and transfer to a chilled 1.5-ml tube. Centrifuge at 18,000 × *g* for 20–30 s at 1°C (see Note 2).
4. Wash the cells two more times with 1 ml of ice-cold water. Resuspend the cell pellet in water to a total volume of 50 μl and keep on ice.

### 3.3. Electroporation of Mini-λ DNA into DH10B Cells Containing the BAC

1. Mix 1 μl of mini-λ DNA (25–50 ng) with 50 μl of electrocompetent DH10B cells containing the BAC and keep on ice for 5 min.
2. Transfer into a 0.1-cm gap prechilled cuvette. Set the Gene Pulser at 1.8 kV, 25 μF capacitance, and 200 Ω resistance. Electroporate the DNA into the *E. coli* and immediately add 1 ml of SOC medium. Transfer the cells into a 15-ml falcon tube.
3. Grow cells at 32°C for 1 h. Spin down the cells and resuspend in 200 μl of "Superbroth special" medium.
4. Plate the cells on an LB agar plate containing tetracycline (12.5 μg/ml) and incubate overnight at 32°°C.
5. Pick isolated colonies for recombineering. Make glycerol stocks and freeze at −80°C for future use.

### 3.4. "Hit" Step: Insertion of a 20-mer Heterologous Sequence at the Site Where the Point Mutation Needs to Be Generated

#### 3.4.1. Generation of a Targeting Vector for the "Hit" Step

A targeting vector for the "Hit" step is generated to insert 20 bases of heterologous sequence in the BAC at the site where a point mutation needs to be generated. The targeting vector is synthesized by PCR using two 100-mer oligonucleotides with overlapping 20 nucleotides at their 3′ ends. The resulting targeting vector contains two homology arms that are 80 bases in length, flanking a 20-mer heterologous sequence in the middle (Fig. 1b). This heterologous sequence (5′-GGATCCTAGAATTCCTCGAG-3′) can be the same for all "Hit" targeting vectors. This heterologous sequence contains multiple restriction sites that can be used to confirm correct targeting, by restriction enzyme digestion of the DNA amplified from the targeted region:

1. Set up the following reaction using the Expand High Fidelity (HiFi) PCR System (Roche). Use 6 μl of each 100-mer oligonucleotide (10 μM) and 10 μl of 2 mM dNTPs, 2 μl of HiFi Taq Polymerase (3.5 U/μl) in a 100-μl PCR. PCR cycle includes an initial denaturation at 94°C for 1 min, followed by 30 cycles of 94°C for 30 s, 55–60°C for 30 s, and 72°C for 30 s, and a final extension at 72°C for 2 min.
2. Examine 1 μl of the reaction on a 1.0–1.5% agarose gel in 1× TAE buffer to confirm the amplification of a 180-bp PCR product.
3. Purify the targeting vector using a Qiagen PCR Purification Kit and elute in 30 μl of Qiagen Elution Buffer or ethanol-precipitate and dissolve in 20–30 μl $dH_2O$.
4. Denature the PCR product by heating at 94°C for 5 min and immediately chill on ice to obtain single-stranded 180-mer oligonucleotides.

### 3.5. Induction of the Lambda Recombination Genes and Preparation of Electrocompetent Cells

1. Inoculate DH10B cells containing the BAC and the mini-λ from an isolated single colony into 3 ml "Superbroth special" medium. Culture overnight at 32°C in a shaking incubator.
2. Add 1 ml of the overnight culture to 30 ml of "Superbroth special" medium in a 250-ml baffled Erlenmeyer flask.
3. Grow cells at the 32°C until the $OD_{600}$ is between 0.55 and 0.6.
4. Transfer 10 ml of the culture in a 50-ml Oak Ridge tube and store on ice. These cells will be used as uninduced control (see Note 3).
5. Transfer 10 ml of the culture into a 50-ml Erlenmeyer flask and place in a 42°C shaking water bath. Shake for 15 min at 200 rpm to induce the lambda recombination genes.
6. Immediately chill the flask on ice/water slurry with gentle swirling for 10–15 min.
7. Transfer the culture into a prechilled 50-ml Oak Ridge tube.
8. Centrifuge both the induced and uninduced cultures for 10 min at 4,600 × *g* at 4°C.
9. Make cells electrocompetent, as described in Subheading 3.2.

### 3.6. Electroporation of the Targeting Vector into DH10B Cells Containing the BAC

1. Chill two 0.1-cm electroporation cuvettes in a freezer.
2. In a 0.5-ml tube, mix appropriate volume of DNA (200–300 ng of salt-free PCR fragment) with 50 μl of electrocompetent induced or uninduced cells. Leave the tubes on ice for 5 min.
3. Electroporate the DNA into the cells using 1.8 kV, 25 μF capacitance, and 200 Ω resistance. Immediately add 1 ml SOC medium to the cuvette and transfer the electroporation mix to sterile 15-ml culture tubes. Incubate the tubes with shaking at 32°C for 1.5 h.

4. Serially dilute the cell suspension in "Superbroth special" medium and plate 200 μl of $10^{-3}$ and $10^{-4}$ dilutions onto a 15-cm LB agar plate containing an appropriate antibiotic (chloramphenicol at 12 μg/ml). Use sterile glass beads instead of a bacteriological spreader to achieve uniform distribution of colonies throughout the plate. Incubate agar plates for 18–22 h at 32°C.

### 3.7. Identifying the Recombinant Clones by Colony Hybridization

1. Pick the plate with approximately 3,000–6,000 colonies. Avoid using plates with too few or too many colonies.
2. Cover the plate with a circular, charged nylon membrane (Hybond) and mark three edges by punching holes with a 17-gauge needle.
3. Carefully lift up the membrane with two forceps. Autoclave the membrane for 1 min in dry cycle to denature the bacterial DNA. Cross-link DNA on the membrane by using a UV cross-linker and perform colony hybridization.
4. Generate $\gamma$-$^{32}$P-labeled probe using a 20-mer oligonucleotide complementary to the heterologous sequence. Set up the labeling reaction in a 20-μl volume by mixing 2 μl of 10× Polynucleotide Kinase buffer, 1 μl of 10 μM oligonucleotide, 5 μl of [γ-32P] ATP (5,000 Ci/mmole, 10 mCi/ml), and 1 μl of 10 U/μl T4 polynucleotide kinase. Incubate at 37°C for 45 min.
5. Inactivate the enzyme at 68°C or by boiling for 1 min. Add 30 μl TE to bring the total volume of reaction mix up to 50 μl.
6. Purify the labeled oligonucleotides using Sephadex G-25 columns (Quick Spin Columns, Roche). Measure the activity of 1 μl of the product in a scintillation counter. The typical activity is usually around 100,000–200,000 cpm. Use one third of the labeled probe for the hybridization. The rest can be stored at –20°C and used for the subcloning step (see below).
7. Hybridize for 2–3 h at 50°C. Hybridizing for more than 3 h will result in high background. Wash and expose to X-ray film to identify positive colonies.
8. After positive colonies are identified, they should be subcloned to obtain a pure recombinant clone that does not contain the original nonrecombinant BAC by a second round of hybridization. Repeat steps 2–7 of Subheading 3.7.
9. Confirm correct targeting by testing for the presence of the restriction sites present in the heterologous sequence (see Note 4).
10. Select the recombinant clone for presence of the mini-λ by streaking the colonies on LB/Tet plate. Colonies that are tetracycline resistant have retained the mini-λ and can be used for next "Fix" step (see Note 5).

### 3.8. "Fix" Step: Replace the Heterologous Sequence with the Desired Mutation

To replace the heterologous sequence inserted in the "Hit" BAC DNA with a sequence having the desired mutation, a second round of recombineering will be performed. "Fix" targeting cassette is composed of the same homology arms as the corresponding "Hit" vector, while the heterologous middle region of the "Hit" vector is replaced with the final sequence including a desired mutation. At this point, a 20-mer oligonucleotide encompassing this region can serve as a probe to differentiate the "Fix"-recombinant from the "Hit"-intermediate clones (see Note 6):

1. Once the correctly targeted pure "Hit" colonies are identified, repeat the targeting step (Fix targeting) to replace the 20-nucleotide heterologous sequence with the desired mutation exactly as described above in Subheadings 3.4–3.7.
2. Check the integrity of the BAC DNA by digesting the BAC with a few restriction enzymes (e.g., *Bam*HI, *Eco*RI, *Hin*dIII, *Eco*RV) and comparing the restriction pattern with the original BAC clone by running the two samples in parallel on a 0.8% agarose gel.
3. Confirm the mutation by sequencing. Make a glycerol stock and freeze at –80°C.

## 4. Notes

1. The yield of mini-λ DNA is considerably lower compared to the yield of high-copy plasmids. Resuspending the DNA pellet in 30–50 μl ensures that the DNA is not too dilute. DNA concentration of 25–50 ng/μl is desirable.
2. Remove tubes from the centrifuge promptly. Because the pellet is very soft, care should be taken not to dislodge it, especially when processing multiple tubes.
3. It is very important to have a negative control when screening for recombinant clones. Uninduced control should not yield any positive clone. Signals on the autoradiograph from uninduced cells should be considered as background signal.
4. To confirm the presence of restriction sites, use two PCR primers outside the homology arms of the targeting vector. Digest the PCR product for the presence of *Bam*HI, *Eco*RI, or *Xho*I sites present in the heterologous sequence.
5. Because the attachment sites *attL* and *attR* are present in mini-λ DNA, the prophage DNA circle can be excised out and either it can reintegrate or be lost from the cell during heat induction. If no tetracycline-resistant colony is obtained, the mini-λ DNA should be electroporated in one of the "Hit" clones, as described in Subheadings 3.2 and 3.3.

6. Although a 20-mer oligonucleotide works well as a probe, occasionally, a longer oligonucleotide gives more specific signal. If the 20-mer oligonucleotide probe either results in very weak signal or very high background signal, 35–40-mer oligonucleotide should be tried.

## Acknowledgments

The research was sponsored by the Center for Cancer Research, National Cancer Institute, US National Institutes of Health.

### References

1. Copeland NG, Jenkins NA, Court DL (2001): Recombineering: a powerful new tool for mouse functional genomics. Nat Rev Genet 2:769–79.
2. Yu D, Ellis HM, Lee EC, Jenkins NA, Copeland NG, Court DL (2000): An efficient recombination system for chromosome engineering in *Escherichia coli*. Proc Natl Acad Sci USA 97:5978–83.
3. Zhang Y, Buchholz F, Muyrers JP, Stewart AF (1998): A new logic for DNA engineering using recombination in *Escherichia coli*. Nat Genet 20:123–8.
4. Zhang Y, Muyrers JP, Testa G, Stewart, AF (2000): DNA cloning by homologous recombination in *Escherichia coli*. Nat. Biotechnol18: 1314–1317.
5. Datta S, Costantino N, Zhou X, Court DL (2008): Identification and analysis of recombineering functions from Gram-negative and Gram-positive bacteria and their phages. Proc Natl Acad Sci USA 105:1626–1631.
6. Warming S, Costantino N, Court DL, Jenkins NA, Copeland NG (2005): Simple and highly efficient BAC recombineering using galK selection. Nucleic Acids Res 33:e36.
7. Court DL, Swaminathan S, Yu D, Wilson H, Baker T, Bubunenko M, Sawitze J, Sharan SK (2003): Mini-l: a tractable system for chromosome and BAC engineering. Gene 315:63–69.
8. Datta S, Costantino N, Court, DL (2006): A set of recombineering plasmids for gram-negative bacteria Gene 379:109–115.
9. Chan W, Costantino N, Li R, Lee SC, Su Q, Melvin D, Court DL, Liu P. (2007): A recombineering based approach for high-throughput conditional knockout targeting vector construction. Nucleic Acids Res 35:e64.
10. Yang Y, Sharan SK (2003): A simple two-step, 'hit and fix' method to generate subtle mutations in BACs using short denatured PCR fragments. Nucleic Acids Res 31:e80.
11. Swaminathan S, Ellis HM, Waters LS, Yu D, Lee EC, Court DL, Sharan SK (2001): Rapid engineering of bacterial artificial chromosomes using oligonucleotides. Genesis 29:14–21.
12. Ellis HM, Yu D, DiTizio T, Court DL (2001): High efficiency mutagenesis, repair, and engineering of chromosomal DNA using single-stranded oligonucleotides. Proc Natl Acad Sci USA 98:6742–6.
13. Sharan SK, Thomason LC, Kuznetsov SG, Court DL (2009): Recombineering: a homologous recombination-based method of genetic engineering. Nat Protoc 4:206–23.

# Chapter 10

## Using Recombineering to Generate Point Mutations: *galK*-Based Positive–Negative Selection Method

**Kajal Biswas, Stacey Stauffer, and Shyam K. Sharan**

### Abstract

Recombineering is a recombination-based highly efficient method of genetic engineering. It can be used to manipulate the bacterial chromosomal DNA as well as any episomal DNA. Recombineering can be used to insert selectable or nonselectable DNA fragments and subclone DNA fragments without the use of restriction enzymes and also to make precise alterations including single nucleotide changes in the DNA. Here we describe a *galactokinase* (*galK*)-based two-step method to generate point mutations in the bacterial artificial chromosome (BAC) insert using the recombineering technology. It takes advantage of the ability to select and also counterselect for the presence of *galK*.

**Key words:** Recombineering, Bacteriophage lambda recombination genes, *GalK*, Point mutation, Bacterial artificial chromosome, Oligonucleotides

### 1. Introduction

Recombineering, genetic engineering using recombination proteins, is a powerful system for engineering bacterial chromosomes and episomes in vivo by homologous recombination using PCR products and synthetic oligonucleotides as substrates (1). Francis Stewart and colleagues in 1998 made an important advance in the field of genetic engineering by describing the use of recombination systems encoded by the *recE* and *recT* genes of the Rac prophage and an analogous λ red system to manipulate genes (2). Since then, several different systems have been developed utilizing different recombination machinery (1–3). This highly efficient technique can be exploited in various ways to manipulate the genome, including the construction of chromosomal gene knockouts, point mutations,

Jean Peccoud (ed.), *Gene Synthesis: Methods and Protocols*, Methods in Molecular Biology, vol. 852,
DOI 10.1007/978-1-61779-564-0_10, © Springer Science+Business Media, LLC 2012

deletions, small insertions, in vivo cloning, mutagenesis of bacterial artificial chromosomes, and genomic libraries (4–8).

Bacterial artificial chromosome (BAC) is an ideal vector for cloning and manipulating large fragments of DNA (9). BACs are maintained in *Escherichia coli* (*E. coli*) cells that are *recA*⁻ to ensure the stability of the insert. However, this hinders the manipulation of BACs by homologous recombination in the bacterial cells. To perform BAC recombineering, a bacterial strain that expresses a bacteriophage recombination system is required. The recombination functions are provided by three genes of bacteriophage λ red locus: *exo*, *bet*, and *gam*. Exo is a 5′–3′ endonuclease that degrades 5′ ends of linear DNA. Bet binds to the single-stranded DNA and promotes annealing to the complementary DNA, whereas Gam inhibits the recBCD exonuclease and protects the DNA from degradation. The defective λ prophage system encoding the *exo*, *bet*, and *gam* genes is commonly used for providing recombination functions. The phage recombination systems in the defective λ prophage are under control of bacteriophage λ temperature-sensitive *c*I857 repressor. At low temperatures (30–34°C), the recombination genes are not expressed. By shifting the bacterial cultures to 42°C, the recombination genes are expressed at high levels from the λ *pL* promoter.

Here we describe the generation of point mutations, deletions or insertions in BAC DNA using *galK*-based positive–negative selection method developed by Warming et al. (10). This two-step selection allows the modification of BAC DNA without introducing a selectable marker at the modification site (Fig. 1). This system uses a bacterial strain containing the λ prophage recombineering system and defective in the utilization of galactose as a carbon source due to the deletion of the galactokinase gene (*galK*) from the galactose operon. The rest of the genes of the galactose operon are intact. So, when the *galK* function is added *in trans*, the ability of *galK*⁻ *E. coli* to grow in a media containing galactose as the carbon source is restored. The *galK* selection scheme is a two-step system; the first step, a positive selection step, involves targeting the region of interest with the *galK* cassette containing homology to a specified position in a BAC. The recombinant bacteria are selected on the minimal plate containing galactose as the carbon source. In the second step, a DNA fragment containing the particular mutation of interest replaces the *galK* cassette. Cells are selected for the loss of *galK* in the presence of 2-deoxy-galactose (DOG) on minimal plates with glycerol as the carbon source. DOG is harmless unless phosphorylated by functional *galK*. Phosphorylation of DOG by galactokinase turns DOG into 2-deoxy-galactose-1-phosphate, a nonmetabolizable intermediate toxic to the cells. This is negative selection as it selects the bacteria for the absence of *galK* cassette. This positive–negative selection for BAC manipulation is highly efficient (60–80%) even with 50 bases of homology on both sides.

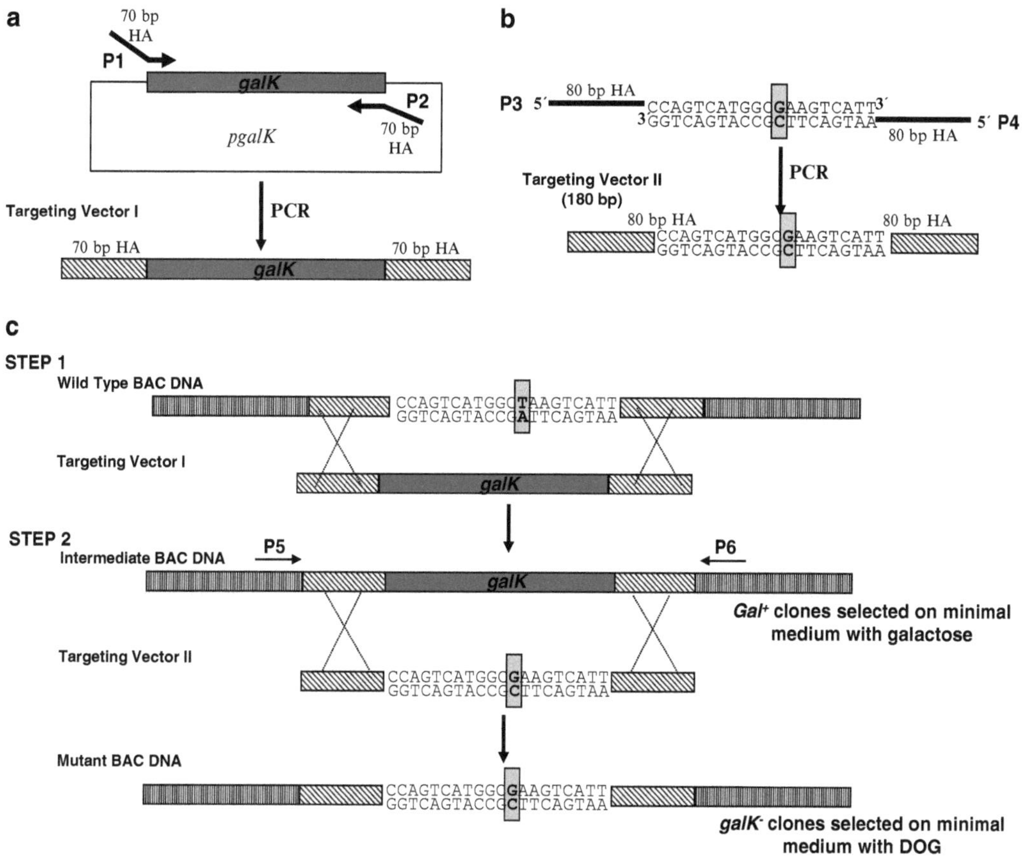


Fig. 1. Schematic representation of *galK*-based recombineering to generate point mutations. (**a**) Generation of targeting vector by PCR to insert the *galK* cassette using primers P1 and P2. 3′ end of each P1 and P2 anneal to the 5′ and 3′ ends of the *galK* cassette, respectively. The 5′ ends of the primers have 70 bases of homology to the target site. (**b**) The targeting vector to replace the *galK* cassette with desired mutation (*marked in bold in the box*) is generated by PCR using two 100-mer oligonucleotides, P3 and P4. The two oligonucleotides have complementary 3′ ends where they anneal together and amplify the homology arms (HA) to generate the targeting vector. (**c**) The two steps depicting the recombineering procedure to generate point mutation. The base to be replaced in the wild-type BAC and the mutated base in mutant BAC are marked in *bold letters in the box.* P5 and P6 mark the two PCR primers located outside the homology arms that can be used for screening the correctly targeted clones in both steps of recombineering.

## 2. Materials

### 2.1. Bacteria and Plasmids

1. *E. coli* SW102: A modified *E. coli* Dy380 strain (11) containing the defective λ prophage and a fully functional *gal* operon, except for a deletion of *galK*. Loss of *galK* allows for efficient BAC modification using *galK* positive–negative selection. This strain is tetracycline resistant (12.5 μg/ml). This strain is temperature sensitive and must be grown at 32°C.
2. *pGalK* plasmid: This plasmid is used as a template to amplify the *galK* cassette by PCR. SW102 and *pGalK* can be obtained from the NCI-Frederick recombineering resource (http://web.ncifcrf.gov/research/brb/recombineeringInformation.aspx).

### 2.2. Reagents

1. Ampicillin (Sigma).
2. Chloramphenicol (Sigma).
3. Sterile distilled $H_2O$, chilled on ice.
4. Plasmid DNA isolation reagents (Qiagen maxiprep and miniprep kits).
5. PCR purification kit (Qiagen).
6. Gel extraction kit to purify DNA from agarose gels (Qiagen).
7. Expand High Fidelity Taq polymerase (Roche).
8. Restriction enzymes (New England Biolabs).
9. Standard Taq polymerase (Invitrogen).
10. dNTP mixture, 10 mM each, PCR grade (Invitrogen).
11. Agarose (SeaKem LE, ISC Bioexpress).
12. Primers for PCR amplification of recombineering substrates, 25 pmol/μl in $H_2O$ (Invitrogen).

### 2.3. Equipment

1. Constant temperature bacterial incubator set at 30–34°C (2005 Low-Temperature Incubator, VWR).
2. Two shaking $H_2O$ baths (200 rpm) set at 30–32°C and at 42°C.
3. Spectrophotometer and cuvettes.
4. Electroporator.
5. Electroporation cuvettes with 0.1 cm gap, labeled and prechilled.
6. Floor model low-speed centrifuge at 4°C.
7. Refrigerated microcentrifuge at 4°C.
8. Thermal cycler and accessories for PCR.
9. Agarose gel electrophoresis apparatus.
10. Sterile 125- and 250-ml Erlenmeyer flasks, preferably baffled.
11. Sterile 35–50-ml centrifuge tubes.
12. 1.5-ml microfuge tubes.
13. 0.2-ml flat-cap PCR tubes.
14. Insulated ice buckets.
15. Sterile glass culture tubes (16 × 150 mm) for overnight growth of bacterial cultures.
16. Stainless steel closures for culture tubes.
17. Pipettes of various volumes with aerosol-resistant sterile tips.
18. Petri plates, 100 × 15 mm.

### 2.4. Media

All media were sterilized by autoclave after preparation and stored at room temperature.

1. *M9 salt solution*: For 1 L solution 3.0 g $KH_2PO_4$, 12.8 g $Na_2HPO_4 \cdot 7H_2O$, 1.0 g $NH_4Cl$, 0.5 g NaCl.

2. *LB (Luria broth)*: For 1 L broth 10 g bacto-tryptone, 5 g yeast extract, 5 g NaCl, pH 7.2.
3. *"Superbroth special"*: For 1 L broth 35 g bacto-tryptone, 20 g yeast extract, 5 g NaCl.
4. *SOC medium*: For 1 L media 20 g bacto-tryptone, 5 g yeast extract, 2 ml of 5 M NaCl, 2.5 ml of 1 M KCl, 10 ml of 1 M $MgCl_2$, 10 ml of 1 M $MgSO_4$, 20 ml of 1 M glucose.
5. *M63 minimal plates*: For 1 L 5× M63 10 g $(NH_4)_2SO_4$, 68 g $KH_2PO_4$, 2.5 mg $FeSO_4{\cdot}7H_2O$. Adjust to pH 7 with KOH.

#### 2.5. Other Reagents

1. 0.2 mg/ml D-biotin (sterile filtered) (1:5,000).
2. 20% galactose (autoclaved) (1:100).
3. 20% 2-deoxy-galactose (autoclaved) (1:100).
4. 20% glycerol (autoclaved) (1:100).
5. 10 mg/ml L-leucine (1%, heated, then cooled down, and sterile filtered).
6. 25 mg/ml chloramphenicol in EtOH (1:2,000).
7. 1 M $MgSO_4{\cdot}7H_2O$ (1:1,000).
8. Agar plates: Autoclave 15 g agar in 800 ml $H_2O$ in a 2-L flask. Let it cool down a little. Add 200 ml autoclaved 5× M63 medium and 1 ml 1 M $MgSO_4{\cdot}7H_2O$. Adjust volume to 1 L with $H_2O$ if necessary. Let it cool down to 50°C ("touchable hot"). Add 10 ml carbon source (final conc. 0.2%), 5 ml biotin (1 mg), 4.5 ml leucine (45 mg), and 500 μl chloramphenicol (final conc. 12.5 μg/ml). Pour the plates, 25–40 plates/L.
9. MacConkey indicator plates:
   Prepare MacConkey agar plus galactose according to manufacturer's instructions (Difco). After autoclaving and cooling to 50°C, to 1 L add 500 μl chloramphenicol (final conc. 12.5 μg/ml), and pour the plates, 25–40 plates/L.

## 3. Methods

#### 3.1. Preparation of BAC DNA

Identify a BAC containing the gene of interest using a genome browser (such as http://genome.UCSC.edu/) and order the BAC from a commercial supplier or Children's Hospital Oakland Research Institute's BACPAC Resource (http://bacpac.chori.org/). Before proceeding further, make sure the BAC contains the gene of interest by PCR analysis and check the integrity of the BAC insert by comparing the restriction digestion pattern (using restriction enzymes such as *Eco*RI, *Bam*HI, *Pst*I, or *Spe*I) of 2 or 3 overlapping BACs.

1. Inoculate a single colony of *E. coli* containing the BAC of interest into 10 ml of "superbroth special" medium with 12.5 μg/ml of chloramphenicol and grow overnight at 32°C.
2. Pellet down the cells at 6,000 × *g* for 5 min and remove the supernatant.
3. Dissolve the pellet in 200 μl of buffer P1 (Qiagen miniprep kit) and transfer to an eppendorf tube.
4. Add 200 μl of buffer P2 (Qiagen miniprep kit) and mix the tubes by inversion. Incubate at room temperature for 5 min.
5. Add 200 μl of buffer P3 (Qiagen miniprep kit) and mix vigorously. Incubate on ice for 5 min.
6. Clear the supernatant by centrifuging two times at 12,000 × *g* for 10 min in a tabletop centrifuge. Transfer supernatant to a new tube each time.
7. Precipitate DNA by adding 600 μl isopropanol to the supernatant and incubating on ice for 10 min followed by centrifugation at 12,000 × *g* for 10 min.
8. Wash the pellet in 70% ethanol and dissolve the air-dried pellet into 50 μl sterile distilled water. From that, 40 μl (approximately 1 μg) can be used for restriction analysis in a 50-μl reaction, and 1 μl can be used as template for PCR analysis or for transformation into electrocompetent SW102 bacteria.

### 3.2. Electroporation of BAC DNA into E. coli SW102 Strain

The first step of BAC manipulation is to introduce the BAC into SW102 cells that harbor the defective λ prophage and have deletion of the galactokinase (*galK*) gene.

#### 3.2.1. Preparation of Electrocompetent Cells

1. Inoculate 5 ml overnight culture in "superbroth special" with an isolated colony of SW102. Grow the culture at 32°C overnight. Use of antibiotics is optional (SW102 cells are resistant to tetracycline, and BACs are chloramphenicol resistant).
2. Next morning, dilute the o/n culture 1:50, i.e., transfer 1 ml into an autoclaved 250-ml Erlenmeyer baffled flask with 50 ml "superbroth special" and grow at 32°C to an $O.D_{600}$ 0.50–0.60. Place a 50-ml tube containing sterile $ddH_2O$ in the ice/water slurry.
3. After cooling down the flask containing the bacteria in an ice/water bath slurry for 1 or 2 min, transfer 10 ml of culture into precooled Oak Ridge centrifuge tube and spin down at 6,000 × *g* in prechilled rotor for 10 min at 1°C.
4. Discard the supernatant and invert the tube on a paper towel. Add 1 ml cold $ddH_2O$ while keeping the tube in the ice water. Resuspend the pellet in the $ddH_2O$ by gently shaking the tube in the ice/water bath (this can take a while the first time, around 5 min). When resuspended, fill up to 10 ml with ice-cold $ddH_2O$, invert a couple of times, and spin again for 5 min.

5. Pour off supernatant and resuspend the pellet in 1 ml cold $ddH_2O$ (resuspension will be easy this time) and transfer to a chilled 1.5-ml tube. Spin at 12,000×*g* for 30 s at 1°C.
6. Wash the cells one more time with 1 ml ice-cold water.
7. Gently remove all supernatant by inverting the tube on paper towel (be careful not to lose the pellet). Resuspend the cell pellet in 50 μl ice-cold water and keep on ice.

#### 3.2.2. Electroporation

1. Mix 500 ng of BAC DNA (1–5-μl volume) with 50 μl freshly prepared electrocompetent SW102 cells in a precooled 1.5-ml tube. Let it sit for 5 min on ice and then transfer into a precooled 0.1-cm cuvette.
2. Electroporate the BAC DNA (1.8 kV voltage, 25-μF capacitance, and 200-Ω resistance) in cells. Add 1 ml SOC medium immediately after electroporation and transfer the cells into a 1.5-ml tube.
3. Grow the cells for 1 h at 32°C. Spin down cell for 30 s in a microcentrifuge. Discard the supernatant and resuspend the pellet in 200 μl LB medium.
4. Plate the transformed bacteria in a 10-cm LB plate containing 12.5 μg/ml chloramphenicol. Incubate at 32°C for 18–24 h. Ten to one hundred chloramphenicol-resistant colonies may be obtained.

#### 3.2.3. Identification of SW102 Clones Containing BAC DNA

Isolate the BAC DNA from eight to ten individual BAC colonies using alkaline lysis method as described in Subheading 3.1 and perform restriction digestion to confirm the integrity of BAC DNA after electroporation into SW102 cells. Compare the restriction pattern with the original BAC DNA. Most of the BAC DNA from SW102 cells should be identical to the original DNA. Freeze the aliquot of the SW102 cells containing BAC in 15% glycerol at –70°C.

### 3.3. Manipulation of BAC DNA Using Positive–Negative Selection

Modifications like single base changes or deletions or insertions in the BAC DNA can be generated by using *galK*-based positive–negative selection (10). To perform those steps, the first step is to design the primers with the homology arms to generate the targeting vectors.

#### 3.3.1. Generating the Targeting Vector for Step I

1. Design the first set of primers to amplify the *galK* gene. These primers will have 50–70-bp homology to an area flanking the desired site to be mutated. The 3′ end of these primers anneal to the *galK* cassette (Fig. 1a). For example, to generate a single base change, the homology arms should be the 50–70 bases on either side of this nucleotide. This will result in a deletion of that nucleotide and insertion of the *galK gene* in the first step. The primers should be as follows:

   *Forward*: 5′-70 bp homology-CCTGTTGACAATTAATCA TCGGCA-3′

*Reverse*: 5′-70 bp homology of complementary strand-TCAGCACTGTCCTGCTCCTT-3′

2. Amplify the *galK* cassette using the primers from step 1 and a Taq polymerase with proofreading activity (e.g., Expand High Fidelity Taq from Roche). Use 1–2 ng of *pGalK* plasmid as template. PCR steps are 94°C for 30 s, 55°C 30 s, 72°C 1.5 min, for 35 cycles. After the PCR, add 1–2 μl *Dpn*I per 50 μl reaction and incubate at 37°C for 1 h. This step removes any plasmid template as *Dpn*I digests methylated plasmid but not the PCR products that are not methylated. Gel purify the *Dpn*I-digested PCR product, preferably overnight at low voltage. Extract the DNA from agarose gel using a PCR purification kit and elute in 20 μl $ddH_2O$ (see Note 1).

#### 3.3.2. Induction of Bacteriophage λ Recombination System in SW102 Cells

1. Start an overnight culture of SW102 cells containing the BAC from a single colony and grow at 32°C in "superbroth special" containing chloramphenicol (12.5 μg/ml).
2. Next morning, dilute 1 ml of overnight culture in 50 ml "superbroth special" containing chloramphenicol (12.5 μg/ml) in a 250-ml baffled conical flask and grow at 32°C to an $OD_{600}$ of 0.55–0.6. This will take 3–4 h. During that time, turn on the 42°C shaking water bath and make an ice/water slurry. Chill a 50-ml tube containing sterile $ddH_2O$ in ice/water slurry.
3. Once the SW102 culture reaches the $OD_{600}$ of 0.55–0.6, transfer 10 ml of culture to an Oak Ridge tube and place on ice. This will be the uninduced control. Transfer another 10 ml of culture to another baffled 50-ml conical flask. Heat shock at 42°C for exactly 15 min in a shaking water bath. Stop the induction immediately by placing the flask into ice/water slurry for 15 min with intermittent shaking.

#### 3.3.3. Targeting the galK Cassette

1. Prepare the electrocompetent cells from both induced and uninduced cells, as described in Subheading 3.2.1 (steps 3–7).
2. Electroporate 300 ng of the targeting vector containing the *galK* cassette (from Subheading 3.3.1) in both uninduced and induced electrocompetent cells as described in Subheading 3.2.2 (steps 1 and 2).
3. After the 1-h growth at 32°C, spin down the bacteria in 1.5-ml tube at 12,000×*g* for 30 s and remove the supernatant with a pipette. Resuspend the pellet in 1 ml M9 salts, and spin again. This washing step is repeated once more.
4. After the second wash, discard the supernatant and resuspend the pellet in 1 ml M9 salts. Plate serial dilutions in M9 (100 μl, 100 μl of a 1:10 dilution, and 100 μl 1:100) onto M63

containing biotin and leucine minimal media plates with galactose as the carbon source to select the *galK*$^+$ colonies (see Note 2).

5. Incubate the plates for 3 days at 32°C.

#### 3.3.4. Identifying galK-Positive Recombinants

1. To screen for *Gal*$^+$ colonies, streak eight to ten colonies from above (Subheading 3.3.3) onto MacConkey indicator plates containing galactose and chloramphenicol to obtain single colonies. Incubate the plates at 32°C overnight (see Note 3).
2. Pick eight to ten single red colonies to perform colony PCR to confirm the integration of *galK* cassette at the desired site. Use the primers flanking the targeted region (but not included in the homology arms of the targeting vector) to amplify BAC DNA. The size of the PCR product should show the presence of *galK* cassette into the desired site (see Note 4).
3. Grow the correct clones in 5 ml of "superbroth special" media containing chloramphenicol at 32°C overnight. Use an aliquot of the culture to freeze in 15% glycerol at −70°C and one correct clone to replace the *galK* cassette with desired mutation.

#### 3.3.5. Replacing the galK Cassette with Desired Mutation (Step II)

1. Design the PCR primers to generate the targeting vector II that will be used to replace the *galK* cassette with the desired mutation. For this, each primer should have 80-bp homology at the 5′ end, and the bases need to be inserted/replaced at the 3′ end (Fig. 1b). The two oligonucleotides should have 20-bp complementary bases at the 3′ end that will help to anneal to each other and extend the two by PCR to generate a 180-bp targeting vector (Fig. 1b).
2. Amplify the targeting vector to replace the *galK* cassette using the primers from step 1 and a Taq polymerase with proofreading activity. PCR steps are 94°C 15 s, 55°C 20 s, 72°C 20 s, for 30 cycles. Purify the PCR products using PCR purification kit (Qiagen).
3. Follow the steps described in Subheadings 3.3.2 and 3.3.3 for induction of recombination system and electroporating the targeting vector. After the second wash, discard the supernatant and resuspend the pellet in 1 ml M9 salts. Make serial dilutions in M9 and plate (100 μl, 100 μl of a 1:10 dilution, and 100 μl 1:100) onto M63 containing biotin and leucine minimal containing 2-deoxy-galactose (DOG) and chloramphenicol (12.5 μg/ml) plates with glycerol as a carbon source to select *galK*$^-$ colonies.
4. Incubate the plates for 3 days at 32°C.
5. Screen for the correct clones for the replacement of *galK* cassette by colony PCR using the same primers from step 2 of Subheading 3.3.4. The size of the PCR band should show the

absence of *galK* cassette at the desired site. Purify the PCR product and sequence using the same flanking primers in two separate reactions to confirm the presence of the desired mutation (see Note 5).

6. Confirm the integrity of the BAC by examining the restriction digestion pattern of BAC miniprep DNA (see Subheading 3.3.1). Discard clones with rearrangements that may have occurred during the BAC manipulation. Include the parent BAC clone as a control. Select the clones that show a digestion pattern identical to the parental clone.

## 4. Notes

1. Removal of the plasmid DNA is essential to reduce the background *galK*$^+$ colonies. Efficiency of incorporation of plasmid DNA is much higher compared to the homologous recombination efficiency, and the presence of even picogram quantity of plasmid DNA will give thousands of nonrecombinant colonies.
2. Prior to selection on minimal media, washing in M9 salts is important to remove any rich media from the bacteria. The uninduced samples routinely have a higher degree of lysis/bacterial death after electroporation, and some bacteria will be lost, so the uninduced sample is diluted in 0.25–0.75 ml of M9 salts in the final step to make up for the difference. Plate 100 μl of the uninduced sample as a control.
3. The colonies that appear after the 3 days of incubation are likely to be *gal*$^+$, but they are often mixed with nonrecombinant *galK*$^-$ cells. In order to get rid of any *gal*$^-$ contaminants, it is important to obtain single, bright red colonies before proceeding to the second step. On the MacConkey agar plates, *galK*$^-$ colonies will be white/colorless and the *gal*$^+$ colonies will be bright red due to pH change resulting from fermented galactose. Streak the SW102 cells containing the BAC of interest for comparison.

   Alternatively, the colonies can be first screened by PCR using the primers outside the homology arms. After identifying the PCR positive clones, streak the colonies on the MacConkey agar plates to obtain single *gal*$^+$ colonies. It is important to remove any nonrecombinant *galK*$^-$ cells before proceeding to step 2 of targeting.
4. It is important to check the integrity of the BAC by examining the restriction digestion pattern. Occasionally, BACs undergo large deletions. Such BAC clones should be discarded. BACs

that show a restriction digestion pattern very similar (a few fragments may be different due to insertion of *galK* cassette) to the parental BAC clone should be used for the next step.

5. Rarely, 70–80 bases of homology are not sufficient to obtain correctly targeted clones. In such cases, increasing the length of homology by using an additional set of 100-mer oligonucleotides with 20 bases of homology to the first set of primers is helpful.

## Acknowledgments

The research was sponsored by the Center for Cancer Research, National Cancer Institute, US National Institutes of Health.

### References

1. Muyrers JP, Zhang Y, Testa G, Stewart AF (1999): Rapid modification of bacterial artificial chromosomes by ET-recombination. Nucleic Acids Res 27:1555–7.
2. Zhang Y, Buchholz F, Muyrers JP, Stewart AF (1998): A new logic for DNA engineering using recombination in Escherichia coli. Nat Genet 20:123–8.
3. Yu D, Ellis HM, Lee EC, Jenkins NA, Copeland NG, Court DL (2000): An efficient recombination system for chromosome engineering in Escherichia coli. Proc Natl Acad Sci USA 97:5978–83.
4. Copeland NG, Jenkins NA, Court DL (2001): Recombineering: a powerful new tool for mouse functional genomics. Nat Rev Genet 2:769–79.
5. Ellis HM, Yu D, DiTizio T, Court DL (2001): High efficiency mutagenesis, repair, and engineering of chromosomal DNA using single-stranded oligonucleotides. Proc Natl Acad Sci USA 98:6742–6.
6. Sarov M, Schneider S, Pozniakovski A, Roguev A, Ernst S, Zhang Y, Hyman AA, Stewart AF (2006): A recombineering pipeline for functional genomics applied to Caenorhabditis elegans. Nat Methods 3:839–44.
7. Swaminathan S, Ellis HM, Waters LS, Yu D, Lee EC, Court DL, Sharan SK (2001): Rapid engineering of bacterial artificial chromosomes using oligonucleotides. Genesis 29:14–21.
8. Yang Y, Sharan SK (2003): A simple two-step, 'hit and fix' method to generate subtle mutations in BACs using short denatured PCR fragments. Nucleic Acids Res 31:e80.
9. O'Connor M, Peifer M, Bender W (1989): Construction of large DNA segments in Escherichia coli. Science 244:1307–12.7.
10. Warming S, Costantino N, Court DL, Jenkins NA, Copeland NG (2005): Simple and highly efficient BAC recombineering using galK selection. Nucleic Acids Res 33:e36.
11. Lee EC, Yu D, Martinez de Velasco J, Tessarollo L, Swing DA, Court DL, Jenkins NA, Copeland NG (2001): A highly efficient Escherichia coli-based chromosome engineering system adapted for recombinogenic targeting and subcloning of BAC DNA. Genomics 73:56–65.

# Chapter 11

## Assembling Large DNA Segments in Yeast

**Héloïse Muller, Narayana Annaluru, Joy Wu Schwerzmann, Sarah M. Richardson, Jessica S. Dymond, Eric M. Cooper, Joel S. Bader, Jef D. Boeke, and Srinivasan Chandrasegaran**

### Abstract

As described in a different chapter in this volume, the *u*racil-*s*pecific *e*xcision *r*eaction (USER) fusion method can be used to assemble multiple small DNA fragments (~0.75-kb size) into larger 3-kb DNA segments both in vitro and in vivo (in *Escherichia coli*). However, in order to assemble an entire synthetic yeast genome (Sc2.0 project), we need to be able to assemble these 3-kb pieces into larger DNA segments or chromosome-sized fragments. This assembly into larger DNA segments is carried out in vivo, using homologous recombination in yeast. We have successfully used this approach to assemble a 40-kb chromosome piece in the yeast *Saccharomyces cerevisiae*. A lithium acetate (LiOAc) protocol using equimolar amount of overlapping smaller fragments was employed to transform yeast. In this chapter, we describe the assembly of 3-kb fragments with an overlap of one building block (~750 base pairs) into a 40-kb DNA piece.

**Key words:** Synthetic yeast, USER fusion, DNA assembly, Large DNA fragments

## 1. Introduction

Our long-term goal is to use the model eukaryote *Saccharomyces cerevisiae* as the basis for a cell with a synthetic genome "Sc2.0" that can be used to answer a wide variety of profound questions about fundamental properties of chromosomes, genome organization, gene content, the function of RNA splicing, the extent to which small RNAs play a role in eukaryotic biology, the distinction between prokaryotes and eukaryotes, and the intimate relationship between genome structure and evolution. The availability of a fully synthetic genome will allow the direct testing of evolutionary questions that are not otherwise approachable. The eventual "synthetic yeast" that will be designed and refined is likely to play an important

Jean Peccoud (ed.), *Gene Synthesis: Methods and Protocols*, Methods in Molecular Biology, vol. 852,
DOI 10.1007/978-1-61779-564-0_11, 

practical role as well. Experiments are underway in our lab using an iterative replacement of wild-type genomic segments of yeast with designed and synthesized large DNA segments in order to achieve, initially, the total synthesis of a functional yeast chromosome and ultimately to create a fully designed synthetic yeast organism as a model eukaryote.

In chapter 7, we utilized the *u*racil-*s*pecific *e*xcision *r*eaction (USER) fusion method to assemble multiple smaller DNA fragments, called building blocks (BBs) (1, 2) (~0.75-kb size), into larger 3-kb DNA segments both in vitro and in vivo (in *Escherichia coli*) (3). The next step is to assemble these 3-kb pieces into larger DNA segments or chromosome-sized fragments. This assembly strategy into larger DNA segments is carried out in vivo, by using homologous recombination in yeast (4–6), which is the focus of this chapter. Because of very high homologous recombination rate in yeast, it is the organism of choice to assemble multiple DNA fragments in vivo into larger DNA segments or chromosome-sized fragments.

In the example we use in this chapter to describe our method, we assemble 17 sets of 3-kb fragments into a 40-kb chromosome piece. The recombined assembly strategy aims to link together all sets with a yeast plasmid backbone in order to be able to maintain the plasmid in yeast as well as a bacterial artificial chromosome (BAC) backbone for the easy recovery of the large-size-assembled plasmid from bacteria. Figure 1 depicts schematically all fragments that are to be made and cotransformed into yeast for assembly. This chapter describes all the steps that are needed to make each of these fragments (Subheadings 3.1–3.4), followed by the cotransformation technique (Subheadings 3.5 and 3.6) and the screening and analyzing methods to determine integrity of the construct (Subheadings 3.7–3.11).

## 2. Materials

### 2.1. Reagents

1. *Plasmids*:
   pBeloBAC11 (NEB) and pRS416 (NEB).
2. *Strains*:
   Yeast strains of *S. cerevisiae* BY4741, *E. coli* strains DH5α, K12 ER2420 containing pBeloBAC11 (NEB E4154S), and TransforMax™ EPI300™ (EPICENTRE #EC300105).
3. *Media*:
   Luria–Bertani (LB) with carbenicillin or LB with chloramphenicol: For 1 L of medium, add 10 g of sodium chloride (NaCl), 10 g of Bacto™ Tryptone, and 5 g of Bacto™ Yeast Extract. Dissolve in dd$H_2O$, adjusting the volume to 1 L. Autoclave and allow medium to cool down below

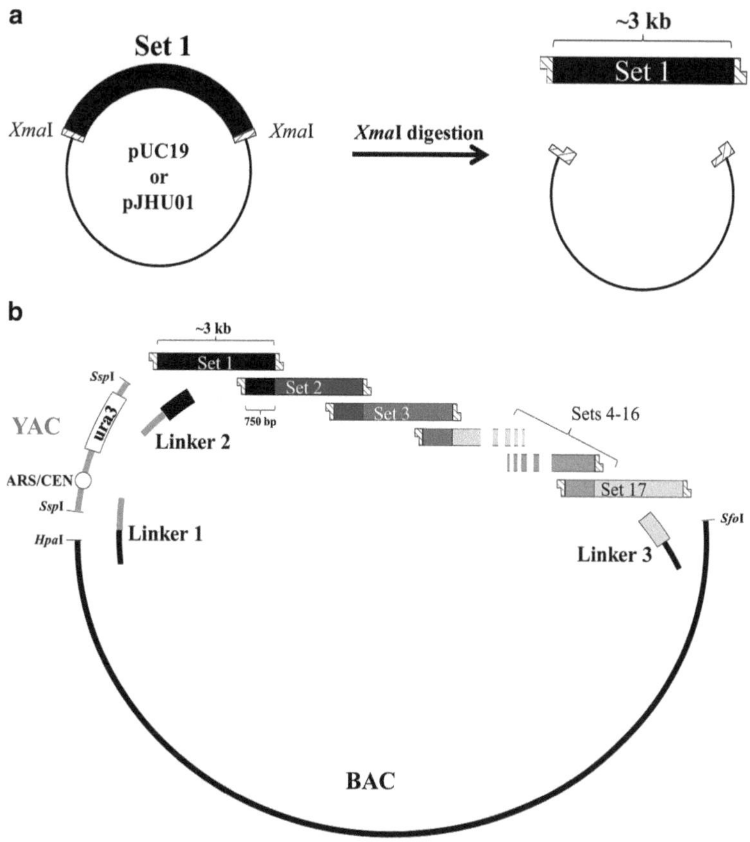


Fig. 1. DNA fragments to be cotransformed into yeast for assembly. Pieces to be cotransformed are the following: 17 DNA sets from the previous assembly step; two plasmids backbones, one for maintenance of assembly in yeast (YAC), and one for maintenance of assembly in bacteria (BAC); and linkers enabling recombination between DNA segments that have not been designed with overlap. (**a**) DNA sets are 3-kb fragments that were previously assembled and cloned in a bacterial plasmid (*black line*, pUC19, or pJHU1). Each 3-kb set was released from the plasmid backbone by restriction enzyme digestion (*Xma*I), resulting in 1 bp + 4 bp 5′ overhangs at the ends of each set. For example, set 1 is represented as a *black box*, and *hatched boxes* at the ends to represent restriction enzyme site overhangs after digestion. (**b**) DNA sets are represented as *grayscale boxes* and numbered 1–17; *hatched boxes* at the ends represent restriction enzyme site overhangs after digestion. Each 3-kb set was designed to overlap its neighbor by ~750 bp (one BB), enabling homologous recombination between the two fragments. YAC (*gray line*) is the *Ssp*I restriction fragment of pRS416 containing ARS/CEN sequence and the marker gene *URA3*. BAC (*black line*) is the *Hpa*I–*Sfo*I large restriction fragment of pBeloBAC11 containing the Cm$^R$ resistance marker gene, *redF′*, Ori2, *repE*, *sopA*, *sopB*, and *sopC*. Linker fragments were designed to enable recombination between consecutive DNA fragments that have no overlap. They contain 200–500-bp overlap with adjoining fragments. In our assembly, three linkers have been designed: linker 1, to enable recombination between BAC and YAC; linker 2, to enable recombination between YAC and the first set of our assembly; and linker 3, to enable recombination between the last set of our assembly and BAC. For further explanation regarding linker preparation, see Fig. 2.

55°C before adding any antibiotic: 1 ml of 100 mg/ml carbenicillin or 1 ml of 20 mg/ml of chloramphenicol for 1 L of medium. Petri dish or liquid media can be stored at 4°C for 3–4 weeks after addition of the antibiotic.

SOC: Complement 1 L of LB medium with 10 ml of 1 M $MgSO_4$ (10 mM final), 10 ml of 1 M $MgCl_2$ (10 mM final), and 18 ml of 20% dextrose (0.36% w/vol).

YPD: For 1 L of medium, add 20 g of Bacto™ Peptone, 10 g of Bacto™ Yeast Extract, and 20 g of dextrose. Dissolve in $ddH_2O$, adjusting the volume to 1 L. Sterilize by autoclaving.

SC-Ura: For 1 L of medium, add 6.7 g of yeast nitrogen base w/o amino acids, 20 g of dextrose, and 1.92 g of yeast synthetic drop-out medium supplement without uracil. Dissolve in $ddH_2O$, adjusting the volume to 1 L, and autoclave.

To make agar plates, add 20 g/L Bacto™ Agar before autoclaving.

4. *Chemicals*:

Ethanol 100% and 70%.

Ethidium bromide 10 mg/ml, diluted to 0.5 μg/ml in agarose gel for direct staining of DNA (Note 1).

Isopropanol.

Polyethylene glycol 3350.

EDTA.

NaOH.

Potassium acetate.

5. *Buffers*:

Plasmid purification buffers P1, P2, P3, N3, PE, and EB from Qiagen.

EDTA 0.5 M (pH 8): Dissolve 186.1 g $Na_2EDTA \cdot 2H_2O$ in 800 ml of $ddH_2O$. Adjust pH to 8.0 with NaOH (~20 g of NaOH pellets). EDTA will dissolve at pH 8.0. Adjust volume to 1 L with $ddH_2O$. Sterilize by autoclaving.

Tris–HCl 1 M (pH 7.5 and pH 8).

Dithiothreitol (DTT) 1 M: Dissolve 0.77 g of DTT in $ddH_2O$, adjusting the volume to 5 ml.

Diethanolamine (pH 9.0).

SDS 20% w/vol.

Solution TE 10×: Add 50 ml of 1 M Tris–HCl (pH 7.5) (100 mM final) and 10 ml of 0.5 M EDTA (pH 8.0) (10 mM final). Adjust volume to 500 ml with $ddH_2O$ and autoclave.

Solution LiOAc 10×: lithium acetate 1 M (pH 7.5): Dissolve 6.6 g of lithium acetate in $ddH_2O$, adjusting the volume to 100 ml and autoclave.

Solution PEG-3350: polyethyleneglycol PEG-3350 50% w/vol. Do not autoclave. Use filtration unit for sterilization. Keep aliquots at −20°C.

Solution TE/LiOAc: In a sterile environment, add 10 ml of TE 10× and 10 ml of LiOAc 10× and adjust volume to 100 ml with sterile $ddH_2O$.

Solution TE/LiOAc/PEG: In a sterile environment, add 0.1 ml of TE 10×, 0.1 ml of LiOAc 10×, and 0.8 ml of PEG-3350 50% (40% final) to make 1 ml of solution.

Solution yeast miniprep I: Add 5 ml of 0.5 M EDTA (pH 8.0) (25 mM final), 5 ml of 1 M Tris–HCl (pH 8.0) (50 mM final), and 2 ml of 1 M dithiothreitol (DTT) (20 mM). Adjust volume to 100 ml with $ddH_2O$.

Solution yeast miniprep II: Add 20 ml of 1 M diethanolamine (pH 9.0) (200 mM final), 16 ml of 0.5 M EDTA (pH 8.0) (80 mM final), and 5 ml of 20% SDS (w/vol) (1% final). Adjust volume to 100 ml with $ddH_2O$.

Solution yeast miniprep III: potassium acetate 5 M: Dissolve 245 g of potassium acetate in $ddH_2O$, adjusting the final volume to 500 ml. Keep at 4°C.

6. *Enzymes*:

Zymolyase (100 T), 100 mg/ml.

Taq polymerase.

Taq DNA ligase.

*PfuTurbo*® Cx Hotstart DNA polymerase (Agilent Technologies 600410).

USER™ Enzyme (NEB).

All restriction enzymes and buffers are from New England Biolabs.

7. *Miscellaneous*:

QIAprep Spin Miniprep Kit (Qiagen 27104).

PureLink™ Quick Gel Extraction Kit (Invitrogen K210012).

Carrier DNA: Deoxyribonucleic acid, single stranded, from salmon testes (10 mg/ml).

dNTPs: 2.5 mM solution of each of the nucleotide dATP, dCTP, dGTP, and dTTP.

Acid-washed glass beads.

Glass beads 4 mm.

Parafilm.

Petri dishes.

COREX® 30-ml glass tube.

Sterile centrifuge tubes.

Gene Pulser cuvettes.

Molecular weight marker: 1-kb ladder.

Specific amplicons: Two sets of oligonucleotides pairs have been designed all along *S. cerevisiae* chromosomes. First set, native-specific amplicons, is specific to the wild-type sequence of the yeast strain BY4741 and will allow amplification of PCR product from endogenous chromosomes of yeast. The other set, synthetic-specific amplicons, is specific to the sequence of the synthetic yeast Sc2.0 and will allow amplification of PCR products only from the newly assembled synthetic structure (7–9). Stock solutions of oligonucleotides (primers) have a concentration of 25 μM.

10× Loading dye: Dissolve 0.24 g of bromophenol blue and 0.42 g of xylene cyanol FF in $ddH_2O$, add 30 ml of 30% glycerol, and adjust the volume to 100 ml with $ddH_2O$.

### 2.2. Equipment

Vortexer.

Centrifuge.

Table-top centrifuge.

Water bath (or heat block) set up to 50°C, 42°C, 65°C, or 70°C.

Incubators at 30°C and 37°C.

Electroporation unit (Bio-Rad).

Gel electrophoresis box.

Microscope.

Malassez cell.

Thermocycler.

NanoDrop or spectrophotometer.

Speed-Vac.

UV light box and camera.

## 3. Methods

### 3.1. YAC Preparation

What we refer to as a yeast artificial chromosome (YAC) in this chapter is a DNA fragment containing the necessary genetic elements for replicating, segregating, and screening in yeast. The YAC used in this study is the *Ssp*I restriction fragment of the pRS416 plasmid containing the ARS/CEN sequence and the *URA3* gene marker (Fig. 1b).

1. pRS416 plasmid DNA is prepared from 4 ml of a culture of the *E. coli* strain DH5α carrying the plasmid in LB + carbenicillin (100 μg/ml), using Qiagen QIAprep Spin Miniprep Kit according to manufacturer's instructions.
2. Digest ~5 μg of pRS416 with *Ssp*I restriction enzyme in 50 μl of dd$H_2O$.

| | |
|---|---|
| DNA | 10.0 μl (5 μg) |
| Enzyme buffer 10× | 5.0 μl |
| Enzyme (*Ssp*I-HF) | 1.0 μl (20 units) |
| dd$H_2O$ | 34.0 μl |
| Total | 50.0 μl |

   Incubate at 37°C for 2–3 h.
3. Add 6 μl of 10× loading dye, load on agarose gel, and migrate at 80 V for ~1 h.
4. Purify the 2-kb band containing ARS/CEN and *URA3* gene from *S. cerevisiae* with the PureLink™ Quick Gel Extraction Kit according to manufacturer's instructions.
5. Determine precise concentration of the YAC by using $OD_{260}$ with a UV spectrophotometer or a NanoDrop.

### 3.2. BAC Preparation

In order to be able to recover the large plasmid containing the assembled DNA segment from bacteria, we also introduced a BAC sequence into the assembled construct. We chose pBeloBAC11 that includes all necessary elements for maintaining the large recombined plasmid (over 10 kb) in bacteria and carries the chloramphenicol resistance gene marker. pBeloBAC11 DNA is prepared from 100 ml of a culture of *E. coli* strain K12 ER2420, carrying the plasmid in LB + chloramphenicol (20 μg/ml) as follows:

1. Harvest cells by centrifugation 15 min at 6,000 × *g*.
2. Resuspend bacterial pellet in 4 ml of buffer P1.
3. Add 4 ml of buffer P2, invert tube four to six times, and incubate at room temperature for 5 min.
4. Add 4 ml of buffer P3, invert four to six times, and incubate on ice for 15 min.
5. Centrifuge 30 min at 20,000 × *g* at 4°C, remove and keep supernatant.
6. Centrifuge supernatant again at 20,000 × *g* for 15 min at 4°C and transfer supernatant promptly in a COREX® 30-ml glass tube.

7. Precipitate DNA by adding 0.7 volume (8.4 ml) of room temperature isopropanol to the lysate, cover the tube with parafilm, and mix by inverting four to six times.
8. Centrifuge for 30 min at 20,000 × *g* at 4°C and carefully decant the supernatant.
9. Let the pellet dry for 10 min at room temperature, dissolve in 100 μl buffer TE, and transfer to a microcentrifuge tube.

   Then follow the manufacturer's instruction for QIAprep Spin Miniprep Kit Handbook:
10. Add 250 μl of buffer P1.
11. Add 250 μl of buffer P2 and mix thoroughly by inverting four to six times.
12. Add 350 μl of buffer N3 and mix immediately and thoroughly by inverting the tube four to six times.
13. Centrifuge for 10 min at 15,000 × *g* in a table-top centrifuge.
14. Apply the supernatant to a QIAprep spin column by decanting or pipetting.
15. Centrifuge for 1 min, discard flow-through.
16. Wash column by adding 750 μl of buffer PE and centrifuge for 1 min.
17. Discard flow-through and centrifuge for an additional 1 min to remove residual wash buffer.
18. Place the QIAprep column in a clean 1.5-ml microcentrifuge tube. To elute DNA, add 50 μl of buffer EB that has been pre-heated to 65°C to the center of the column. Let it stand for 1 min and centrifuge in a table-top centrifuge for 1 min at 15,000 × *g*.

The BAC is then linearized by enzymatic digestion using restriction enzymes *Hpa*I and *Sfo*I, which remove the loxP site of the pBeloBAC11 that might possibly interfere with our construct. The 6,850-bp *Hpa*I–*Sfo*I fragment is then gel-purified with the PureLink™ Quick Gel Extraction Kit according to manufacturer's instructions.

### 3.3. Preparation of the Set of 3-kb DNA Fragments

The 3-kb DNA fragments, derived from assembly of four BBs (3), are cloned in the pUC19 or the pJHU1 vector between two restriction sites. Plasmid DNA of each set of cloned 3-kb fragments is prepared using QIAprep Spin Miniprep Kit, and separation of the 3-kb fragment from the plasmid backbone is performed using restriction enzyme digestion (Fig. 1a).

1. Use restriction enzymes as advised by the manufacturer to digest ~5 μg of plasmid, in a final volume of 50 μl. In our case, all fragments are cloned between two *Xma*I restriction sites.

| | |
|---|---|
| DNA | 10.0 μl (5 μg) |
| Enzyme buffer 10× | 5.0 μl |
| BSA | 0.5 μl |
| Enzyme (*Xma*I) | 1.0 μl (10 units) |
| $ddH_2O$ | 33.5 μl |
| Total | 50.0 μl |

Incubate at 37°C for 2–3 h.

2. At the end of incubation time, inactivate the enzyme according to manufacturer's instructions; here, we used 20 min at 70°C for *Xma*I.
3. Determine accurate concentration of all DNA sets using $OD_{260}$ with a UV spectrophotometer or a NanoDrop.

Gel purification of the 3-kb inserts is not always necessary, but you may choose to do so if contaminating products might interfere with the assembly reaction. In this case, perform a gel purification of the 3-kb fragment of interest using PureLink™ Quick Gel Extraction according to manufacturer's instructions. When choosing to purify fragments, performing a double digestion with an additional enzyme that cuts in the plasmid only might be necessary for ease of purification. This allows for a better separation between the plasmid bands and the fragment of interest, which might otherwise migrate too close to the desired 3-kb fragment on agarose gel.

### 3.4. Linker Fragments Preparation

Linker fragments, which enable recombination between DNA segments that have not been designed to overlap, need to be synthesized. In our current construct, we need three linker fragments: a first to link the BAC with the YAC, a second to link the YAC with the first BB of the intended 40 kb insert, and a third to link the last BB of the intended 40-kb insert with the BAC (Fig. 1b). Linkers were obtained by USER fusion in vitro assembly of two U-containing PCRs of each extremity, designed to have no identity with the fragment that follows; see chapter on USER fusion for details of the procedure (3).

1. Perform PCR amplification from the extremities of each fragment using one regular oligonucleotide and one U-containing oligonucleotide; here, U is being placed 7–15 bp from the termini in order to make compatible ends between the two overlapping parts of the linker fragments (Fig. 2).

   Polymerase PfuTurbo® Cx has to be used at this step to allow amplification with the U-containing oligonucleotides and to prevent the addition of untemplated A at the 3′ end of the product.

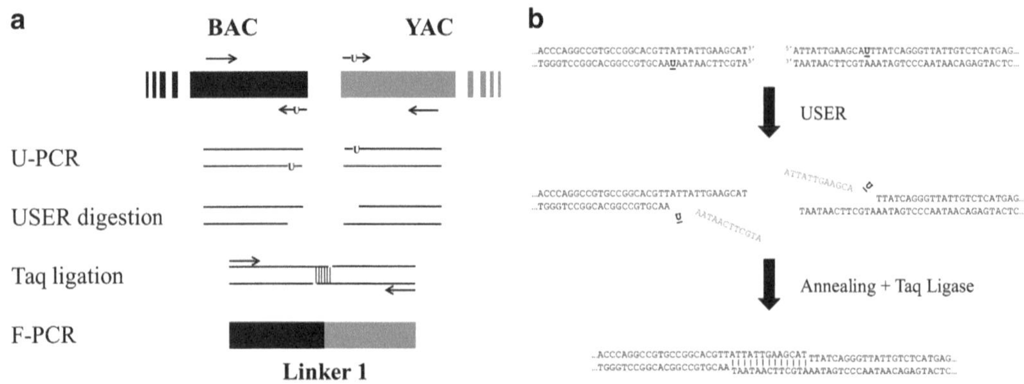


Fig. 2. Construction of linkers. Example for linker 1. Linkers are generated using the USER fusion in vitro method (3). (**a**) Construction of linkers is achieved in four steps: First, segments to be linked are amplified by PCR with one regular oligonucleotide and one U-containing oligonucleotide on the side where the linkage will occur. Second, PCR products are digested with the USER enzyme, leaving 3′ complementary overhangs on each side. Third, both ends are ligated using Taq ligase under cycling condition. In a final step, the two outmost (regular) primers amplify the full-length linker, which is then gel-purified before yeast transformation. (**b**) Sequences at junction of BAC and YAC linker segments before and after USER digestion and Taq ligation.

2. Mix together 500 ng of each parts of the linker and digest with USER enzyme overnight at 37°C.
3. Ligate the USER digestion products using Taq DNA ligase under cycling conditions.
4. The linker is obtained by a final PCR step using the two outermost regular primers. Linkers are then gel-purified with the PureLink™ Quick Gel Extraction Kit according to manufacturer's instructions.

### *3.5. Preparation of Controls and Samples to Be Transformed*

To perform the DNA assembly step, a mixture of all the 3-kb sets, plasmid, and linker DNAs will be cotransformed in yeast (Fig. 1b). For this, the following four samples, including controls, have to be prepared:

1. Positive control: 1 μg of a supercoiled yeast plasmid (e.g., pRS416).
2. Background control: containing only YAC, as prepared for the assembly and in the same amount as used in the assembly sample 4.
3. Sample containing YAC, all 3-kb sets (either gel-purified or not), and all linkers, but omitting the BAC.
4. Sample containing the YAC, the BAC, all 3-kb sets (either gel-purified or not), and all linkers.

During transformation, a negative control, without any DNA to be transformed, is added as well.

For samples that contain several DNA fragments, it is important that they are present in equimolar amounts of between 50 and 100 fmol. If using non-gel-purified fragments derived from a plasmid (pUC19 or pJHU1), remember that the plasmid backbone is taken into account in the concentration measurement; therefore, the fragment concentration needs to be adjusted according to the plasmid size. If using a gel-purified fragment, only the size of the purified fragment is taken into account.

The following is an example calculation for a non-gel-purified 3-kb fragments (set 2 of our assembly) to obtain 70 fmol of the 3-kb fragment that has been cloned into pJHU1:

- Concentration of the DNA (DNA of set 2 + DNA of pJHU1) in tube after digestion: 185 ng/μl.
- Size of the DNA construct before digestion: 3 kb from set 2 + 2 kb from pJHU1 = 5 kb.
- Molecular weight of a 5-kb DNA fragment: 5,000 × 660 (approximate molecular weight of a bp) = 3,300,000 g/mol.
- The weight of 70 fmol will be $3{,}300{,}000 \times (70 \times 10^{-15}) = 2.3 \times 10^{-7}$ g = 230 ng.
- With a concentration of 185 ng/μl, 230 ng is present in 230/180 = 1.2 μl.

After mixing of all DNA fragments for each sample, the volume may be too large to perform direct transformation (i.e., >10 μl). In this case, the samples are lyophilized for 1–2 h in a Speed-Vac, and the pellet is resuspended in 5–10 μl of $ddH_2O$.

### 3.6. Yeast Transformation (Lithium Acetate)

This protocol was adapted from Gietz et al. (10).

Phase I: Making competent cells

1. Pregrow cells in 10-ml YPD and incubate overnight at 30°C with shaking.
2. Dilute to $4 \times 10^6$ cells/ml in 50-ml YPD and grow at 30°C with shaking for 3–4 h, until concentration reaches $2\text{–}3 \times 10^7$ cells/ml (approximately 3 generations).
3. Harvest $10^8$ cells per transformation to be performed ($5 \times 10^8$ for the 5 transformation samples in this example) into a sterile centrifuge tube.

   Remark: The volume of cell culture can be adjusted if more than 10 transformations are planned.
4. Centrifuge culture 5 min at 3,000 × *g* and at 4°C.

5. Wash cells with 20 ml of sterile TE/LiOAc.
6. Centrifuge 5 min at $3{,}000 \times g$ and at 4°C.
7. Resuspend pellet in TE/LiOAc in order to yield $2 \times 10^9$ cells/ml, use 50 μl per transformation (take into account the volume of the pellet).
8. Incubate at 30°C for 15 min without shaking.

Phase II: Transformation

9. Prepare one microcentrifuge tube per transformation (include a negative control without DNA).
10. Add 300 μl of fresh TE/LiOAc/PEG (made on the same day).
11. Add 50 μg of carrier DNA (5 μl) that has been denatured previously for 3 min at 95°C and rapidly transferred to ice.
12. Add the DNA to be transformed (1–10 μl). Mix with vortex.
13. Add 50 μl of competent cells per tube ($10^8$ cells/transformation) and mix carefully by pipetting up and down.
14. Incubate at 30°C for 30 min without shaking.
15. Heat shock for 20 min at 42°C.
16. Centrifuge briefly and wash cells in 1 ml of sterile $ddH_2O$.
17. Centrifuge briefly and resuspend the cells in 100 μl of sterile $ddH_2O$.
18. Plate cells with glass beads on selective media. Prepare two petri dishes per transformation, and plate 10% (10 μl) of the transformed cells onto the first dish and the remaining 90% (90 μl) onto the second one.

After 2 days of growth, count and record the colony number for each transformation. The negative control should have no colonies. The positive control gives the transformation efficiency, which should be around of $10^5$ transformants/μg of DNA. Various controls omitting one or more fragments used in the assembly can be used to determine the background of the experiment. For each experiment that included the complete set of fragments (YAC, BAC, 3-kb fragments, and linkers), pick 12 clones and restreak them on selective medium for subcloning. Subclone also one transformant from the background control transformation (containing YAC only) in order to obtain a negative control for further PCR analysis.

### 3.7. Yeast Total DNA Extraction

Since the assembly is performed using a yeast centromeric plasmid, the full construct can be extracted using standard yeast total DNA isolation techniques, along with the other yeast chromosomes.

In order to screen positive yeast transformant clones, total DNA minipreps are performed for one of the subclones of each of the 12 that were chosen from each transformation that contained the complete set of fragments and for one subclone obtained from the background control transformation. The protocol is as follows:

1. The day before yeast DNA extraction, grow each colony in 3 ml of appropriate selective media. Incubate overnight at 30°C with shaking.
2. Transfer 2 ml of the culture into a microcentrifuge tube and centrifuge at 15,000 × *g* for 1 min at room temperature (repeat this step twice with 1 ml of the culture if only 1.5-ml microcentrifuge tubes are available).
3. Discard supernatant by tube inversion or aspiration.
4. Resuspend pellet in 200 μl of yeast miniprep solution I. Add 2 μl of zymolyase (make sure zymolyase is resuspended before adding it to the tube). Incubate for 45 min to 1 h at 37°C.
5. Add 200 μl of yeast miniprep solution II. Mix by inversion (do not vortex). Incubate for 30 min at 65°C, place the tubes on ice, and let them cool down before opening.
6. Add 100 μl of cold yeast miniprep solution III. Mix gently by inversion several times (do not vortex). A white precipitate appears. Leave for 30 min to 2 h on ice.
7. Microcentrifuge for 20 min at 4°C at 15,000 × *g*.
8. Transfer supernatant into a new microcentrifuge tube.

   Optional: If the sample does not look clean and clear after the first centrifugation, repeat centrifugation and transfer the supernatant into a new tube.
9. Add 300 μl of isopropanol and mix solution gently.
10. Microcentrifuge for 10 min at 15,000 × *g*.
11. Discard supernatant (be careful not to lose the pellet).
12. Rinse pellet with 1 ml of cold 70% ethanol and invert tube several times for washing.
13. Microcentrifuge briefly and discard supernatant.
14. Dry the pellet for 5 min in a Speed-Vac or for 30 min at 37°C (with tubes opened).
15. Dissolve pellets in 50-μl buffer TE (1×) and incubate for 10 min at 65°C.
16. DNA can be quantified on an agarose gel. Store at −20°C.

### 3.8. Specific Amplicon Analysis

In the example described, DNA to be assembled corresponds to a synthetic version of the yeast chromosome III. Since this assembly is done in yeast as well, we need to check for any unwanted recombination that may have occurred between the endogenous chromosome III and the synthetic chromosome construct. Therefore, we check integrity of them with specific amplicons. Native-specific-amplicons and synthetic-specific-amplicons must both be amplified from the DNA prep at this stage. People willing to assemble DNA from another organism will need to check their assembly only. As a first step, transformants are screened using one native-specific amplicon primer pair (WT) and three synthetic-specific amplicon primer pairs (SYN) that are scattered along the construct. For one primer pair, set up PCR as follows:

| | |
|---|---|
| DNA | 0.1 μl |
| PCR buffer, 10× | 2.5 μl |
| dNTPs | 2.0 μl |
| Oligo 1 | 1.0 μl |
| Oligo 2 | 1.0 μl |
| Taq polymerase | 0.1 μl (0.5 units) |
| $ddH_2O$ | 18.3 μl |
| Total | 25.0 μl |

Cycle:

| | | |
|---|---|---|
| 94°C | 4 min | (Initial denaturation step) |
| ↓ | | |
| 30 Cycles: | | |
| 95°C | 30 s | |
| 55°C | 30 s | |
| 72°C | 30 s | |
| ↓ | | |
| 72°C | 10 min | (Final elongation step) |
| ↓ | | |
| 4°C | | Forever |

After the PCR, add 2.5 μl of 10× loading dye to each tube and run a 10-μl aliquot on a 1% agarose gel with EtBr, along with a molecular weight marker (1-kb ladder). Take a picture using a UV light box and a camera to check for presence of the product and its size. For clones that give positive results for all four specific amplicons (one WT and three SYN), all the specific amplicons (WT and SYN) along the construct will be checked for presence and size

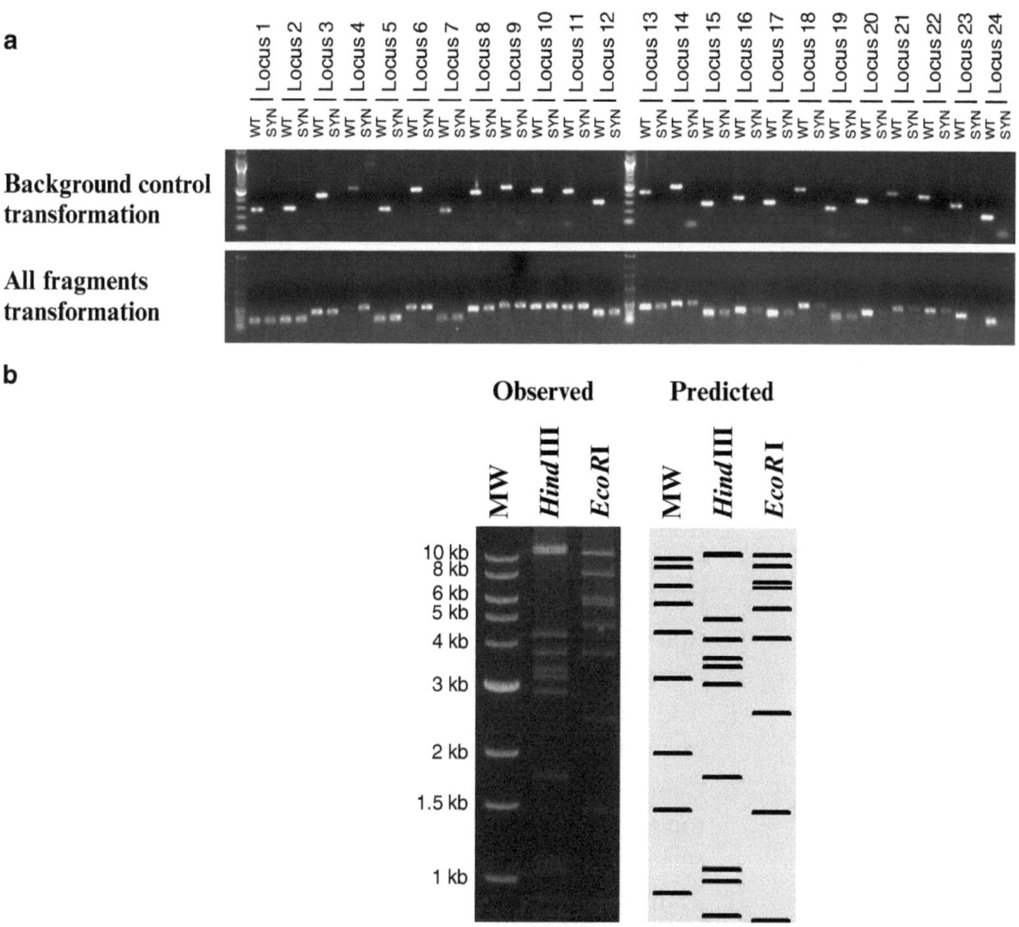


Fig. 3. Native-specific and synthetic-specific amplicons and RFLP analysis of the assembled construct. Along our ~40-kb segment are scattered 22 specific amplicons primer pairs (loci 1â€“22); each of which has a native-specific version (WT), amplifying only endogenous chromosomes of *Saccharomyces cerevisiae*, and a synthetic-specific version (SYN), amplifying only the synthetic DNA of our construct. Loci 23 and 24 correspond to amplicons of a subsequent portion of the synthetic chromosome and serve as a negative control. (**a**) Agarose gel electrophoresis of the amplicons are shown for one clone from the background control transformation, with the YAC only (*top gel*), and one clone from the transformation including all fragments, using non-gel-purified 3-kb sets (*bottom gel*). (**b**) RFLP profile of the ~40-kb construct cloned in the YACâ€“BAC vector after recovery from bacteria. The *picture on the left* shows the bands observed after *Hind*III or *EcoRI* digestion. The *picture on the right* shows the predicted bands that would be generated for the in silico construct generated from pDRAW32 (v1.1.107, Acaclone Software) after either *Hind*III digestion (predicted band sizes: 12,348, 11,388, 4,389, 3,805, 3,375, 3,215, 2,904, 1,778, 1,123, 1,057, 893, 798 bps) or *EcoRI* digestion (predicted band sizes: 11,563, 8,370, 6,239, 5,922, 4,789, 3,848, 2,483, 1,487, 874, 854, 765 bps). *MW* molecular weight, 1-kb ladder, sizes are indicated on the *left*.

(Fig. 3a). DNA extracted from the clone that corresponds to the background control transformation, which contains only YAC, is used as a positive control for the native-specific amplicons and a negative control for the synthetic-specific amplicons.

### 3.9. Plasmid Isolation from Yeast Transformants

In transformants that contain all specific amplicons along the assembled construct, it is important to check for potential rearrangements or small deletions within the construct that do not include any of the amplicons. This analysis will be performed by Restriction Fragment Length Polymorphism (RFLP). In order to isolate the plasmid DNA for such analysis, the plasmid is purified from positive yeast clones, followed by its transformation into bacteria for amplification and easy recovery of the clean plasmid assembly. Plasmid isolation from yeast is performed using a user-modified protocol using Qiagen columns from the QIAprep Spin Miniprep Kit:

1. Inoculate a single colony into 2–5 ml of appropriate selective media and grow the culture for 16–24 h at 30°C.
2. Harvest cells by centrifugation for 1 min at 20,000 × *g* and resuspend cells in 250 μl of buffer P1 containing 0.1 mg/ml RNase A. Transfer the cell suspension to a 1.5-ml microcentrifuge tube.
3. Add 50–100 μl of acid-washed glass beads and vortex for 5 min. Let it stand to allow the beads to settle. Transfer supernatant to a fresh 1.5-ml microcentrifuge tube.
4. Add 250 μl of lysis buffer P2 to the tube and invert gently four to six times to mix. Incubate at room temperature for 5 min.
5. Add 350 μl of neutralization buffer N3 to the tube and invert immediately but gently four to six times.
6. Centrifuge the lysate for 10 min at maximum speed in a tabletop microcentrifuge (15,000 × *g*). Meanwhile, place a QIAprep spin column in a 2-ml collection tube.
7. Transfer the cleared lysate from step 6 to the QIAprep spin column by decanting or pipetting.
8. Microcentrifuge for 30–60 s at 15,000 × *g*. Discard flow-through.
9. Wash QIAprep spin column by adding 0.75 ml of buffer PE and microcentrifuging for 30–60 s at 15,000 × *g*.
10. Discard flow-through and microcentrifuge for an additional 1 min at 15,000 × *g* to remove residual wash buffer.

    Residual wash buffer will not be completely removed unless the flow-through is discarded before this additional centrifugation. Residual ethanol from buffer PE may inhibit subsequent enzymatic reactions.
11. Place QIAprep spin column into a clean 1.5-ml microcentrifuge tube. To elute DNA, add 25 μl of buffer EB (10 mM Tris–HCl, pH 8.5) that has been preheated to 65°C to the center of each QIAprep spin column, and let it stand for 1 min and microcentrifuge for 1 min.

### 3.10. Recovery of BAC in Bacteria

Use 3 μl of the plasmid DNA extract from each yeast assembly reaction to transform TransforMax™ EPI300™ bacteria by electroporation. Follow the manufacturer's instructions and plate on selective media, LB + chloramphenicol if using the pBeloBAC11. Grow cells overnight and then extract plasmid DNA of bacterial transformants using method described in Subheading 3.1.

### 3.11. RFLP Analysis

Choose two or three restriction enzymes that will generate fragments ranging in size from 500 bp to 10 kb. For example, we chose to use *Eco*RI and *Hind*III restriction enzymes to analyze final assembled construct (Fig. 3b). Digest 1 μg of DNA using each restriction enzyme according to manufacturer's instructions in 50 μl of $ddH_2O$ (see Subheadings 3.1 and 3.4 for examples of digestion). Add 5 μl of 10× loading dye, load onto a 1% agarose gel with EtBr along with a molecular weight markers (1-kb ladder), and run at ~80 V. When separation is completed, take a picture using a UV light box and a camera. Analyze the size of each fragment by comparison with the molecular weight markers and record any difference with the predicted RFLP profile.

## 4. Notes

1. Ethidium bromide is a mutagen. Always wear gloves when handling gels and any solution containing EtBr.

## Acknowledgments

This work was supported by grants from National Science Foundation (MCB0718846) to JDB, JSB, and SC; from Microsoft to JSB and JDB; and from National Institutes of Health (GM077291) to SC. HM was a recipient of a fellowship from the Fondation pour la Recherche Medicale (FRM).

### References

1. Dymond, JS, Scheifele, L., Richardson, S, Lee, P, Chandrasegaran, S, Bader, JS, Boeke, JD (2009) Teaching synthetic biology, bioinformatics and engineering to undergraduates: the interdisciplinary Build-a-Genome course. *Genetics* 18:13–21.
2. Cooper E, Book chapter 20. THE BUILD A GENOME CLASS.
3. Chandrasegaran S, Book chapter 7. USER FUSION.
4. Gibson, DG, Benders, GA, Andrews-Pfannkoch, C, Denisova, EA, Baden-Tillson, H, Zaveri, J, Stockwell, TB, Brownley, A, Thomas, DW, Algire, MA, Merryman, C, Young, L, Noskov, VN, Glass, JI, Venter, JC, Hutchison, CA 3rd, Smith, HO (2008) Complete chemical synthesis, assembly, and cloning of a Mycoplasma genitalium genome. *Science* 319:1215–20.
5. Gibson, DG, Benders, GA, Axelrod, KC, Zaveri, J, Algire, MA, Moodie, M, Montague, MG, Venter, JC, Smith, HO, Hutchison, CA 3rd (2008) One-step assembly in yeast of 25 overlapping DNA fragments to form a complete synthetic Mycoplasma genitalium genome. *Proc Natl Acad Sci USA* 105:20404–9.
6. Gibson D, Book chapter 14. SYNTHESIS AND TRANSPLANTION OF BACTERIAL GENOMES.
7. Richardson, SM, Wheelan, SJ, Yarrington, RM, Boeke, JD (2006) GeneDesign: rapid, automated

design of multikilobase synthetic genes. *Genome Res* 16:550–6.

8. Richardson, SM, Nunley, PW, Yarrington, RM, Boeke, JD, Bader, JS (2010) GeneDesign 3.0 is an updated synthetic biology toolkit. *Nucleic Acids Res* 38:2603–6.
9. Richardson S, Book chapter 18 GENE DESIGN.
10. Gietz, RD, Schiestl, RH, Willems, AR, Woods, RA (1995) Studies on the transformation of intact yeast cells by the LiAc/SS-DNA/PEG procedure. *Yeast* 11:355–60.

# Chapter 12

# Recursive Construction of Perfect DNA Molecules and Libraries from Imperfect Oligonucleotides


## Gregory Linshiz, Tuval Ben Yehezkel, and Ehud Shapiro



## Abstract

Making faultless complex objects from potentially faulty building blocks is a fundamental challenge in computer engineering, nanotechnology, and synthetic biology. We developed an error-correcting recursive construction procedure that attempts to address this challenge. Making DNA molecules from synthetic oligonucleotides using the procedure described here surpasses existing methods for de novo DNA synthesis in speed, precision, and amenability to automation. It provides for the first time a unified DNA construction platform for combining synthetic and natural DNA fragments, for constructing designer DNA libraries, and for making the faultless long synthetic DNA building blocks needed for de novo genome construction.



**Key words:** Synthetic biology, DNA synthesis, Recursion, Error correction, Automation


## 1. Introduction

Making faultless complex objects from potentially faulty building blocks is a fundamental challenge in computer engineering (1), nanotechnology (2, 3), and synthetic biology (4, 5). We address this challenge with the development of an error-correcting recursive construction procedure for the generation of DNA molecules from synthetic oligonucleotides. This method surpasses existing methods for de novo DNA synthesis (6–11) in speed, precision, and amenability to automation.

Complex mathematical objects such as functions (12), fractals (13), natural and formal languages (14, 15), as well as computer data structures (16) are typically described using recursion. Although the promise of recursion to physical construction has been recognized (3), its application in engineering has been scarce (17, 18). Here, we show how recursion can be used as the basis of an error-correcting procedure for constructing faultless complex physical objects from potentially faulty building blocks. The physical


Jean Peccoud (ed.), *Gene Synthesis: Methods and Protocols*, Methods in Molecular Biology, vol. 852,
DOI 10.1007/978-1-61779-564-0_12,

objects are DNA molecules and their libraries, and the building blocks are short synthetic oligonucleotides.

Long DNA molecules encoding novel genetic elements are broadly needed in biological and biomedical research (4, 6, 19, 20); however, only short oligonucleotides (<100 nt) can be made quickly and cheaply by machines (21). Such oligonucleotides are used as building blocks to construct longer DNA molecules using one of two basic construction strategies, namely, polymerase cycling assembly of multiple overlapping synthetic oligonucleotides (PCA) (7) and ligation of synthetic oligonucleotides (8).

The major challenges facing DNA synthesis are fidelity, automation-level robustness of the construction process, and compliance with diverse construction needs such as de novo molecules, specified de novo libraries, and integration of natural and synthetic DNA. Our new procedure attempts to address all these challenges with one unified methodology for the first time.

The utility of synthetic DNA constructs in biology depends on their being free of sequence errors (4–6), yet the synthetic oligonucleotides serving as their building blocks are inherently error prone (about one sequence error per 160 nt) (4, 6). Therefore, all DNA construction protocols struggle from the labor-intensive time-consuming task of cloning and sequencing synthetic DNA fragments, seeking an error-free fragment. If none is found, a clone with sufficiently few errors that can be patched without undue effort using site-directed mutagenesis (22) is used.

Our procedure for constructing error-free DNA molecules integrates recursive construction and error correction. It uses "Divide & Conquer" (D&C) (16, 23), the quintessential recursive problem-solving technique, to construct long DNA molecules from short oligonucleotides and to then error-correct the resulting molecules until an error-free molecule is obtained.

D&C solves a problem (in our case, the construction of a particular single-stranded (ss)DNA molecule) by dividing it into two smaller subproblems (in our case, the construction of two shorter ssDNA molecules, as shown in the top part of Fig. 1a). Each subproblem is then solved recursively using D&C, and the solutions to the subproblems are combined into a solution to the original problem (in our case, combining the two ssDNA molecules into the desired longer ssDNA molecule, as shown in Figs. 1a and 2). If the problem is sufficiently small (in our case, the ssDNA molecule is quite short), it is not divided further, but solved directly (in our case, synthesized as an oligo). Our D&C DNA synthesis procedure performs the "Divide" phase in silico, dividing the desired sequence of the target DNA molecule further and further, until it is so short that it can be synthesized directly (Fig. 1). It then performs the "Conquer" phase in vitro, systematically composing pairs of smaller molecules into ever-larger molecules until the target molecule is constructed, as shown in Fig. 1a.

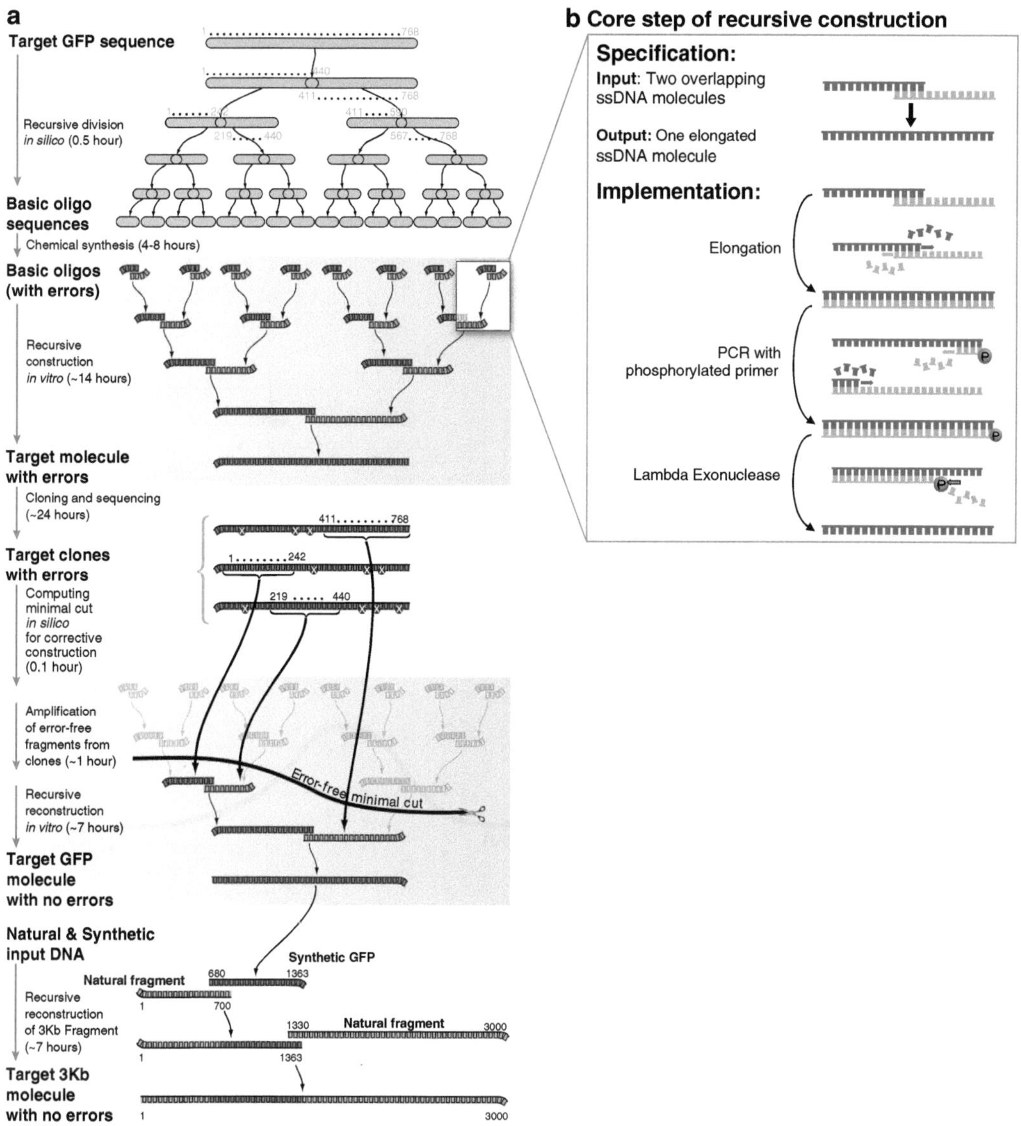


Fig. 1. Recursive DNA construction protocol. (**a**) Recursive construction of error-free DNA molecules from error-prone oligonucleotides. The target sequence is recursively divided in silico into basic oligo sequences using D&C. Each node in the tree of the divided sequence (top tree, basic oligo sequences) represents two overlapping DNA sequences, with overlaps delimited in the center of each node. Basic DNA building blocks from in silico division are obtained (with errors) by conventional oligonucleotide synthesis and used as inputs (basic oligos for recursive construction in vitro). We performed recursive construction from these oligonucleotides until full-length molecules (expected to be erroneous) were formed. *Light* and *dark blue* correspond to sense and antisense strands of recursively combined DNA. Erroneous target molecules were cloned and sequenced, and errors (marked with *red*) were identified. A minimal cut from error-free sequences found in clones (*nonfaded*, below minimal cut line) of the original construction process was computed in silico from an ensemble of three error-prone clones (*center*). Fragments constituting this minimal cut (*black brackets* on clones) were amplified from the clones and used as inputs to recursive reconstruction in vitro according to the minimal cut. This resulted in a target clone with no errors. Recursive construction can utilize synthetic DNA fragments and DNA from a natural (bacterial plasmid) source as inputs (natural fragment) for the recursive construction process. (**b**) The core step of recursive construction. In the core step of recursive construction, two overlapping ssDNA molecules are received as input and produce the elongated ssDNA molecule as output as follows. The overlapping ssDNA molecules hybridize and prime each other for an overlap extension elongation reaction to form a dsDNA molecule (elongation), which is then amplified by PCR with one of the two primers phosphorylated at its 5′ end (PCR with phosphorylated primers). The phosphate-labeled PCR strand is then degraded with lambda exonuclease, yielding an elongated ssDNA molecule as output (lambda exonuclease).

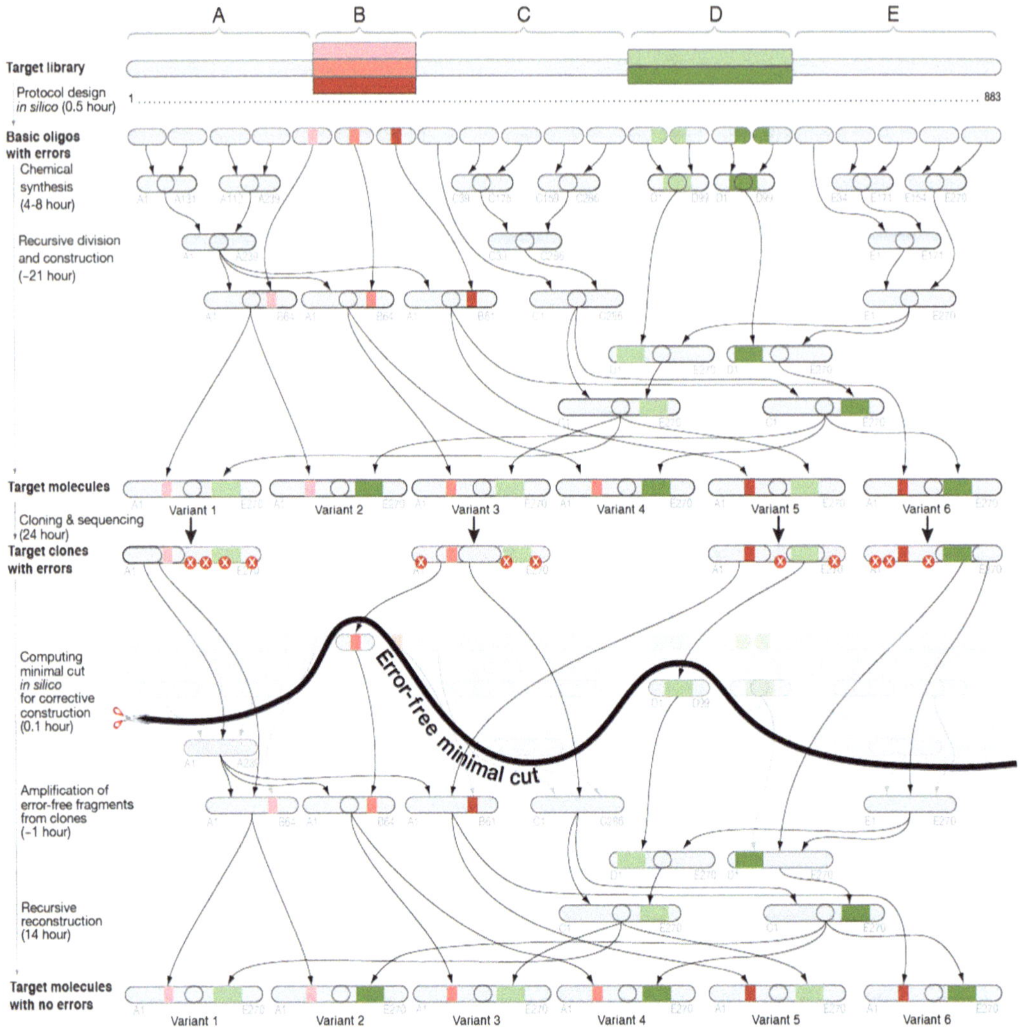


Fig. 2. Recursive construction and error correction of a simple combinatorial library of six p53 variants. The top of the figure shows a diagram of the target library describing shared (**a**, **c**, **e**) and unique (**b**, **d**) components of the target p53 combinatorial DNA library. Below is a graph (from and below basic oligos with errors) depicting the library construction process corresponding to the D&C-based in silico division. *Arrows* show the full-length variants that were created from the oligos. Target library sequences were recursively divided into shared and unique components and then further divided into basic oligonucleotides using D&C. These oligo sequences were made (with errors) by conventional chemical oligo synthesis. Using in vitro recursive construction with these oligos as input, we composed error-prone molecules of all six target p53 library members (target molecules with errors). These p53 variants were cloned and sequenced, and errors were identified (marked in *red* on top of clones). An error-free minimal cut (the *nonfaded* part of the graph below the minimal cut black line) of the original library construction plan was computed from only four error-prone clones (variants' numbers 1, 3, 5, and 6) for all six library variants. Error-free segments out of these four clones (delimited) were used as input for the error-free recursive reconstruction of all six target library members (target molecules with no errors) across clones and across variants according to the computed minimal cut graph. Expected optimal times for each step using standard equipment are shown on the *left*.

Recursive construction can be extended to produce error-free combinatorial DNA libraries with prespecified and/or randomized members. Most construction methods deliver combinatorial libraries in "one pot," which poses a limitation on the methods that can be used for their screening. Our library construction protocol can deliver each library member separately, say, in a separate well of a plate, which may facilitate a richer set of screening methods. In addition, the starting material for the libraries can be either natural or synthetic DNA.

In the application of library construction, our error correction method becomes even more efficient as we only need to find one error-corrected instance of fragments that are shared between several library members (Fig. 3).

## 2. Materials

### 2.1. Phosphorylation

1. Phosphorylation buffer: 70 mM Tris–HCl, 10 mM $MgCl_2$, 7 mM dithiothreitol, pH 7.6 at 37°C, and 1 mM ATP.
2. T4 polynucleotide kinase (NEB).

### 2.2. Overlap Extension Elongation Between Two ssDNA Fragments

1. Overlap extension buffer: 25 mM TAPS pH 9.3 at 25°C, 2 mM MgCl2, 50 mM KCl, and 1 mM β-mercaptoethanol 200 μM each of dNTP.
2. Thermo-Start DNA polymerase (ABgene).

### 2.3. PCR Amplification

1. PCR buffer: 25 mM TAPS pH 9.3 at 25°C, 2 mM $MgCl_2$, 50 mM KCl, and 1 mM β-mercaptoethanol 200 μM each of dNTP.
2. AccuSure DNA polymerase (BioLINE).

### 2.4. Lambda Exonuclease Digestion

1. Lambda exonuclease buffer: 25 mM TAPS pH 9.3 at 25°C, 2 mM $MgCl_2$, 50 mM KCl, 1 mM β-mercaptoethanol, and 5 mM 1,4-dithiothreitol.
2. Lambda exonuclease (Epicentre).

### 2.5. Chemical Oligonucleotide Synthesis

Oligonucleotides for all experiments were ordered from commercial providers (Sigma-Genosys and IDT) with standard desalting.

### 2.6. DNA Purification Kits

QIAGEN's QIAquik 96-well PCR purification kit and QIAGEN's MinElute PCR purification kit.

### 2.7. Cloning System

pGEM-T Easy Vector System and JM109 competent cells from Promega.

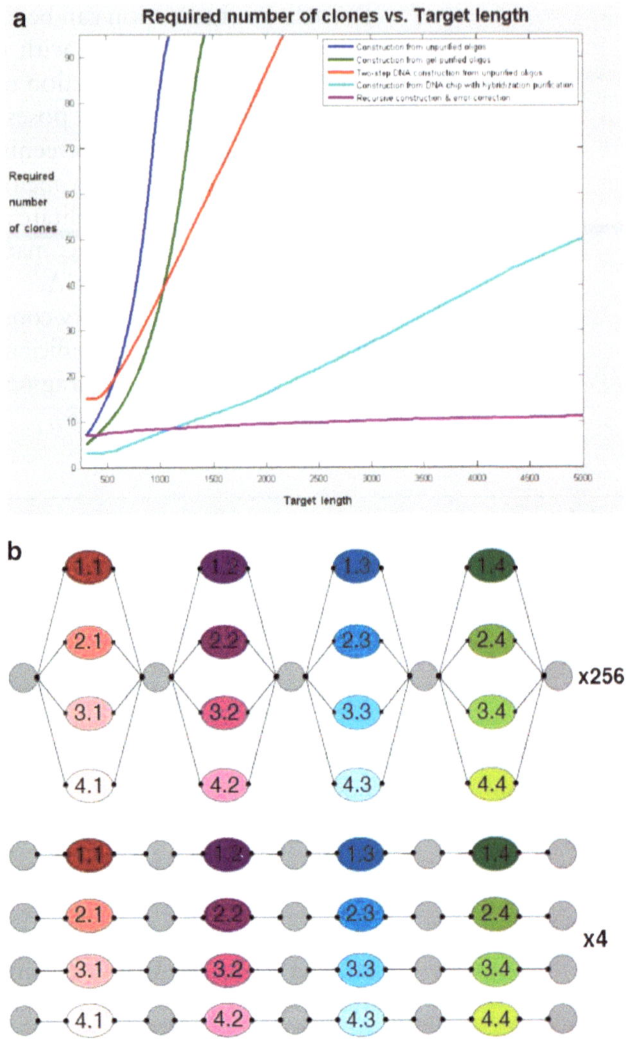


Fig. 3. Comparative analysis of error correction methodologies. (**a**) Error correction for single sequences. The number of clones that are actually sequenced (i.e., the required number of clones) in order to obtain an error-free synthetic DNA fragment is plotted against fragment length for different methods of construction, from left to right: naive construction from synthetic oligos with no error correction, construction from gel-purified oligos, two-step DNA construction, two-step construction from oligos purified by hybridization (6), and recursive construction with iterative error correction. (**b**) Error correction of libraries. A graph representing a DNA library with four variable sites, each containing four variants, totaling 256 possible library members (*top*). Using our protocol, we can first construct and error-correct a representative set of only four library members, which constitute a minimal cut through the construction graph of the entire library. A subsequent iteration of the protocol can use error-free fragments obtained from these four library members to efficiently construct the entire 256 strong library. This dramatically economizes the error correction of libraries compared to the correction of single sequences presented in panel (**a**).

#### 2.8. Liquid Handling Robot

Tecan Freedom 200.

#### 2.9. PCR Machine

Biometra TRobot.

## 3. Methods

### 3.1. Core Recursive Construction Steps

The core recursive construction step (Fig. 1b) requires four basic enzymatic reactions: phosphorylation, elongation, PCR, and lambda exonucleation. These are described in the order of execution by the following protocol:

#### 3.1.1. Phosphorylation

All PCR primers used in the recursive construction protocol are phosphorylated beforehand simultaneously according to the following protocol: 300 pmol of 5′ DNA termini are incubated in a 50-μl reaction containing phosphorylation buffer, 10 U of T4 polynucleotide kinase at 37°C for 30 min, followed by inactivation at 65°C for 20 min.

#### 3.1.2. Overlap Extension Elongation Between Two ssDNA Fragments

1–5 pmol of 5′ DNA termini of each progenitor are placed in a reaction containing overlap extension buffer and 4 U of Thermo-Start DNA polymerase. The thermal cycling program is as follows: enzyme activation at 95°C for 15 min, slow annealing of 0.1°C/s from 95 to 62°C, followed by elongation at 72°C for 10 min.

#### 3.1.3. PCR Amplification of the Elongation Product with Two Primers, One of Which Is Phosphorylated

1–0.1 fmol of template, 10 pmol of each primer in a 25-μl reaction containing PCR buffer, and 1.9 U of AccuSure DNA polymerase.

The thermal cycler program is as follows: 20 cycles of enzyme activation at 95°C for 10 min, denaturation at 95°C, annealing at Tm of primers, extension at 72°C, each for 1.5 min/kb.

#### 3.1.4. Lambda Exonuclease Digestion of the Above PCR Product to Regenerate ssDNA

1–5 pmol of 5′ phosphorylated DNA termini are digested in a reaction containing lambda exonuclease buffer and 5 U of lambda exonuclease. The thermal cycler program is 37°C for 15 min, 42°C for 2 min, followed by enzyme inactivation at 70°C for 10 min.

### 3.2. Automated DNA Purification

Automated DNA Purification was performed with QIAGEN's QIAquik 96-well PCR purification kit using standard protocols adapted to work with Tecan Freedom 200 and a vacuum manifold.

### 3.3. Preparation of Reactions

The preparation of all reactions listed above including QC sampling for capillary and gel electrophoresis was performed automatically by a Tecan Freedom 200 liquid-handling robot controlled with in-house developed software.

### 3.4. Protocol Automation

The steps of the protocol that were executed automatically were performed by a Tecan Freedom 200 robot mounted with a Biometra TRobot PCR block controlled with in-house developed software. Some of the steps were not performed robotically owing to lack of required automation equipment in our lab. Some of the DNA purifications were performed manually using a tabletop microcentrifuge owing to the lack of an automated plate centrifuge. Transfer of capillary electrophoresis and RT-PCR plates from the robot to their slots in the corresponding machinery was also performed manually due to lack of a robotic arm in our lab.

### 3.5. DNA Purification

Manual DNA purification was performed with QIAGEN's MinElute PCR purification kit using standard procedures.

### 3.6. Cloning

Fragments were cloned into the pGEM-T Easy Vector Systeml from Promega. Vectors containing cloned fragments were transformed into JM109 competent cells and sequenced.

### 3.7. Divide and Conquer Algorithm for DNA Molecule Synthesis

#### 3.7.1. Goal

The algorithm searches for an optimal protocol for the construction of the target molecule under a set of constraints and a set of cost parameters.

#### 3.7.2. General Description

The algorithm starts with the target molecule provided by the user and searches for an optimal and valid division point. Such a point fulfills a set of constraints such as the existence of specific primers and overlap specificity for the division point in addition to the existence of two valid protocols which may build the two divided parts. Once a valid division point is found, the algorithm recursively searches a protocol to build its two parts. The recursion stops when the target molecule is sufficiently short to be produced by an oligo synthesizer. A cost function is computed for each subprotocol based on the number of oligos and their lengths, the number of reactions, and the number of protocol tree levels required to build its target, and the smallest cost protocol is selected as the optimal protocol. Since the protocol space is very large, a dynamic programming algorithm is used to keep previously computed subprotocols in a cache and is reused when needed in a different search path. In addition, a branch and bound algorithm is used to trim the search space when the intermediate costs show that the current best cost for a protocol cannot be improved.

#### 3.7.3. Algorithm

Input

TARGET_SEQUENCE and CONFIGURATION with parameter values.

Output

A protocol to construct the target molecule including:

- Sequence list: a list of all oligos and intermediate target products.
- Reaction list: a list of reactions; each reaction has two sequences from the sequence list as its input and one sequence as its product, thus describing the protocol tree.
- Oligos and primers list: a list of oligos that are needed to be synthesized as building blocks for the protocols.

*3.7.4. Algorithm Pseudocode*

1. Preprocessing:
   1.1. Find best overlap for each point and compute its range of specificity.
   1.2. Find best primer for each point and compute its range of specificity.
2. Divide & Conquer (TARGET_SEQUENCE):
   2.1. If TARGET_SEQUENCE is in cache, then return the protocol from cache.
   2.2. If TARGET_SEQUENCE is shorter than MAX_OLIGO_SIZE, then return the OLIGO as the current protocol.
   2.3. Set CURRENT_PROTOCOL to NO_PROTOCOL and CURRENT_BEST_COST to Inf.
   2.4. For each division point, check the following:
      2.4.1. Valid overlap exists and complies with overlap constraints.
      2.4.2. Primers for division exist and comply with primers' constraints.
      2.4.3. Check primer dimers for each subtarget.
      2.4.4. If any of the checks 2.4.1–2.4.3 failed, then continue to next point.
      2.4.5. Compute lower bound on the cost of the protocol if LOWER_BOUND_COST > CURRENT_BEST_COST, then continue to next point.
      2.4.6. Divide & Conquer (LEFT_SUBTARGET).
      2.4.7. Divide & Conquer (RIGHT_SUBTARGET).
      2.4.8. Merge protocols and compute the CURRENT_COST.
      2.4.9. If CURRENT_COST < BEST_CURRENT_COST, set CURRENT_BEST_PROTOCOL = CURRENT_PROTOCOL (update cache).
   2.5. Return CURRENT_BEST_PROTOCOL.

#### 3.7.5. Pseudocode of Recursive Cost Function

In case of division
CURRENT_TARGET_COST = LEFT_SUBTARGET_COST + RIGHT_SUBTARGET_COST + LEVELCOST + REACTIONCOST

In case of oligo
oligo_constant_cost + oligo_nuc_cost × oligo_length

The following are the parameters used to evaluate the specificity and affinity of primers and elongation overlaps and parameters used to compute the cost function:

- max_oligo_len: 80 maximal oligo length.
- min_oligo_len: 30 minimal oligo length.
- max_primer_Tm: 70 maximal primer melting temperature.
- min_primer_Tm: 60 minimal overlap melting temperature.
- min_primer_len: 14 minimal primer length.
- max_primer_len: 30 maximal primer length.
- min_overlap_Tm: 60 minimal overlap melting temperature.
- min_overlap: 15 minimal overlap length.
- max_overlap: 70 maximal overlap length.
- levelcost: 50 Cost of additional level in the protocol.
- reactioncost: 10 Cost of additional reaction in the protocol.
- oligo_constant_cost: 2 Constant cost for as single Oligo.
- oligo_nuc_cost: 0.2500 length dependent cost for an Oligo.

The specificity of fragments in elongation reactions and of PCR primers was evaluated using sequence alignment algorithms and Tm formulas from the MATLAB bioinformatics toolbox.

### 3.8. Divide and Conquer Algorithm for DNA Combinatorial Library Synthesis

#### 3.8.1. Goal

The algorithm searches using Divide & Conquer approach for a protocol to construct a combinatorial library described by the user with an efficient utilization of the library shared sequences.

#### 3.8.2. General Description

The algorithm receives a library description with variable regions separated by shared regions. Each variable region may have two or more variants of different sequence and size. Using the D&C approach, the algorithm finds the optimal library protocol to construct the library from its shared and variable regions with minimal number of reactions considering that intermediate product may be a multiplication of the sizes of two variable regions. The algorithm then finds a specific valid overlap within the shared regions that is suitable for synthesizing the two adjacent regions with all their variants. The overlap defines the building blocks of the library, both the shared and variant fragments. Each building block is then planned using a D&C algorithm for a single molecule (described above). The protocols of the building blocks are merged, and additional reactions are added according to the libraries' optimal construction protocol previously calculated.

#### 3.8.3. Minimal Cut

A *cut* in a tree is a set of nodes that includes a single node on any path from the root to a leaf.

Let T be a recursive construction protocol tree and S a set of strings. We say that S covers T if there is a set of strings C such that every string in C is a substring of some string in S and C is a cut C of T. In such a case, we also say that S covers T with C.

*Claim*: If S covers T, then there is a unique minimal set C such that S covers T with C.

*Proof*: Easy.

Error-free reconstruction algorithm: Given an RC protocol T and a set of sequences (of molecular clones) S, find a minimal C such that S covers T with C.

Then, we lift C with PCR and do the recursive construction starting with C.

#### 3.8.4. Computing the Minimal Cut

We use a recursive approach for computing the minimal cut of a protocol tree. Each node in the tree represents a biochemical process with a product and two precursors. The algorithm starts with the root of the tree (target molecule) and for each node checks whether its product sequence exists with no errors in one of the clones. If such a clone exists, this product is marked as a new basic building block for reconstruction of the target molecule, and its primer pair and relevant clone (as template) are registered as its generating PCR. If there is no clone that contains an error-free sequence of the node product, the reaction is registered as existing reaction in the new protocol, and the algorithm is recursively executed on the two precursors of the product. The output of such a protocol is a tree of reactions which comprises a minimal cut of the original tree. It contains leaves for which error-free products exist and that all its internal nodes are error-free in the clones that contain them. An automated program that utilizes these new error-free building blocks for recursive construction of the target molecule is generated for the robot.

#### 3.8.5. Computing the Required Number of Clones

For a fragment of size $L$ under mutation rate $R$, the probability of having an error-free fragment in a single clone is taken from a Poisson distribution with lambda $= L \times R$ (the probability to have 0 errors when the expected errors are $L \times R$). To find the smallest number of clones required to get an error-free fragment with a probability larger than 95%, we use a binomial distribution and compute the probability of having at least one error-free fragment out of N clones.

In the D&C approach, the length of the pure fragment can be reduced to the size of an oligo (~80 bp) at the expense of having to perform more steps during reconstruction. Thus, in order to guarantee that we have full error-free coverage of the target sequence molecule, the probability of having a pure fragment of size *L* in *N* clones—*P* Success (*L*, *N*) is multiplied by itself, the number of fragments of size *L* that are required to construct the target molecule. (The first part is error-free, and the second part is error-free, etc.) We compute this number after considering the overlap which reduces the contribution of each oligo to be smaller than its actual size (~55 bp). Then, we find the smallest number of clones which satisfies the requirement that the total probability of having a minimal cut will exceed 95%.

## References


1. John Von Neumann, R. S. P. (1952) *Lectures on probabilistic logics and the synthesis of reliable organisms from unreliable components*, California Institute of Technology, Pasadena.
2. Drexler, K. E. (1992) *Nanosystems: molecular machinery, manufacturing, and computation*, Wiley, New York.
3. Merkle, R. C. (1997) Convergent assembly, *Nanotechnology 8*, 18–22.
4. Forster, A. C., and Church, G. M. (2006) Towards synthesis of a minimal cell, *Mol Syst Biol 2*, 45.
5. Carr, P. A., Park, J. S., Lee, Y. J., Yu, T., Zhang, S., and Jacobson, J. M. (2004) Protein-mediated error correction for de novo DNA synthesis, *Nucleic Acids Res 32*, e162.
6. Tian, J., Gong, H., Sheng, N., Zhou, X., Gulari, E., Gao, X., and Church, G. (2004) Accurate multiplex gene synthesis from programmable DNA microchips, *Nature 432*, 1050–1054.
7. Stemmer, W. P., Crameri, A., Ha, K. D., Brennan, T. M., and Heyneker, H. L. (1995) Single-step assembly of a gene and entire plasmid from large numbers of oligodeoxyribonucleotides, *Gene 164*, 49–53.
8. Au, L. C., Yang, F. Y., Yang, W. J., Lo, S. H., and Kao, C. F. (1998) Gene synthesis by a LCR-based approach: high-level production of leptin-L54 using synthetic gene in Escherichia coli, *Biochem Biophys Res Commun 248*, 200–203.
9. Xiong, A. S., Yao, Q. H., Peng, R. H., Li, X., Fan, H. Q., Cheng, Z. M., and Li, Y. (2004) A simple, rapid, high-fidelity and cost–effective PCR-based two-step DNA synthesis method for long gene sequences, *Nucleic Acids Res 32*, e98.
10. Gao, X., Yo, P., Keith, A., Ragan, T. J., and Harris, T. K. (2003) Thermodynamically balanced inside-out (TBIO) PCR-based gene synthesis: a novel method of primer design for high-fidelity assembly of longer gene sequences, *Nucleic Acids Res 31*, e143.
11. Smith, H. O., Hutchison, C. A., 3rd, Pfannkoch, C., and Venter, J. C. (2003) Generating a synthetic genome by whole genome assembly: phiX174 bacteriophage from synthetic oligonucleotides, *Proc Natl Acad Sci USA 100*, 15440–15445.
12. Rogers, H. (1967) *Theory of recursive functions and effective computability*, McGraw-Hill, New York.
13. Mandelbrot, B. B. (1982) The Fractals Book, *Observatory 102*, 151–151.
14. Chomsky, N. (1964) *Syntactic structures*, Mouton, The Hague.
15. Hopcroft, J. E., and Ullman, J. D. (1979) *Introduction to automata theory, languages, and computation*, Addison-Wesley, Reading, Mass.
16. Aho, A. V., Hopcroft, J. E., and Ullman, J. D. (1983) *Data structures and algorithms*, Addison-Wesley, Reading, Mass.; London.
17. SLONING, BIOTECHNOLOGY, and GMBH. (2006) DE NOVO ENZYMATIC PRODUCTION OF NUCLEIC ACID MOLECULES.
18. Knight, T. (2003) Idempotent Vector Design for Standard Assembly of Biobricks, *MIT Synthetic Biology Working Group.*

19. Heinemann, M., and Panke, S. (2006) Synthetic biology – putting engineering into biology, *Bioinformatics 22*, 2790–2799.
20. Ryu, D. D., and Nam, D. H. (2000) Recent progress in biomolecular engineering, *Biotechnol Prog 16*, 2–16.
21. Caruthers, M. H. (1985) Gene synthesis machines: DNA chemistry and its uses, *Science 230*, 281–285.
22. Hutchison, C. A., 3rd, Phillips, S., Edgell, M. H., Gillam, S., Jahnke, P., and Smith, M. (1978) Mutagenesis at a specific position in a DNA sequence, *J Biol Chem 253*, 6551–6560.
23. Alsuwaiyel, M. H. (1999) *Algorithms: design techniques and analysis*, World Scientific, Singapore; New Jersey.

# Chapter 13

# Cloning Whole Bacterial Genomes in Yeast

Gwynedd A. Benders

## Abstract

Many bacterial and archaeal genomes are of a similar size to molecules that have been cloned in the yeast *Saccharomyces cerevisiae* and thus might be clonable as single, circular episomes in this host. Yeast offers a variety of efficient tools for the manipulation and study of cloned DNA. One strategy to clone a genome in yeast is to cotransform yeast spheroplasts with the genome of interest and a linear yeast vector whose termini are homologous to a spot in the genome. Clones are selected on auxotrophic medium and then screened for completeness and size; they may also be sequenced.

**Key words:** Cloning, Bacteria, Genome, *Saccharomyces cerevisiae*, Yeast, Yeast centromeric plasmid, YAC, Synthetic biology

## 1. Introduction

Recently, a new frontier in microbiology has been opened by the realization that, in some environments, the vast majority of microbes are not cultivatable. Many microbes that can be cultured in the lab cannot be grown easily, cheaply, or safely and may lack genetic tools. Genomes from such organisms could be more easily studied if cloned in a convenient host.

Yeast can stably maintain cloned sequences of at least 2 Mb (1), which is a size that includes many bacterial and archaeal genomes. (By contrast, the largest episomes in *Escherichia coli* are generally about 300 kb.) Yeast is a well-studied model organism that has a repertoire of well-defined, efficient tools that can be used for manipulation of cloned DNA.

Large sequences can be cloned circularly in yeast as yeast centromeric plasmids (YCps) in the presence of a yeast centromere (*CEN*) and at least one sequence that functions as a yeast replication origin (autonomously replicating sequence (*ARS*)). A YCp is

Jean Peccoud (ed.), *Gene Synthesis: Methods and Protocols*, Methods in Molecular Biology, vol. 852,
DOI 10.1007/978-1-61779-564-0_13, © Springer Science+Business Media, LLC 2012

maintained at chromosomal copy number (1–2 copies per cell), replicates autonomously, and segregates mitotically (2, 3).

To clone a sequence as a YCp, a yeast vector containing a *CEN*, ARS, and a yeast selectable marker (often an auxotrophic marker such as *HIS*) is inserted into the sequence. This can be accomplished in several ways (4). The yeast vector can be transformed into the bacterium, where it integrates into the genome by a designed strategy such as transposition or homologous recombination. Bacterial genomes containing the vector are then isolated and transformed into yeast. Alternatively, the bacterial genome and the linear vector can be cotransformed into yeast, where they will recombine by homologous recombination. In this case, the termini of the vector are homologous to a spot in the genome. This protocol outlines the latter strategy.

Large molecules are transformed into yeast by incubation of yeast spheroplasts with the DNA in the presence of polyethylene glycol (5). The transforming DNA must be isolated carefully to avoid breakage (this can be done in agarose), and the spheroplasts must be handled gently to avoid lysis.

Clones are auxotrophically selected in top agar and screened for completeness by multiplex PCR. Those that amplify all PCR products are screened for size by restriction digestion and either pulsed-field or field inversion gel electrophoresis (PFGE or FIGE), followed if necessary by Southern blot. Clones can also be sequenced (6), although it is likely to be difficult to isolate significant quantities of clone DNA separately from yeast chromosomal DNA.

Thus far, yeast has been used almost exclusively to clone eukaryotic sequences. Sequences from a few bacteria have been cloned (7–9), and we have stably cloned and modified several complete bacterial genomes in yeast (4, 6, 10–13). To date, we have cloned whole genomes from mycoplasmas.

## 2. Materials

### 2.1. PCR Amplification of Vector

1. YCp vector such as pTARBAC3 (BACPAC Resources, Children's Hospital and Research Center at Oakland).
2. Pair of 88-mer primers (Integrated DNA Technologies, ultramer quality or PAGE-purified).
3. TaKaRa LA Taq™ DNA polymerase (TaKaRa Bio Inc., catalog no. RR002A) or Phusion® DNA polymerase (New England BioLabs, catalog F530), with respective reaction buffers.

### 2.2. PCR-Amplified Vector Purification from Agarose

1. Low-melting-point agarose (Invitrogen, catalog no. 16520050).
2. 1× TAE buffer (diluted from Invitrogen 10×, catalog no. 15558-042).

3. 1-kb DNA ladder (Invitrogen, catalog no. 15615-024).
4. Ethidium bromide, 10 mg/ml (20,000×; Invitrogen, catalog no. 15585-011) or SYBR® Gold, 10,000× (Invitrogen, catalog no. S-11494). Caution: Ethidium bromide is toxic.
5. Dark Reader transilluminator (Clare Chemical Research).
6. 3 M sodium acetate in 10× TAE, filter-sterilized.
7. Beta-agarase I (New England BioLabs, catalog no. M0392).
8. Buffer-saturated phenol (Invitrogen, catalog no. 15513-039). Caution: Phenol is toxic and corrosive.
9. Isopropanol.
10. GlycoBlue™ (Invitrogen, catalog no. AM9515).
11. 70% ethanol.
12. TE buffer (10 mM Tris–HCl, 1 mM EDTA, pH 8.0; Invitrogen, catalog no. AM9858).
13. Quantitative DNA ladder (High DNA Mass Ladder, Invitrogen, catalog no. 10496-016).

### 2.3. Isolation of Bacterial Genomes in Agarose

1. Medium specific to the bacterium whose genome will be cloned.
2. Chloramphenicol.
3. CHEF Bacterial Genomic DNA Plug Kit (Bio-Rad, catalog no. 170-3592). Alternatively, the solutions in this kit can be prepared as listed in the Bio-Rad Manual for the CHEF-DR® II or CHEF-DR® III under the heading "Preparation of Agarose Embedded Bacterial DNA." These solutions are listed as Subheading 2.3 items 4–7.
4. Cell suspension buffer: 10 mM Tris, pH 7.2, 20 mM NaCl, and 50 mM EDTA.
5. Lysozyme buffer: 10 mM Tris, pH 7.2, 50 mM NaCl, 0.2% sodium deoxycholate, 0.5% sodium lauryl sarcosine, and 1 mg/ml lysozyme. Add lysozyme fresh to the buffer immediately before using.
6. Proteinase K reaction buffer: 100 mM EDTA, pH 8.0, 0.2% sodium deoxycholate, 1% sodium lauryl sarcosine, and 1 mg/ml proteinase K. Add proteinase K powder fresh to the buffer immediately before using.
7. Wash buffer: 20 mM Tris-HCl, pH 8.0, and 50 mM EDTA.
8. Low-melting-point agarose (Invitrogen, catalog no. 16520050).
9. Phenylmethylsulfonyl fluoride (PMSF), 1 mM (caution: toxic).
10. Plug molds (Bio-Rad, catalog no. 170-3622).
11. Screened caps for 50-ml Falcon tubes (Bio-Rad, catalog no. 170-3711).

#### 2.4. Electrophoretic Analysis of Bacterial Genome Intactness

1. Pulsed-field gel electrophoresis apparatus, such as Bio-Rad's CHEF-DR® II or CHEF-DR® III.
2. 0.5× TBE buffer (diluted from Invitrogen 10×, catalog no. AM9863).
3. Pulsed-field certified agarose (Bio-Rad, catalog no. 162-0137).
4. Yeast chromosome PFG marker (New England BioLabs, catalog no. N0345S).
5. Low-range PFG marker (New England BioLabs, catalog no. N0350S).
6. SYBR® Gold, 10,000× concentrate stock solution (Invitrogen, catalog no. S-11494).

#### 2.5. Restriction Digestion of Agarose-Embedded Bacterial Genomes

1. Restriction endonuclease and buffer that will cleave the bacterial genome at, or near, the vector insertion site.
2. Materials listed in Subheading 2.4.

#### 2.6. Preparation of DNA for Yeast Transformation

1. TE buffer (10 mM Tris–HCl, 1 mM EDTA, pH 8.0; Invitrogen, catalog no. AM9858).
2. Beta-agarase I (New England BioLabs, catalog no. M0392).
3. Wide-bore (genomic) pipette tips, 200 μl (Molecular BioProducts ART tips, catalog no. 2069G).
4. Carrier DNA, e.g., calf thymus (Invitrogen, catalog no. 15633-019).

#### 2.7. Yeast Spheroplast Transformation

All solutions can be either autoclaved or filter-sterilized and stored at room temperature, except as noted.

1. Yeast strain VL6-48 (14) (ATCC number MYA-3666).
2. YPD plates: 1% Bacto™ yeast extract, 2% Bacto™ peptone, 2% dextrose, and 2% Bacto™ agar.
3. YPAD medium: 1% Bacto™ yeast extract, 2% Bacto™ peptone, 2% dextrose, and 0.006% adenine sulfate. For optimal yeast growth, make this medium from components, not a premixed YPD powder, and filter-sterilize.
4. Sterile distilled water.
5. Sorbitol solution: 1 M.
6. SPE solution: 1 M sorbitol, 10 mM sodium phosphate, 10 mM $Na_2EDTA$, pH 7.5.
7. Beta-mercaptoethanol (caution: toxic).
8. Zymolyase® solution: 10 mg/ml Zymolyase® 20T (MP BioMedicals, catalog no. 32092) in 25% glycerol, 50 mM Tris–HCl, pH 7.5. Filter-sterilize and store in aliquots at –20°C.
9. SDS solution: 2%. Sterilization not necessary.

10. SOS solution: 1 M sorbitol, 6.5 mM $CaCl_2$, 0.25% Bacto™ yeast extract, 0.5% Bacto™ peptone.
11. STC solution: 1 M sorbitol, 10 mM Tris–HCl, 10 mM $CaCl_2$, pH 7.5.
12. Wide-bore (genomic) pipette tips, 200 μl (Molecular BioProducts ART tips, catalog no. 2069G).
13. PEG solution: 20% PEG8000, 10 mM $CaCl_2$, 10 mM Tris–HCl, pH 7.5. Filter-sterilize. This solution will oxidize with time, lowering the pH. To prolong its lifetime, store at 4°C and warm to room temperature before use. Prepare a new solution every several months.
14. Wide-bore (genomic) pipette tips, 1,000 μl (Molecular BioProducts ART tips, catalog no. 2079G).
15. Selective sorbitol plates and top agar (SD-HIS + sorbitol): 1 M sorbitol, 2% dextrose, 0.17% yeast nitrogen base (without amino acids), 0.5% $(NH_4)_2SO_4$, 0.006% adenine (hemisulfate salt), 0.01% L-aspartic acid, 0.002% L-arginine (HCl), 0.01% L-glutamic acid (monosodium salt), 0.006% L-leucine, 0.003% L-lysine (mono-HCl), 0.002% L-methionine, 0.005% L-phenylalanine, 0.0375% L-serine, 0.02% L-threonine, 0.004% L-tryptophan, 0.003% L-tyrosine, 0.015% L-valine, 0.002% uracil, and 2% Bacto™ agar. This medium can be purchased as a premixed powder (without the agar or sorbitol) from TEKnova, Inc. (catalog no. C7112). TEKnova, Inc., also sells this medium in liquid and agar plate forms. If the vector used contains a different auxotrophic marker than HIS, this selective medium should be altered by the inclusion of 0.002% L-histidine and the exclusion of the relevant amino acid.

### 2.8. Culture of Transformants for Analysis

1. SD-HIS agar plates (see Subheading 2.7 item 15; omit sorbitol).
2. SD-HIS liquid medium (see Subheading 2.7 item 15; omit sorbitol and agar).
3. 48-well deep well plates (e.g., VWR Scientific, catalog no. 82004-674).
4. AirPore Tape Sheets (Qiagen, catalog no. 19571).

### 2.9. Crude DNA Isolation from Yeast Cells for PCR Template

1. SPE solution (see Subheading 2.7 item 6).
2. Beta-mercaptoethanol (caution: toxic).
3. Sterile distilled water.
4. SDS solution, 2%.
5. Sodium acetate, 5 M.
6. Isopropanol.

#### 2.10. Screening Clones for Completeness by Multiplex PCR

1. Multiplex PCR primers specific to the target bacterial genome.
2. Qiagen Multiplex PCR Kit (catalog no. 206143).
3. 100-bp DNA ladder (Invitrogen, catalog no. 15628-019).

#### 2.11. Screening Clones for Size by Electrophoresis

1. SD-HIS liquid medium (see Subheading 2.7 item 15; omit sorbitol and agar).
2. CHEF Yeast Genomic DNA Plug Kit (Bio-Rad, catalog no. 170-3593). Alternatively, the solutions in this kit can be prepared as listed in the Bio-Rad Manual for the CHEF-DR® II or CHEF-DR® III under the heading "Preparation of Agarose Embedded Yeast DNA." (These solutions are the same as those listed as Subheading 2.3 items 4, 6, and 7, with the replacement of Subheading 2.3 item 5 with the material listed as Subheading 2.11 item 4).
3. Cell suspension buffer (see Subheading 2.3 item 4).
4. Zymolyase® buffer: 10 mM Tris, pH 7.2, 50 mM EDTA, 1 mg/ml Zymolyase®. Add Zymolyase® fresh to the buffer immediately before using.
5. Zymolyase®, 10 mg/ml (see Subheading 2.7 item 8).
6. Proteinase K reaction buffer (see Subheading 2.3 item 6).
7. Wash buffer (see Subheading 2.3 item 7).
8. Low-melting-point agarose (Invitrogen, catalog no. 16520050).
9. Phenylmethylsulfonyl fluoride (PMSF), 1 mM (caution: toxic).
10. Plug molds (Bio-Rad, catalog no. 170-3622).
11. Screened caps for 50-ml Falcon tubes (Bio-Rad, catalog no. 170-3711).
12. 1× TAE buffer (diluted from Invitrogen 10× buffer, catalog no. 15558-042).
13. *Not*I and reaction buffer (New England BioLabs, catalog no. R0189).
14. Pulsed-field gel electrophoresis apparatus, such as Bio-Rad CHEF-DR® II or CHEF-DR® III.
15. 0.5× TBE buffer (diluted from Invitrogen 10× buffer, catalog no. AM9863).
16. Pulsed-field certified agarose (Bio-Rad, catalog no. 162-0137).
17. Yeast chromosome PFG marker (New England BioLabs, catalog no. N0345S).
18. Low-range PFG marker (New England BioLabs, catalog no. N0350S).
19. SYBR® Gold, 10,000× (Invitrogen, catalog no. S-11494).

## 3. Methods

The following methods assume that the bacterial genome being cloned has been sequenced. It is very helpful to visualize and analyze the genome of interest using software such as CLC DNA Workbench or Invitrogen's Vector NTI® software. This will aid in determining restriction maps, choosing a vector insertion site, and designing PCR primers.

If a genome does not clone in its entirety, the cloning of large sections of the genome can next be attempted. In this strategy, the genome can be digested with a restriction enzyme that produces a few large fragments. Fragment-specific vectors, with ends that span the termini of a fragment, can then be mixed separately or together with the digested genome for cotransformation into yeast.

### *3.1. Vector Insertion Site Determination*

1. The vector is designed to recombine with the bacterial genome by homologous recombination. There are several considerations when choosing the vector insertion site:
   (a) There is a much higher (~20×) rate of recombination of the vector with the genome if there is a double-stranded break at the site of recombination (4, 15, 16). Therefore, the most straightforward way to recombine the YCp vector into the bacterial genome of interest is at, or near, a unique restriction site. If no unique sites are present, a rare site can be used, and the genome can be incompletely digested.
   (b) It may be beneficial to leave certain genome regions intact. For instance, vector insertion within an essential gene or in a region of interest might be undesirable; if this occurs, the vector can be moved within the genome after cloning (13).

### *3.2. Vector Amplification Design*

1. Design primers to amplify a YCp vector such that 60-bp termini that are homologous to the target insertion site of the bacterial genome to be cloned are added. Each primer should have 60 nt of bacterial genome target sequence, followed by the 8 nt *Not*I site, followed by about 20 nt of homology to the YCp vector. Thus, these are 88-mer primers. The *Not*I sites will be used to cleave the vector from the insert for sizing of clones. Be sure to orient the bacterial genome sequences correctly (Fig. 1). One bacterial primer sequence should be in the same orientation as the reference strand of the genome sequence, and the other should be the reverse complement of the reference strand. Thus, for example, if the vector insertion site were 5′-GACTT.. AGCTA**CCCGGG**TACAG..GTAGG-3′, with cleavage at the SmaI site (in bold), the bacterial portion of one primer sequence would be 5′-GACTT..AGCTA**CCC**-3′ and of the other would be 5′-CCTAC..CTGTA**CCC**-3′.

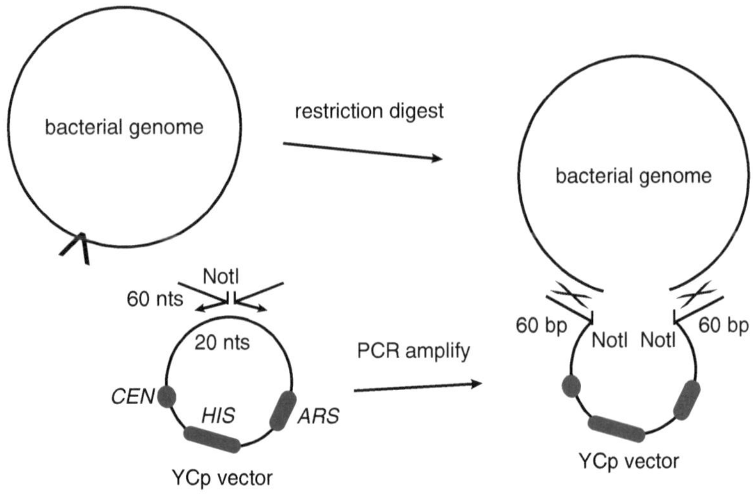


Fig. 1. Schematic for cloning a bacterial genome in yeast by homologous recombination of a yeast centromeric plasmid (YCp) vector with the genome at a double-stranded break created by restriction digestion at a unique site (*caret*). The vector contains a yeast centromere (*CEN*), replication origin (*ARS*), and selectable auxotrophic marker (*HIS* or other) and is prepared by PCR amplification. *Not*I sites in the amplification primers allow clones to be sized by *Not*I digestion and electrophoresis.

2. If amplifying the BAC-YCp portion of pTARBAC3, the following primer sequences can be used (*Not*I site is in bold): 5′-60 nt bacterial genome sequence-**GCGGCCGC**ATTGCATCAACGCATATAGC-3′ and 5′-60 nt bacterial genome sequence-**GCGGCCGC**AGAGCCTTCAACCCAGTCAG-3′. This will amplify about 9 kb of this 14-kb vector.
3. If mutations are undesirable at the site of homologous recombination of the vector with the target genome, the primers should be ordered purified.

#### 3.3. PCR Amplification of Vector

1. The vector amplified may be on the order of 10 kb. To produce such a large amplicon, we have had success both with TaKaRa LA Taq™ and Finnzymes' Phusion® DNA polymerases. The vector may amplify better if linearized. If the whole vector is amplified, it may be helpful to phosphatase the vector template so that it does not produce vector-only yeast transformants. If only part of the vector is amplified, phosphatase treatment is unnecessary since the template will be left behind during gel purification of the amplicon.
2. If amplifying pTARBAC3, the following final reaction concentrations may be used: 1× LA Taq™ polymerase buffer II ($Mg^{2+}$ plus), 1.0 mM $MgCl_2$ (in addition to the $MgCl_2$ in the LA buffer), 0.4 mM each dNTP, and 0.5 μM each of 2 primers. Make a 100 μl reaction and use 5 U TaKaRa LA *Taq*™ polymerase and 10 ng AatII-linearized pTARBAC3. Cycling conditions

are 94°C for 1 min; followed by five cycles of 94°C for 20 s, 55°C for 30 s, and 72°C for 12 min; followed by 25 cycles of 94°C for 20 s, 60°C for 30 s, and 72°C for 12 min; followed by a terminal extension of 72°C for 8 min.

### 3.4. PCR-Amplified Vector Purification from Agarose

1. Cast a small 0.8% gel using low-melting-point agarose and 1× TAE buffer. Use a gel comb that has been taped to create one large well adjacent to one single well. The gel will set much faster if poured in a cold room or refrigerator.
2. Load the single well with a 1-kb ladder. Load the PCR-amplified vector into the large well.
3. Run the gel in 1× TAE until good ladder separation is achieved.
4. Stain the gel with ethidium bromide or SYBR® Gold. Stain with ethidium bromide if you may need to run the gel further poststaining. Sometimes, running a gel after SYBR® Gold staining will cause distorted bands.
5. View the gel on a Dark Reader transilluminator (blue light box) and excise the band of interest with a razor blade. (The higher energy of a UV box will damage the DNA.) Be careful not to scratch the top of the transilluminator. Place the gel fragment in an Eppendorf or other conveniently sized tube.
6. Spin down the gel fragment to estimate its volume in the tube. Or weigh the tube with the gel and subtract the tube's weight. Or weigh the gel slice on plastic wrap or weigh paper.
7. Add 0.1× volume of 10× TAE buffer containing 3 M sodium acetate (pH ~9).
8. Heat the gel slice at 42–45°C for 10 min in a heat block or water bath. If the agarose is at a concentration greater than 1%, consider to diluting it to 1% to aid in digestion.
9. Melt the gel slice by incubating it at 70°C for 5 min.
10. Equilibrate the gel slice to 42–45°C by incubation at this temperature for 10 min.
11. Add beta-agarase I (New England BioLabs, 1 U/μl) at 1 μl per 100 μl of agarose. If the gel was run in TBE instead of TAE, double the amount of agarase.
12. Incubate at 42–45°C for 1–2 h. The agarose will gel below 42°C, and the agarase will be inactivated above 45°C.
13. During this time, remove buffer-saturated phenol from the refrigerator to warm to room temperature (RT). (Do not use phenol–chloroform.) The phenol should be colorless, not pink or brown.
14. Extract with an equal volume of phenol in a fume hood. Centrifuge at RT several minutes.

15. Remove the aqueous (top) phase to a fresh tube. As it cools, it may become cloudy due to residual phenol. This phenol will be removed by ethanol washes. Dispose of phenol waste in the correct hazardous waste containers.
16. Add an equal volume of isopropanol and 2 μl GlycoBlue™. Mix, then centrifuge 30 min at RT.
17. Decant the supernatant, wash the pellet 2× with 70% ethanol, and resuspend in TE.
18. Quantitate the gel-purified vector both spectrophotometrically and by running 20–50 ng on a gel with a quantitative ladder.

### 3.5. Isolation of Bacterial Genomes in Agarose

This procedure will vary with the bacterium, but in general follow the protocol as listed in the Bio-Rad Manual for the CHEF-DR® II or CHEF-DR® III, under the heading "Preparation of Agarose Embedded Bacterial DNA." This protocol ($5 \times 10^7$ cells per plug) will produce about 200 ng of genomic DNA per plug of a 2 Mb bacterial genome, assuming 2 copies per cell. If possible, scale up the density of the cells in the plug as much as possible to have more DNA for yeast transformation. This protocol is as follows below:

1. Grow a culture of the bacterium of interest. It may be helpful to add chloramphenicol (to a final concentration of 180 μg/ml) to the culture for the last several generations to maximize for fully replicated genomes.
2. Harvest and wash the cells with cell suspension buffer.
3. For each plug to be made, resuspend the cells in cell suspension buffer to a volume of 50 μl. Warm the resuspended cells to 50°C.
4. Mix the warmed, resuspended cells with an equal volume of 2% low-melting-point agarose (kept at 50°C).
5. Fill plug molds with the cells suspended in agarose. Each plug will be about 100 μl. Place in a refrigerator to speed solidification.
6. Add the solidified plugs to lysozyme buffer (5 ml/ml of plugs) in a 50-ml Falcon tube. Incubate for 1 h at 37°C. Lysostaphin can be used for bacteria not sensitive to lysozyme.
7. Using a screened cap to prevent plug loss during decantation, rinse the plugs with 50 ml wash buffer.
8. Add proteinase K reaction buffer (5 ml/ml of plugs). Incubate overnight at 50°C. The plugs can also be left over the weekend.
9. Rinse the plugs several times with 50 ml wash buffer for at least 30 min each time. During the second wash, add PMSF to 1 mM to inactivate residual proteinase K so that it will not interfere with subsequent restriction digestion.

***3.6. Electrophoretic Analysis of Bacterial Genome Intactness***

1. Bacterial genomes isolated in agarose should be checked for intactness before transformation into yeast. Their Mb-scale size requires pulsed-field gel electrophoresis for resolution. Even under these conditions, such large circular molecules will not electrophorese out of the plug into the gel. However, due to breakage and incomplete replication, a fraction of the isolated genomes will be linear and full-length. These should form a distinct band of the correct linear size on the gel, with very little to no smearing (indicates degradation) below this band.
2. Prepare and run an agarose gel per the instructions of the pulsed-field gel apparatus. The following conditions separate a yeast chromosome ladder (200 kb to 2 Mb) well: a 1% agarose gel, 0.5× TBE buffer, 120° angle, 14°C circulating buffer, 6 V/cm, 60–120 s switch time, and 24-h run time (Bio-Rad). Stain the gel with SYBR® Gold and destain. For the clearest detection, destain overnight.

***3.7. Restriction Digestion of Agarose-Embedded Bacterial Genomes***

1. To guide how many units of a restriction enzyme to use for digestion of agarose-embedded genomic DNA, follow New England BioLab's technical reference, "Digestion of Agarose-Embedded DNA: Enzyme-Specific Information."
2. Check for successful digestion by electrophoresis, as described in Subheading 3.6.

***3.8. Preparation of DNA for Yeast Transformation***

1. Dialyze the plug(s) containing digested bacterial genomes against TE buffer (several washes of 1 ml or more TE, pH 8.0 per plug) and remove the buffer.
2. Heat the plug at 42–45°C for 10 min in a heat block or water bath.
3. Melt the plug by incubating it at 70°C for 5 min.
4. Equilibrate the plug to 42–45°C by incubation at this temperature for 10 min.
5. Add beta-agarase I (New England BioLabs, 1 U/μl) at 1 μl per plug.
6. Incubate at 42–45°C for 1–2 h.
7. Using a wide-bore 200-μl pipette tip, or a 20-μl tip with the end cut off with a razor blade, carefully transfer 18-μl liquefied plug to a fresh Eppendorf. Add an equimolar amount of PCR-amplified, gel-purified vector in a volume of 2 μl. If the total amount of DNA in this 20 μl is less than 2 μg, add carrier DNA (e.g., calf thymus) in a volume of 1–2 μl so that the total amount of DNA is 2–3 μg.

### 3.9. Yeast Spheroplast Transformation

This protocol is adapted from Kouprina and Larionov (17) and Burgers and Percival (18).

1. Pick a single large pink colony from a freshly grown YPD plate of yeast strain VL6-48. Inoculate several milliliters of YPAD and grow overnight at 30°C with shaking. This starter culture may be stored at 4°C and used for several weeks.
2. Dilute the VL6-48 starter culture 1:1,000 in 50 ml YPAD in a 500-ml flask and culture overnight at 30°C with shaking.
3. Harvest the cells when the optical density at 600 nm is 4.5–5.0 (about $10^8$ cells/ml). (Take the reading on a 1:10 dilution of the culture, which will give a measurement of 0.45–0.50.) (see Note 1).
4. Pellet the cells at 3,000 × *g* for 5 min at 4°C in a 50-ml Falcon tube. Decant the medium, draining the last drops onto a paper towel.
5. Resuspend the cell pellet in 30 ml sterile distilled water by vortexing, then add another 20 ml sterile distilled water, and mix. Harvest the cells and drain the supernatant as in step 4.
6. Resuspend the cell pellet in 20 ml of 1 M sorbitol by vortexing. Store the suspension at 4°C overnight or proceed immediately.
7. Harvest the cells and drain the supernatant as in step 4.
8. Resuspend the cell pellet in 20 ml SPE. Add 50 μl beta-mercaptoethanol and 200 μl Zymolyase® and mix by inversion (see Note 2).
9. Lay the tube on its side and gently agitate at 30°C.
10. After 10 min, dilute 100–200 μl spheroplasting cells 10× separately in 1.0 M sorbitol and in 2% SDS. Compare their optical densities at 800 nm. Continue these measurements until the difference is 8–10×. Ideally, this should take 15–20 min.
11. Harvest the spheroplasts at 200–300 × *g* for 5 min at 4°C. This will produce a fluffy white pellet. Very gently decant the supernatant.
12. Add 30 ml 1.0 M sorbitol. Very gently invert the tube to resuspend the spheroplasts. If necessary, aid resuspension by gently pipetting with a 50-ml serological pipette. Add 20 ml 1.0 sorbitol and mix by gentle inversion.
13. Harvest the spheroplasts and drain the supernatant as in step 11.
14. Repeat steps 11 and 12.
15. Add 2.0 ml STC to the washed spheroplast pellet and resuspend by gently swirling.
16. Add 200 μl spheroplasts to 20 μl DNA in an Eppendorf and gently pipette twice with a wide-bore (genomic) pipette tip.
17. Incubate for 10 min at room temperature.
18. Add 800 μl PEG, invert to mix, and incubate 10 min at room temperature.

19. Centrifuge the spheroplasts at 250×*g* for 5 min at room temperature.
20. Carefully remove the supernatant by pipetting. Add 800 μl SOS and pipette gently to mix using a 1,000-μl wide-bore tip.
21. Incubate at 30°C for 40–60 min without shaking.
22. Gently invert the tube to resuspend the settled spheroplasts. Pipette the spheroplasts into a 15-ml Falcon tube containing 7 ml top agar (kept in a water bath at 46°C). Draw the tip up through the agar while pipetting out the spheroplasts, then cap and invert the tube several times to mix. Quickly pour the agar onto a prewarmed selective plate and swirl the plate to spread the agar.
23. Incubate the plates 3–5 days at 30°C.

### 3.10. Culture of Transformants for Analysis

1. Transformants should be restreaked for single colonies before definitive analysis. Without this step, a clone may not be pure. If there are only a few transformants to screen, or the cloning efficiency is high, transformants can be restreaked before any analysis is done. If this is not the case, this step can be done after putatively complete clones are identified to save labor and agar plates. In this case, initially make a small streak or patch of each primary transformant. Use selective medium (e.g., SD-HIS).
2. Inoculate 5 ml SD-HIS with either a single colony or a bit of a patch of each transformant. Culture overnight at 30°C. It may be helpful to use 48-well deep well plates for this purpose. These can be sealed with AirPore Tape Sheets to allow for aeration.
3. Remove 1 ml of each clone for DNA isolation for PCR analysis. Remove a small aliquot of each clone for a frozen glycerol stock. (Add sterile glycerol to a final concentration of 15%, mix well, and store at −80°C.) Reserve the rest of the culture at 4°C. It will be used to isolate DNA in agarose for clones that are PCR-positive.

### 3.11. Crude DNA Isolation from Yeast Cells for PCR Template

This protocol is adapted from Kouprina and Larionov (17).

1. Centrifuge 1 ml yeast culture in an Eppendorf 5 min at 3,000×*g*.
2. Decant medium.
3. Add 1 ml SPE with 30 μl 10 mg/ml Zymolyase® and 2 μl beta-mercaptoethanol; vortex to resuspend cell pellet.
4. Incubate at 37°C for 1 h.
5. Centrifuge 5 min at 3,000×*g* and decant medium.
6. Resuspend each spheroplast pellet in 500 μl water through vigorous vortexing.
7. Add 50 μl of 2% SDS per tube and invert vigorously to mix.
8. Incubate at 70°C for 15 min.
9. Add 50 μl of 5 M sodium acetate; invert vigorously to mix.

10. Incubate on ice for 15 min.
11. Centrifuge at max speed for 5 min.
12. Pour supernatant (~500 μl) into a fresh tube; add 500 μl isopropanol; invert to mix.
13. Spin max speed for 30 min.
14. Decant supernatant and let pellet dry.
15. Resuspend pellet in 1 ml TE by vortexing. Before use in PCR, give the resuspension a quick spin to pellet debris. If the pellet is resuspended in a smaller volume of TE, this must be diluted before use in PCR to prevent inhibition of the reaction.

### 3.12. Screening Clones for Completeness by Multiplex PCR

1. Design PCR amplicons evenly spaced around the target genome. Design amplicons in sets of 10, with 100-bp size increments between amplicons, ranging in size from about 100 bp to about 1 kb. A set could consist of 100 bp, 200 bp, etc., or 125 bp, 225 bp, etc., amplicons. Follow the guidelines of the manual for the Qiagen Multiplex PCR Kit in designing the primers. The number of amplicons will determine the resolution of the screen. For instance, two sets of ten amplicons evenly spaced around a 1-Mb genome would spot check for completeness every 50 kb.
2. Perform multiplex PCRs following the manual for the Qiagen Multiplex PCR Kit. This often requires little to no optimization. Use no template and yeast host DNA as negative controls. Use genomic DNA isolated from the bacterium of interest as a positive control. As another positive control, mix together yeast host DNA and genomic DNA isolated from the bacterium of interest.
3. Analyze PCRs on an agarose gel with a 100-bp ladder.

### 3.13. Screening Clones for Size by Electrophoresis

This procedure should selectively remove linear yeast chromosomes from plugs so that a clone can be visualized without being obscured by this background. Alternatively, or in addition, a Southern blot can be performed on the electrophoresed gel (6).

1. To isolate total DNA from yeast clones of interest, follow the protocol as listed in the Bio-Rad Manual for the CHEF-DR® II or CHEF-DR® III, under the heading "Preparation of Agarose Embedded Yeast DNA." As a negative control, isolate DNA from the yeast host strain. This protocol follows the steps of Subheading 3.5, with the following differences:
   (a) Grow the clones in selective yeast medium (e.g., SD-HIS).
   (b) Make plugs at a concentration of about $3 \times 10^8$ cells per plug. Each $OD_{600}$ of a yeast culture is about $3 \times 10^7$ cells/ml, so this corresponds to, for instance, 1 plug per 5 ml of a culture grown to an $OD_{600}$ of 2.

   (c) Right before mixing the cells with agarose, add lysozyme to a final concentration of 1 mg/ml in the plug.
   (d) Use Zymolyase® buffer in place of lysozyme buffer.
2. Preelectrophorese the plugs to remove linear yeast chromosomes (large circular molecules will remain in the plug). Electrophorese the plugs at 6 V/cm for several hours in a 1% gel with 1× TAE buffer. Remove the plugs from the gel wells (see Note 3).
3. Digest the plugs with *Not*I (overnight may be convenient). This will cleave clones at the vector-genome junctions and at any *Not*I sites in the genome (see Note 4).
4. Prepare and run an agarose gel per the instructions of the pulsed-field gel apparatus. The following conditions separate a yeast chromosome ladder (200 kb to 2 Mb) well: a 1% agarose gel, 0.5× TBE buffer, 120° angle, 14°C circulating buffer, 6 V/cm, 60–120 s switch time, and 24-h run time (Bio-Rad). Run a yeast chromosome and a lambda ladder. If there are no *Not*I sites in the genome, a plug of bacterial genomic DNA can be run as a sizing standard.
5. Stain the gel with SYBR® Gold and destain. For the clearest detection, destain overnight. For the most sensitive detection, scan the gel with an Amersham Typhoon™ 9410 Fluorescence Imager or similar instrument.

## 4. Notes

1. This is a late-log-phase culture. VL6-48 spheroplasts efficiently even at this late growth stage. Other yeast strains may not spheroplast efficiently unless harvested at early mid-log phase (about $10^7$ cells/ml).
2. The amount of Zymolyase® required may vary and should be predetermined for each combination of enzyme supplier and yeast strain.
3. The efficiency of the preelectrophoresis is increased if the yeast chromosomes are digested into smaller pieces. If there are restriction enzymes that do not have recognition sites in the genome of interest or the vector used, they may be used to digest the plugs before preelectrophoresis. Alternately, Plasmid-Safe™ (Epicentre® Biotechnologies) can be used to selectively digest linear yeast chromosomes while leaving circular YCps intact. These strategies will reduce yeast chromosome background on the pulsed-field gel, making clones easier to detect.
4. As an alternative to restriction digestion, circular clones can be linearized by heating plugs at 55°C for 1 h.

## Acknowledgments

The author thanks Vladimir Larionov at the National Institutes of Health for the gift of yeast strains and vectors and Clyde Hutchison for helpful comments. This work was supported by Synthetic Genomics, Inc.

### References

1. Marschall P, Malik N, Larin Z (1999) Transfer of YACs up to 2.3 Mb intact into human cells with polyethylenimine. Gene Ther 6: 1634–1637
2. Clarke L, Carbon J (1980) Isolation of a yeast centromere and construction of functional small circular chromosomes. Nature 287: 504–509
3. Stinchcomb DT, Mann C, Davis RW (1982) Centromeric DNA from *Saccharomyces cerevisiae*. J Mol Biol 158: 157–190
4. Benders GA, Noskov VN, Denisova EA et al (2010) Cloning whole bacterial genomes in yeast. Nucleic Acids Res 38: 2558–2569
5. Hinnen A, Hicks JB, Fink GR (1978) Transformation of yeast. Proc Natl Acad Sci USA 75: 1929–1933
6. Gibson DG, Benders GA, Andrews-Pfannkoch C et al (2008) Complete chemical synthesis, assembly, and cloning of a *Mycoplasma genitalium* genome. Science 319: 1215–1220
7. Kuspa A, Vollrath D, Cheng Y et al (1989) Physical mapping of the *Myxococcus xanthus* genome by random cloning in yeast artificial chromosomes. Proc Natl Acad Sci USA 86: 8917–8921
8. Azevedo V, Alvarez E, Zumstein E et al (1993) An ordered collection of *Bacillus subtilis* DNA segments cloned in yeast artificial chromosomes. Proc Natl Acad Sci USA 90: 6047–6051
9. Heuer T, Burger C, Maass G et al (1998) Cloning of prokaryotic genomes in yeast artificial chromosomes: application to the population genetics of *Pseudomonas aeruginosa*. Electrophoresis 19: 486–494
10. Gibson DG, Benders GA, Axelrod KC et al (2008) One-step assembly in yeast of 25 overlapping DNA fragments to form a complete synthetic *Mycoplasma genitalium* genome. Proc Natl Acad Sci USA 105: 20404–20409
11. Lartigue C, Vashee S, Algire MA et al (2009) Creating bacterial strains from genomes that have been cloned and engineered in yeast. Science 325: 1693–1696
12. Gibson DG, Glass JI, Lartigue C et al (2010) Creation of a bacterial cell controlled by a chemically synthesized genome. Science 329: 52–56
13. Noskov VN, Segall-Shapiro TH, Chuang RY (2010) Tandem repeat coupled with endonuclease cleavage (TREC): a seamless modification tool for genome engineering in yeast. Nucleic Acids Res 38: 2570–2576
14. Kouprina N, Annab L, Graves J et al (1998) Functional copies of a human gene can be directly isolated by transformation-associated recombination cloning with a small 3′ end target sequence. Proc Natl Acad Sci USA 95: 4469–4474
15. Orr-Weaver TL, Szostak JW, Rothstein RJ (1981) Yeast transformation: a model system for the study of recombination. Proc Natl Acad Sci USA 78: 6354–6358
16. Leem SH, Noskov VN, Park JE et al (2003) Optimum conditions for selective isolation of genes from complex genomes by transformation-associated recombination cloning. Nucleic Acids Res 31: e29
17. Kouprina N, Larionov V (2008) Selective isolation of genomic loci from complex genomes by transformation-associated recombination cloning in the yeast *Saccharomyces cerevisiae*. Nat Protoc 3: 371–377
18. Burgers PM, Percival KJ (1987) Transformation of yeast spheroplasts without cell fusion. Anal Biochem 163: 391–397

# Chapter 14

# Production of Infectious Poliovirus from Synthetic Viral Genomes


## Jeronimo Cello and Steffen Mueller



## Abstract

Making use of the nucleotides sequence of the RNA genome (7,440 nt) of poliovirus, synthetic deoxyoligonucleotides, 60–70 nt in length are synthesized. The oligonucleotides that map to adjacent segments in the genome are designed such that they are of plus- and minus-strand polarity with the overlapping complementary sequences at their termini. The oligonucleotides are assembled by asymmetric PCR, and then, the segments are ligated directly into a plasmid. The segments are assembled stepwise via common unique restriction endonuclease cleavage sites to yield a full-length poliovirus complementary DNA (cDNA), carrying a phage T7 RNA polymerase promoter at the (left) 5′ end. Genomic RNA is generated with a phage T7 RNA polymerase. The viral RNA is incubated in a cell-free extract, where it is translated and replicated, resulting in the de novo synthesis of poliovirus (PV). Finally, in vivo and in vitro experiments are carried out to confirm that infectious material isolated from the cell-free extract is indeed infectious PV.

All components of the synthetic PV are generated by biochemical means. No virus-related structure or component that may have been generated previously in vivo is used as template or as building block for the viral particles. Our work shows that it is possible to synthesize an infectious agent in test tube by solely following instructions from a written sequence.



**Key words:** Synthetic virus, Oligonucleotides, Poliovirus, cDNA, Infectious agent, Asymmetric PCR, Synthetic genomes


## 1. Introduction

### *1.1. Genome Organization and Cellular Life Cycle of Poliovirus*

Poliovirus (PV), the causative agent of poliomyelitis (1), is a member of the genus *Enterovirus* of the Picornaviridae. The poliovirus genome consists of a single (+) sense RNA molecule of about 7,440 nt in length. A long 5′ nontranslated region (NTR) of 742 nt is followed by a single open reading frame (ORF) encoding for the viral polyprotein of 2,209 amino acids (aa) and a short 3′ NTR of 70 nt ensued by a virus-encoded poly(A) tract of around 60 adenylate residues in length (2, 3) (Fig. 1a).


Jean Peccoud (ed.), *Gene Synthesis: Methods and Protocols*, Methods in Molecular Biology, vol. 852,
DOI 10.1007/978-1-61779-564-0_14, © Springer Science+Business Media, LLC 2012

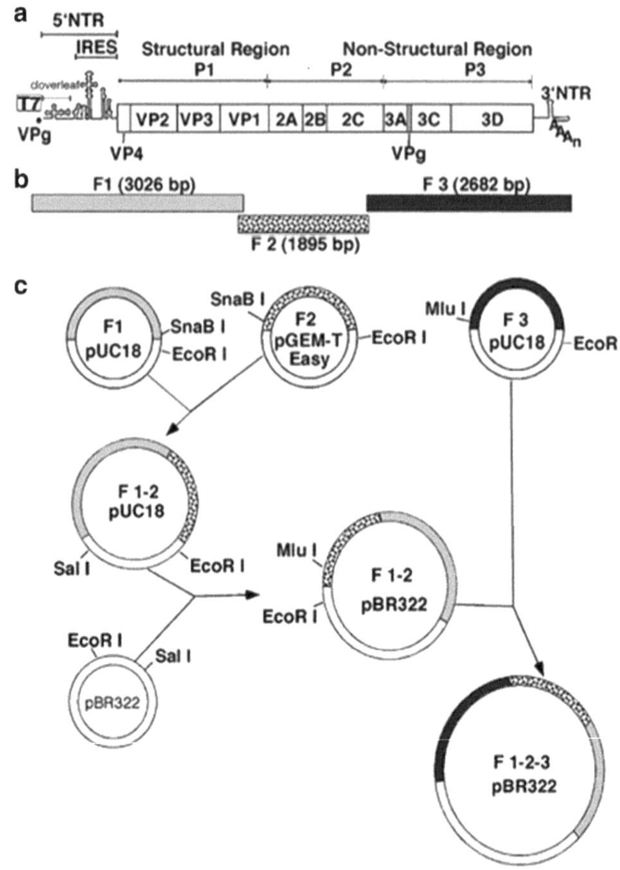


Fig. 1. Genomic structure of PV1 (M) and strategy for the synthesis of its full-length cDNA. (**a**) The 5′ end is terminated with the genome-linked protein VPg, and the 3′ end terminates with the poly (A). In the cDNA, VPg is replaced by the T7 RNA polymerase promoter. The polyprotein contains (N terminus to C terminus) structural (P1) and nonstructural (P2 and P3) proteins that are released from the polypeptide chain by proteolytic processing. (**b**) cDNA carrying a T7 RNA polymerase promoter at the 5′ NTR end is subdivided into three large fragments for the synthesis of full-length sPV1 (M) cDNA. The sizes of the fragments (in base pairs) are depicted above or below each rectangle that represents the respective fragment. (**c**) The three DNA fragments are synthesized, as described in the text. The DNA fragments are assembled stepwise via common unique restriction endonuclease cleavage sites to yield full-length sPV1 (M) cDNA (F1-2-3 pBR322) (Reproduced from ref. 12 with permission).

The virus enters the cell after attaching to the cellular receptor CD155 (4, 5). Immediately after the virus particle uncoats inside the cell, the genomic RNA is released into the cytoplasm of the host cell and translated under the control of internal ribosomal entry site (IRES) into a single highly autocatalytic polyprotein (6). The polyprotein contains structural (P1) and nonstructural (P2 and P3) proteins (Fig. 1a) that are released from the polypeptide chain by proteolytic processing mediated by virally encoded proteinases $2A^{pro}$ and $3C^{pro}/3CD^{pro}$ to ultimately generate 11 mature viral proteins (3). The proteolytic cleavage products serve as capsid precursors (80S particles) and replication proteins. With the aid of viral proteins, most notably, the RNA-dependent RNA polymerase $3D^{pol}$ and the genome-linked protein VPg, along with cellular components, the viral RNA is transcribed into minus-strand copies that serve as templates for the synthesis of new viral genomes (plus-strand RNA). Newly synthesized plus-strand RNA can serve as messenger RNA for more protein synthesis, involve further in RNA replication, or be encapsidated by capsid proteins (3).

Encapsidation of VPg-linked positive-strand RNA molecules constitutes the final steps in the cellular life cycle of PV. $3CD^{pro}$ cleaves the P1 precursor polypeptide, thereby giving rise to proteins VP0, VP1, and VP3, which assemble to form a protomer (7). Five protomers then aggregate, thereby generating a pentamer (8), of which 12 ultimately assemble to constitute the procapsid (9). At this point, the VPg-linked positive-strand virus RNA is encapsidated (9). Cleavage of VPO into VP2 and VP4 finalizes virus assembly by stabilizing the capsid and thereby converting the provirion into a mature, infectious virus particle (10). The mature virus capsid is an icosahedron composed of 60 copies each of VP1–VP4 and exhibiting five-, three-, and twofold axes of symmetry. The outer surface of mature virus capsid is formed by capsid proteins VP1–VP3, while VP4 is found internally (11). In cell culture, the entire replication cycle lasts approximately 6–8 h and yields $10^4$–$10^5$ progeny virions per cell.

Five developments were particularly relevant to pursue the chemical synthesis of infectious poliovirus (12): (1) the elucidation of the primary sequence and gene organization of the poliovirus (2) (Fig. 1a); (2) all properties required for poliovirus proliferation in nature are encoded in the viral genome (3); (3) the test-tube generation of infectious poliovirus RNA by transcription of cDNA with phage T7 RNA polymerase (13); (4) any DNA molecule of any length could be assembled from oligonucleotides (14); and (5) the generation of a cell-free system, consisting of an extract of uninfected HeLa cells, in which poliovirus can be synthesized de novo (15).

Having in mind this information, we considered it important that an example should be set of the de novo biochemical synthesis of a human pathogen. Therefore, we ask ourselves whether poliovirus might be resurrected from pure genomic information. We entertained—correctly—the idea that this achievement could be a wake-up call leading to the public discussion of the potential dangers associated with the misuse of new developments in genetics, genomics, and other areas of biomedical sciences.

The total synthesis of a sequence of poliovirus 1 (Mahoney) [PV1 (M)] is divided into three stages: synthesis and assembly of full-length poliovirus-specific cDNA, transcription of cDNA into RNA, and generation of virions in a cell-free extract (Fig. 2).

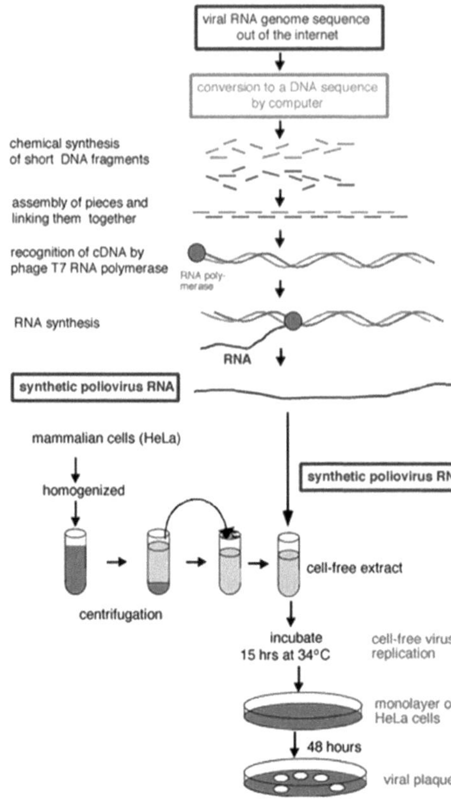


Fig. 2. Synthesis of PV in the absence of natural template (12). Scheme by which cDNA and viral transcript RNA is synthesized, followed by replication of the synthetic viral RNA in a HeLa cell-free extract to form progeny virions (15) (Reproduced from ref. 17 with permission).

Finally, in vivo and in vitro experiments are carried out to confirm that infectious material isolated from the cell-free extract is indeed PV1 (M).

## 2. Materials

### *2.1. Synthesis and Assembly of Full-Length Poliovirus-Specific cDNA*

1. Gel-purified oligonucleotides are dissolved in distilled water at stock solution of 100 μM.
2. PCR cycles are performed in 50 μl of 10 mM Tris–HCl, 2.5 mM $MgCl_2$, 50 mM KCl, pH 8.3, containing 2.5 U Taq and Pwo DNA polymerases mixture (Expand High Fidelity PCR System, Roche), and the four deoxynucleotide triphosphates (dNTP) at a concentration of 200 μM each.

3. Agarose is dissolved at 1.5 mg/ml in TE Buffer 1× (pH 8.0, 10 mM Tris–HCl containing 1 mM EDTA-$Na_2$). Add ethidium bromide to final concentration of 0.2 mg/ml.
4. QIAquick gel extraction kit (QIAGEN, Valencia, CA).
5. Vectors: pGEM®-T Easy Vector (Promega), pUC18 vector (Invitrogen), and pBR322 (New England Biolabs, Inc.).
6. Restriction enzymes.
7. *Escherichia coli* strain DH5α.
8. QIAprep kit for plasmid DNA purification (QIAGEN).

#### 2.2. *In Vitro* Transcription of cDNA into RNA and Generation of Virions in a Cell-Free Extract

1. Synthetic PV1 cDNA (sPV1 cDNA).
2. *Eco*RI restriction enzyme diluted at final concentration of 1 U/μl in SuRE/Cut Buffer H (Roche Diagnostics GmbH).
3. Reaction mixture for in vitro transcription: 40 mM Tris–HCl (pH 8.0), 6 mM $MgCl_2$, 10 mM dithiothreitol (DTT), 2 mM spermidine, 1 mM each NTP, 40 U of RNase inhibitor, and 40 U of T7 RNA polymerase (see Note 1). Reactives for RNA purification: phenol–chloroform 1:1, 10 M ammonium acetate, pure EtOH, and 70% EtOH.
4. HeLa cell-free extract.
5. Reagents for preparation HeLa cell-free extract: PBS buffer (pH: 7.4), 100 mM EGTA (ethylene glycol tetraacetic acid), 50% glycerol, 100 mM $CaCl_2$, *Staphylococcus aureus* nuclease solution (Boehringer Mannheim, lyophilized, resuspended to 2 mg/ml, 5,000 U/ml in 20 mM K-HEPES, pH: 7.5), hypotonic buffer (10 mM K-HEPES, 10 mM potassium acetate, 1.5 mM magnesium acetate. Adjust pH to with 0.1 M KOH. Add 25 μl of 1 M DTT/10 ml of buffer just before use. Store at −20°C), dialysis buffer (10 mM K-HEPES, 90 mM K Ac, 1.5 mM Mg $(Ac)_2$).
6. Translation mix (7.62×): 1 mM ATP, 63 μM GTP, 26 μg/ml creatine phosphate (Sigma, 10 mg/ml in 10 mM K-HEPES, pH 7.4), 20 mM K-HEPES, 20 mM calf liver tRNA, 13 μM amino acid mix, 263 μM spermidine (in 10 mM K-HEPES, Sigma), and 250 μl distilled water. For analyzing the products of in vitro translation of poliovirus RNA in HeLa cell-free extract, the amino acid mix used for the translation mix is not supplemented with methionine. For production of poliovirus in HeLa cell-free extract, the amino acid mix is supplemented with methionine. Aliquot mix and store at −80°C. Individual components should be stored at −20°C.
7. Salt mix (10×): 5.45 μl Mg $(Ac)_2$ (1 M), 13.1 μl $MgCl_2$, 795 μl K acetate (1 M), 186.5 μl distilled water. All stocks should be filter sterilized and stored at −20°C.

8. $^{35}$S-methionine (11 mCi/ml, ICN).
9. RNase inhibitor (Promega, 400 U/µl).
10. To analyze the products of in vitro translation of polio RNA in HeLa cell-free extract, the master mix for PVM translation (sufficient for ten reactions) is prepared as follows: 50 µl HeLa cell-free extract, 17 µl translation mix (minus methionine), 9.17 µl 10× salt mix, 8.3 µl $^{35}$S-methionine, 1 µl RNase inhibitor, and 6.15 µl distilled water. Master mix for the synthesis of poliovirus in HeLa cell-free extract is prepared as follows: 50 µl HeLa cell-free extract, 17 µl translation mix (supplemented with methionine), 9.17 µl 10× salt mix, 1 µl RNase inhibitor, and 14.45 µl distilled water.
11. RNA gel: 0.4 g agarose in 50 ml of Tris buffer (8 ml 1 M Tris base and 200 µl pure acetic acid in 200 ml distilled water).
12. SDS-12.5% polyacrylamid gel Separation gel: 7.8 ml acrylamide/bisacrylamide (40:1), 9.4 ml 1 ml Tris pH: 8.8, 0.25 ml 10% SDS, 7.5 ml distilled water. Add last 15 µl TEMED and 125 µl 10% ammonium persulfate.

Stacking Gel: 1.25 ml acrylamide/bisacrylamide (40:1), 1.25 ml 1 ml Tris pH: 6.8, 0.1 ml 10% SDS, 7.35 ml distilled water. Add last 5 µl TEMED and 50 µl 10% ammonium persulfate.

### 2.3. In Vivo and In Vitro Characterization of Synthetic Poliovirus

1. HBSS buffer: In 400 ml $H_2O$, add 4.0 g NaCl, 0.20 g KCl, 0.50 g glucose, 0.021 g $Na_2HPO_4$, 0.027 g $KH_2PO_4$, 0.096 g $CaCl_2$, 0.111 g $MgCl_2$, 0.176 g $NaHCO_3$, and qc 500 ml pH 7.1. Filter sterilize.
2. RNase A (final concentration 20 µg/ml) and RNase T1 (final concentration 100 U/ml).
3. Confluent HeLa cell monolayers on 6-well plates.
4. Dulbecco's minimal essential medium (DMEM) containing 1% penicillin/streptomycin.
5. 0.6% (w/v) gum tragacanth.
6. 1% crystal violet (1 g crystal violet dissolved in 1% glutaraldehyde and 99% methanol).
7. Poliovirus receptor-specific monoclonal antibody (Mab) D171.
8. Poliovirus type 1- and 2-specific rabbit hyperimmune serum (anti-PV1 and anti-PV2).
9. Mice transgenic for human poliovirus receptor (CD155 tg mice) susceptible to poliovirus infection.

## 3. Methods

### *3.1. Synthesis of Poliovirus cDNA Fragments*

1. The strategy of synthesizing the genome of poliovirus type 1 Mahoney (Fig. 1a) starts with the assembly of the full-length cDNA carrying a phage T7 RNA polymerase promoter at the (left) 5′ end (Fig. 1a) from three large, overlapping DNA fragments (as example see Fig. 1b).
2. Each segment is obtained by combining overlapping segments of 400–600 bp.
3. The segments of the PV1 genome (GenBank accession no.: NC_002058) are presented as purified synthetic oligonucleotides (purchased from commercial company) of plus- and minus-strand polarity with overlapping complementary sequences at their termini. The overlapping region between two oligonucleotides used in one PCR is on average 20 nt long (see Note 2). The oligonucleotides size should be restricted in size to maximally 60 nt, when possible, in order to reduce the errors of the PCR product. The oligonucleotide representing the 5′ end of the nontranslated region (NTR) should be designed to carry the sequence of T7 RNA promoter (TAATACGACTATAGG).
4. The overall strategy of fragment synthesis is outlined in Fig. 2. In general, 8–12 gel-purified oligonucleotides are assembled by asymmetric PCR (Fig. 3). The ratio of the oligonucleotides pair used in each reaction is 5:1. Products H, I, J, and K are obtained from four PCR assays that contained the oligonucleotides pairs D9/D10 and D13/D14 in ratios of 25 pmol:5 pmol

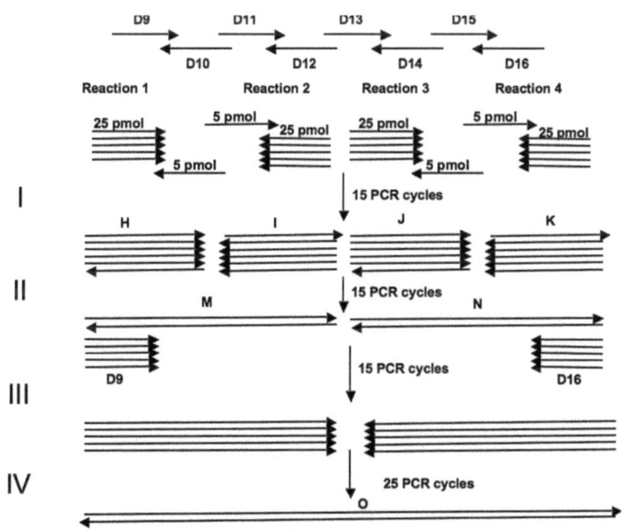


Fig. 3. Flow chart for the synthesis of a polynucleotide of 400–600 bp in length. Eight synthetic oligonucleotides (D9–D16), which overlap their respective adjacent sequences by an average of 20 nt, are mixed pairwise in four reactions at the stoichiometry indicated.

and D11/D12 and D15/D16 in ratios 5 pmol:25 pmol. Fifteen PCR cycles are performed in 50 μl reaction mixture. The typical cycling conditions are 15 s at 94°C (denaturation), 30 s at 55°C (annealing), and 45 s at 72°C (polymerization) and final elongation step 5 min at 72°C (see Note 3). Product M is prepared by combining the products H (24.5 μl of PCR volume from step 1) and I (24.5 μl of PCR volume from step 1) and adding 1 μl of oligonucleotide D9 (final concentration 25 pmol). Product N is prepared by combining the products J (24.5 μl of PCR volume from step 1) and K (24.5 μl of PCR volume from step 1) and adding 1 μl of oligonucleotide D9 (final concentration 25 pmol). Amplify for 15 cycles and then purify products M and N by electrophoresis in 1.5% agarose gel and elute from the gel slice according to QIAquick gel extraction kit (QIAGEN). The final product O is obtained by combining in 46 μl of PCR mixture: 1 μl of purified product M, 1 μl of purified product N, 1 μl of oligonucleotide D9 (25 pmol), and 1 μl of oligonucleotide D16 (25 pmol). Amplify for 25 cycles and then purify product O as described above (see Note 3).

5. The final purified segments of 400–600 bp are ligated directly into a plasmid containing single 3′-T overhangs at the ends (pGEM®-T Easy Vector Systems, Promega) that complement those of the amplified segment. Alternatively, the purified segments are digested with the appropriate restriction enzymes and ligated into similarly cleaved pUC18 plasmid vector.
6. The DNA plasmid is transformed into *E. coli* strain DH5α using the well-established heat-shock method (16). Then, plasmid DNA is purified using QIAprep kit for plasmid DNA purification (QIAGEN) following manufacturer's instructions.
7. Fifteen to 20 clones are sequenced to identify either error-free DNA segments or segments containing small numbers of errors that could be eliminated by combining the error-free portions of segments via an internal cleavage site or by standard site-directed mutagenesis.
8. To obtain three large overlapping DNA fragments (as example see Fig. 1b), all error-free segments are assembled into pGEM®-T Easy or into pUC18 vector or into pBR322 (New England Biolabs, Inc.) via their compatible unique cleavage sites. After completion, the sequence of three fragments should be verified by automated sequence analysis.
9. Finally, the three DNA fragments are assembled stepwise via common unique restriction endonuclease cleavage sites to yield full-length PV1 (M) DNA (as example, see assembly of sPV(M) cDNA, Fig. 1c). The sequence of sPV1 (M) DNA should be confirmed by automated sequence analyses (see Note 4). Full-length PV cDNA is purified as described above (Subheading 3.1 step 6).

#### 3.2. In Vitro Transcription of cDNA into RNA and Generation of Virions in a Cell-Free Extract

1. Using purified phage T7 RNA polymerase, the sPV cDNA that carries a T7 RNA promoter at its 5′ end is transcribed into viral RNA. For that purpose, 5–10 μg of purified full-length PV cDNA is linearized with *Eco*RI (1 U) in 30-μl restriction enzyme buffer for 90 min at 37°C. Then, 1 μg of linearized template DNA is transcribed in a 50-μl transcription reaction mixture for 2 h at 37°C. Two microliters of the transcription reaction and 2 μl of molecular weight marker [(λ-DNA Hind III digest, New England Biolabs)] are loaded onto RNA gel and run for 1 h at 80 V. Full-length PV RNA derived from PV cDNA runs as a 2-kb DNA band.
2. De novo synthesis of poliovirus from transcript RNA derived from sPV1 (M) cDNA is performed in HeLa cell-free extract (see Note 5).
3. Preparation of HeLa cell-free extract: (1) Obtain cells in log-phase spinner culture ($5.2 \times 10^5$ cells/ml). Spin cells in conical flasks at 1,000 rpm for 15 min. Resuspend cells in 10 ml of ice-cold PBS per liter of cells and transfer to a sterile 50-ml Falcon tube. Wash twice more, first with 50 ml/L cells, then with 5 ml/L cells (spin at 1,500 rpm for 10 min) and transfer to 15 ml Falcon tube before last spin. (2) Determine approximately the volume of packed cells and resuspend with same volume ice-cold hypotonic buffer. Swell on ice for 10 min. (3) Homogenize with 10–15 strokes in chilled B dounce homogenizer (Bellco). Dilute 1 μl of homogenate (taken from the bottom side of pestle) in 9 μl of PBS and check cell disruption under phase contrast microscope (intact cell appears highly refractile). Stop when approximately 90% are disrupted. (4) Transfer homogenate to sterile 15 ml tube and cover top with Parafilm. Spin 20 min at 10,000 rpm and 4°C in SS-34 rotor (Sorvall Superspeed RC2-B). (5) Transfer supernatant (cell-free extract) to sterile dialysis tubing (12–14 kD cutoff). Dialyze 2–5 ml of supernatant for 2 h in 1 L dialysis buffer at 4°C under constant stirring. Carefully transfer supernatant (pellet is very soft) to 15 ml Falcon tube and prepare extract in 10% glycerol by adding 200 μl of 50% glycerol to 800 μl supernatant. Aliquot to microfuge tubes (150 μl/tube) and store at −80°C or proceed with nuclease treatment, the last step of the preparation of the cell-free extract. (6) Add 7.5 μl of 100 mM $CaCl_2$ to 1-ml extract in 10% glycerol. Immediately add 7.5 μl of *S. aureus* nuclease solution, vortex gently, and incubate at room temperature for 15 min. Add 30 μl of 100 mM EGTA, vortex gently. Cell-free extract is now ready to use. Alternatively, aliquot 150 μl/microfuge tube and store −80°C (activity drops significantly after 6 h).
4. Transcript RNA (approximately 48 μl) derived from sPV(M) cDNA (Subheading 3.2 step 1) is purified by phenol–chloroform

extraction and ethanol precipitation. One microliter of purified PV RNA resuspended in distilled water (approximately 200 ng, see Note 6) is added to translation reaction mixture prepared by adding 8.8 μl master mix (supplemented with $^{35}$S-methionine) and 2.7 μl distilled water. Incubate the mixture at 34°C in water bath with gently rocking for 15 h. Then, 5 μl of the translation reactions is added to 15 μl distilled water and 20 μl of 2× sample buffer (125 mM Tris–HCl pH 6.8, 4% SDS, 20% glycerol, 10% β mercaptoethanol, trace amount of bromophenol blue), boiled for 5 min and spun for 3 s and run on SDS-12.5% polyacrylamide gel for 3 h at 140 V. For the analysis of the products of in vitro translation, the gel is fixed with a fixing solution (30% methanol and 10% acetic acid) for 30 min at 55°C. Remove the fixing solution and add enhancing solution (En$^3$hance) for 20 min at room temperature. Rinse three times with distilled water and dry the gel for 1 h at 68°C in a gel dryer with vacuum. Expose the dried gel to X-ray film for 24 h at −80°C or for 48–96 h at room temperature. Finally, incubate X-ray film with developer solution for 3 min, rinse with distilled water, add fixer solution for 3 min, and rinse with distilled water. Air-dry the film and analyze the products of in vitro translation and proteolytic processing of PV RNA in HeLa cell-free extract.

5. For the generation of poliovirus in a HeLa cell-free extract, RNA derived from sPV1 (M) is translated as described above (Subheading 3.2 step 4) except that the master mix is supplemented with unlabeled methionine. Four microliters of the purified PV RNA resuspended in distilled water (approximately 1 μg) is added to translation reaction mixture prepared by adding 35.2 μl master mix (supplemented with unlabeled methionine) and 10.8 μl distilled water. Incubate the mixture at 34°C in water bath with gentle rocking. After 15 h, check for the presence of infectious virus particles in the cell-free incubation mixture, as described below.

### *3.3. In Vivo and In Vitro Characterization of Synthetic Poliovirus*

The following experiments are carried out to prove that the infectious material isolated from the HeLa cell-free extract is indeed poliovirus, as designed by oligonucleotides assembly:

1. After 15 h of incubation (Subheading 3.2 step 5), the HeLa cell-free mixture is treated with RNase A (20 g/ml) and RNase T1 (100 U/ml) for 30 min at room temperature, diluted to 300 μl in HBSS, and added to 90% confluent monolayer of HeLa cells. Gently rock for 30 min at room temperature. In parallel, HeLa cell monolayers are infected with wt PV1 (M) as control. The monolayers are washed with HBSS and overlaid with 3 ml of 0.6% (w/v) gum tragacanth. After 48 h at 37°C and 5% $CO_2$, carefully remove the gum tragacanth and stain the cells with 1 ml of 1% crystal violet for 10 min. Rinse off the

excess of stain with running water. The synthetic virus should produce heterogeneous plaques on HeLa cell monolayers similar to those produce by wt PV1 (M).

2. The infectivity of poliovirus is abolished by poliovirus receptor-specific monoclonal antibodies (Mab D171) and type-specific hyperimmune sera. HeLa cells are grown as monolayers in 6-well plates, washed with DMEM, and incubated with 0.5 ml of Mab D171 (final concentration 20 μg/ml) at room temperature for 2 h. Approximately 100 PFU of either sPV1 (M) or wt PV1 (M) is added to the cells and incubated for 1 h at room temperature. Then, the cells are washed three times with DMEM, and the number of plaques is determined after 48 h, as described above. A control plate with unrelated Mab is similarly treated. The treatment of HeLa cells with Mab 171 should completely block infection of the sPV1 (M) and wt PV1 (M) while the unrelated Mab had no effect on the virus infection. For the neutralization test, anti-PV1 and anti-PV2 serum are incubated respectively with approximately 100 PFU of either sPV1 (M) or wt PV1 (M) at room temperature for 2 h. The antibody-virus mixture is added to the HeLa monolayers. Following incubation for 1 h at room temperature, the cells are washed with DMEM, and the plaques stain after 2 days, as described above. No plaque should be observed when sPV1 (M) is incubated with PV type 1-specific antibodies, while serum to PV2 should not inhibit plaque formation. Altogether, these results should confirm that the de novo poliovirus virions chemically synthesized are serotype 1 and require the poliovirus receptor for infection.

3. The sPV1 (M) should be tested to determine whether it expresses a neurovirulent phenotype in mice transgenic for the human poliovirus receptor (CD155 tg mice). To this end, groups of 4–6 CD155 tg mice (equal number of male and females) are inoculated with any given amount of virus ranging from $10^2$ to $10^8$ PFU (30 μl/mouse) intracerebrally for sPV1 (M) and wt PV (M). Mice are examined daily for 3 weeks postinoculation for paralysis and/or death. The virus titer that induced paralysis or death in 50% of the mice ($PLD_{50}$) was calculated by the method of Reed and Muench. Homogenized spinal cord materials are prepared from three diseased mice for each virus tested, and viruses are reisolated from the spinal cord, and the viral RNA is amplified by RT-PCR and sequenced as described above (see Note 7). The sPV1 (M) must cause flaccid paralysis or death in CD155 tg mice resembling the illness produced by wt PV1 (M). Moreover, virus isolated from spinal cord of the paralyzed animals must have the same sequence that the virus inoculated. Confirmation of these data indicates that the synthetic virus is the causative agent of the flaccid paralysis observed in the sPV1 (M)-infected mice.

## 4. Notes

1. The mixture should be prepared fresh every time.
2. To produce the best results in the asymmetric PCR, the overlapping region between two oligonucleotides should have (1) a melting temperature in the range of 52–58°C and (2) GC content of 40–60%.
3. For fragments where 10 or 12 oligonucleotides are required as starting material, purified amplification products corresponding to O and K or N are subjected to amplification steps 3 and 4 to yield the final products (Fig. 3).
4. To ascertain the authenticity of sPV1 (M) and to distinguish it from wt PV1 (M), it is recommended to engineer nucleotide changes into sPV1 (M) genome as genetic markers. Restriction enzyme analysis is a straightforward technique for the authentication of sPV1 (M). Therefore, the nucleotide changes engineered into the synthetic virus genome should be designed, when possible, to create new or abolish restriction sites.
5. De novo synthesis of poliovirus can also be achieved by directly transfecting HeLa cells on a 35-mm-diameter plate with transcript RNA derived from sPV1 (M) cDNA using DEAE-dextran method.
6. Initially, optimal RNA concentration should be found by titrating 50–600 ng/reaction.
7. Confluent HeLa cell monolayers on 35-mm-diameter plates are infected with the inoculated virus and the viruses reisolated from the spinal cord of paralyzed mice at a multiplicity of infection of ten PFU per cell. The cells are incubated at 37°C until they show signs of cytopathic effect (CPE). RNA was isolated from infected cells according to the TRIzol protocol (Life Technologies, Inc.). The region containing a given genetic marker is amplified by the Titan One Tube RT-PCR System (Roche Molecular Biochemicals) using downstream and upstream primers specifically designed for amplification of this region. To exclude the remote possibility that the signals detected in the RT-PCR assays are due to residual template DNA used in the transcriptions reactions, test all of the RNA samples by PCR without reverse transcription. No PCR bands should be observed in the absence of cDNA synthesis, indicating that the signals detected were due to poliovirus RNA. The specific products are analyzed by digestion with the respective restriction enzymes.

## References


1. Landsteiner K, Popper E (1909) Ubertragung der Poliomielitis acuta auf Affen. Z Immunitatstorsch Orig 2:377–390.
2. Kitamura N, Semler BL, Rothberg PG, et al (1981) Primary structure, gene organization and polypeptide expression of poliovirus RNA. Nature 291:547–553.
3. Wimmer E, Hellen CUT, Cao X (1993) Genetics of poliovirus. Annu Rev Genet 27:353–436.
4. Mendelsohn CL, Wimmer E, Racaniello VR (1989) Cellular receptor for poliovirus: molecular cloning, nucleotide sequence, and expression of a new member of the immunoglobulin superfamily. Cell 56:855–865.
5. Koike S, Horie H, Ise I, et al (1990) The poliovirus receptor protein is produced both as membrane-bound and secreted forms. EMBO J 9:3217–3224.
6. Pelletier J, Sonenberg N (1988) Internal initiation of translation of eukaryotic mRNA directed by a sequence derived from poliovirus RNA. Nature 334:320–325.
7. Wetz K (1987) Cross-linking of poliovirus with bifunctional reagents: biochemical and immunological identification of protein neighbourhoods. J Virol Methods 18:143–151.
8. Phillips BA, Fennel R: Polypeptide composition of poliovirions, naturally occurring empty capsids, and 14S precursor particles. J Virol 12:291–29.
9. Jacobson MF, Baltimore D (1968) Morphogenesis of poliovirus. I. Association of the viral RNA with coat protein. J Mol Biol 33:369–378.
10. Holland JJ, Kiehn ED (1968) Specific cleavage of viral proteins as steps in the synthesis and maturation of enteroviruses. Proc Natl Acad Sci USA 60:1015–1022.
11. Hogle JM, Chow M, Filman DJ (1985) Three-dimensional structure of poliovirus at 2.9 A resolution. Science 229:1358–1365.
12. Cello J, Paul A, Wimmer E (2002) Chemical synthesis of poliovirus cDNA: Generation of infectious virus in the absence of natural template. Science 297:1016–1018.
13. Van der Werf S, Bradley J, Wimmer E, et al (1986) Synthesis of infectious poliovirus RNA by purified T7 RNA polymerase. Proc Natl Acad Sci USA 78:2330–2334.
14. Mueller S, Coleman JR, Wimmer E (2009) Putting synthesis into biology: A viral view of genetic engineering through *de novo* gene and genome synthesis. Chem Biol 16:337–347.
15. Molla A, Paul A, Wimmer E (1991) Cell-free *de novo* synthesis of poliovirus. Science 254:1647–1651.
16. Sambrook J, Fritsch EF, Maniatis T (1989). Molecular Cloning A Laboratory Manual. Cold Spring Harbor Laboratory Press 2nd edn, New York, USA.
17. Wimmer E. (2006) The test-tube synthesis of a chemical called poliovirus. The simple synthesis of a virus has far-reaching societal implications. EMBO Rep. 7:S3–9.

# Part III

## Software for Gene Synthesis

# Chapter 15

# In Silico Design of Functional DNA Constructs


## Alan Villalobos, Mark Welch, and Jeremy Minshull



## Abstract

The promise of synthetic biology lies in the creation of novel function from the proper combination of genetic elements. De novo gene synthesis has become a cost-effective method for building virtually any conceptualized genetic construct, removing the constraints of extant sequences, and greatly facilitating study of the relationships between gene sequence and function. With the rapid increase in the number and variety of characterized and cataloged genetic elements, tools that facilitate assembly of such parts into functional constructs (genes, vectors, circuits, etc.) are essential. The Gene Designer software allows scientists and engineers to readily manage and recombine genetic elements into novel assemblies. It also provides tools for the simulation of molecular cloning schemes as well as the engineering and optimization of protein-coding sequences. Together, the functions in Gene Designer provide a complete capability to design functional genetic constructs.



**Key words:** Gene design, Protein expression, Synthetic biology, Gene Designer, Molecular cloning, Molecular biology


## 1. Introduction

Synthetic biology, with its focus on the design of new genetic function, will be enabled by computational tools facilitating the design of new DNA molecules that can encode these functions. Most of today's programs for handling DNA sequence information, however, are primarily oriented toward analysis of existing sequences rather than creation of new ones. This is presumably a legacy of exponentially increasing amounts of genomic and metagenomic sequence information. In consequence, current software is poorly suited to de novo genetic design.

Basic design software should facilitate the design process by providing the designer with the ability to define individual sequence


Jean Peccoud (ed.), *Gene Synthesis: Methods and Protocols*, Methods in Molecular Biology, vol. 852,
DOI 10.1007/978-1-61779-564-0_15, © Springer Science+Business Media, LLC 2012

elements with specific functions, to easily rearrange sequences without error, to organize sequence elements in hierarchical levels of abstraction, and to modify sequence elements thereby modifying their function. The value of these capabilities in a design tool will be demonstrated in this chapter by taking the specific example of designing DNA molecules to express a protein in two different cell types. Expression of a protein requires juxtaposition of appropriate regulatory elements with suitable open reading frames; it is also currently the most widespread use for synthetic genes and therefore probably familiar to most readers.

## 2. Software

In this chapter, we make use of the Gene Designer software application from DNA2.0 (see Note 1). This program is available at no cost to academic and commercial users. Licensing allows for any commercial use except for offering a gene synthesis service which is DNA2.0's main business. Gene Designer is built with the Adobe Flex platform and executes on the Adobe Integrated Runtime (AIR) which enables it to run on multiple platforms and eases installation by leveraging Adobe's broad user base. Gene Designer will run on most of today's desktop systems with at least 1 GB of RAM and on Windows, OSX (Mac), and Linux operating systems (see Note 2).

### 2.1. Graphical User Interface

Users can view, manipulate, and edit genetic constructs in Gene Designer using two different views. In the Icon View, sequence elements are represented as icons that can easily be moved and copied, using a drag-and-drop interface, from libraries of elements to constructs and vice versa. Changes in the arrangement of elements within the Icon View alter the DNA sequence of a construct, which can be viewed in the Sequence View. In Sequence View, the user can see the DNA and encoded amino acid sequences of each element and can access a graphical editing mode for detailed sequence viewing and editing such as manually selecting codons, editing the sequence of an element, and splitting and merging elements. The user can navigate large or circular constructs within the Sequence View using a two-tiered scalable navigator.

Elements can be grouped. This enables the user to keep together a set of DNA sequences that encode a more complex function. A group can easily be moved as a complete unit, while retaining information about all of its component elements. This is an important addition to the 2.0 version of this program and should accommodate increasing levels of abstraction in the design of genetic functions.

A simple demonstration of this will be given here in the design of a vector for expression of proteins in bacteria, and the reuse of elements and functional groups to create a vector for mammalian gene expression.

Gene Designer also includes capabilities that we will not have room to describe in this chapter, including searching for DNA and amino acid motifs, creating reports, importing and exporting different file formats, visualizing restriction enzyme cuts, annotating sequences, and designing sequencing primers. Instead, we will conclude with describing how to use the cloning tool and how to backtranslate a polypeptide sequence to obtain a DNA sequence that encodes it.

### 2.2. Backtranslation

The underlying algorithm that performs backtranslation in Gene Designer 2.0 evaluates hundreds of possible sequences for an open reading frame against several parameters including repeats, unwanted DNA motifs, mRNA secondary structures, and codon bias. Since each amino acid has on average three possible choices for a codon, even a small 300-bp gene will generate a search space of approximately $3^{100} = 5.15 \times 10^{47}$ combinations. This space is too large for all the possible combinations to be evaluated, and the terrain is full of local minima due to the interdependencies between the parameters, so an efficient heuristic must be used to approximate on an optimal solution.

Gene Designer solves the search problem with a genetic algorithm approach (1–3). In this method, we start out with a population of sequences generated randomly but biased toward a codon usage table. Each individual is then evaluated against the set of parameters, and a score is determined based on weights for each parameter. More individuals are then created by crossing individuals from the current population and also by introducing mutations at a given rate. The best individuals are selected to go on to the next round. The process continues until a maximum number of iterations are met, or the user deems that the current best score is good enough. In either case, the best individual is used as the sequence for the construct being backtranslated.

One of the larger challenges regarding the calculation of the critical design parameters for an expressed open reading frame is calculating the number and length of all repeats within the DNA sequence, a process that must be performed hundreds of times. Gene Designer addresses this challenge by using a suffix tree to identify repeats which is constructed via Ukkonen's Online Suffix Tree Construction algorithm (4). This way, it achieves near linear time construction of the suffix tree and therefore identification of the repeated subsequences.

## 3. Methods

### 3.1. Basic In Silico Gene Construction

In the following section, we will illustrate the in silico creation of a genetic design construct. First, we will construct a bacterial expression vector from basic building block components. We will show how use of hierarchical groups simplifies the reuse of sequence elements to create a second vector for expression of proteins in mammalian systems. We will then show how to create a polypeptide-encoding element, clone it into the vector, and backtranslate for optimal expression in a specified host.

Launching Gene Designer opens two windows: a Library Explorer containing an organized set of sequence elements and a Project Window where sequence elements can be arranged to create DNA constructs (Fig. 1). The Project Window consists of two panes: a Navigation Pane which provides an overview of each construct by showing a vertical arrangement of sequence elements and a Workspace Pane in which the elements can be viewed either as icons or as sequences. Moving between these views can be effected using the view selection buttons at the bottom of the screen.

The expression vector to be built in this demonstration will use the T5 promoter (5–8), which is constitutive in *Escherichia coli*. It can be made repressible by flanking it with lac operators, DNA sequences that bind to the lac repressor. To maximize the repression, we use an idealized perfectly symmetrical version of the operator (9) and space two operators around the promoter (10). Some of these elements (the T5 promoter and the lac operator) are present in the library of elements and can be added to the construct by dragging

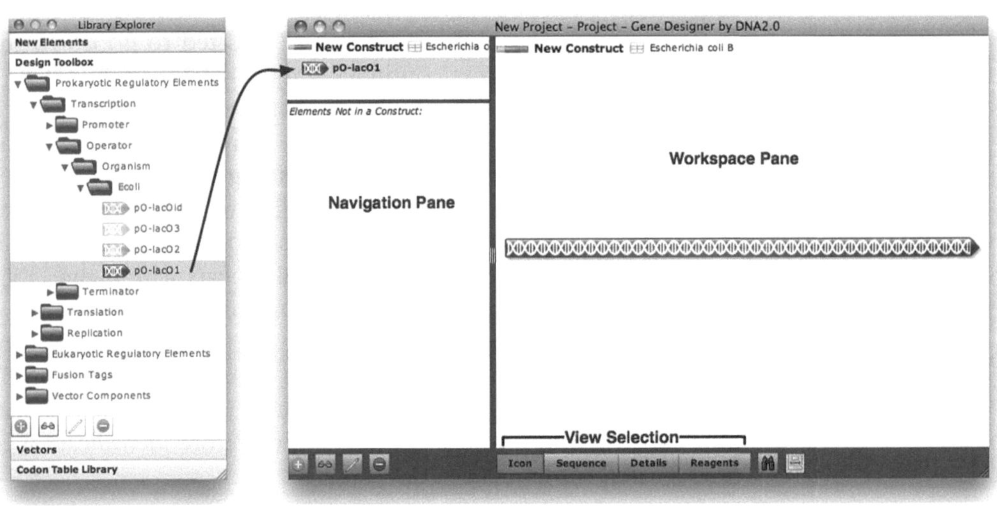


Fig. 1. Library Explorer and Project Window. Elements can be added from the Design Toolbox into the Navigation or Workspace Panes of the Project Window.

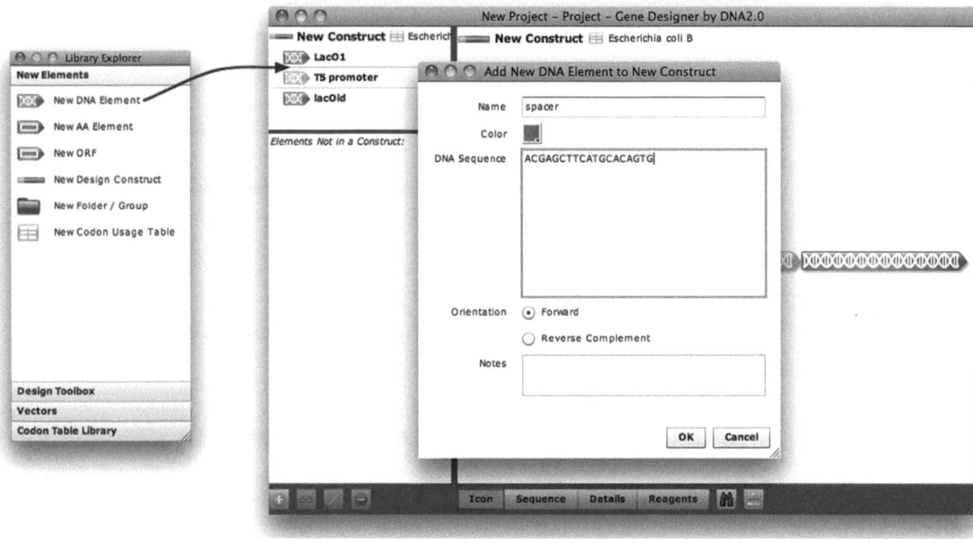


Fig. 2. Creating new sequence elements. Elements that are not preloaded in Gene Designer can be added by dragging a new element from the Library Explorer into the Navigation or Workspace Pane.

across from the Library Explorer (Fig. 1). Two other elements must be added de novo: a sequence to provide the appropriate spacing between the lac operators and an untranslated leader sequence for the mRNA. This is done by dragging in a New DNA Element from the Library Explorer and typing or pasting in a DNA sequence (Fig. 2).

Gene Designer allows manually annotating and editing of sequences. At any point in the in silico construction, it is possible to switch from the Icon View to the Sequence View (Fig. 3). The Sequence View uses two sliders (the Global and Local Sequence Windows) to move through the sequence. Immediately below these is the DNA sequence that results from the current arrangement of all of the sequence elements (the Design Sequence). Below this sequence is a grid for annotations, and below this, each of the sequence elements in the construct is displayed. Below each DNA element, the amino acid sequences encoded in all six possible reading frames are shown; conversely below each amino acid element, all of the possible DNA codons that encode that amino acid, with the actual codon used colored, are shown. The codon for any amino acid can be changed by clicking on one of the other codons.

In the design of the inducible T5 promoter, the spacing between the centers of the lac operators should be 92 bp to allow cooperative binding of the repressor. This can be checked in Sequence View by adding an annotation and, if necessary, editing.

An annotation can be added by highlighting a region of the Design Sequence and selecting "Annotate" from just below the Design Sequence (Fig. 3). Selecting the top strand will add the

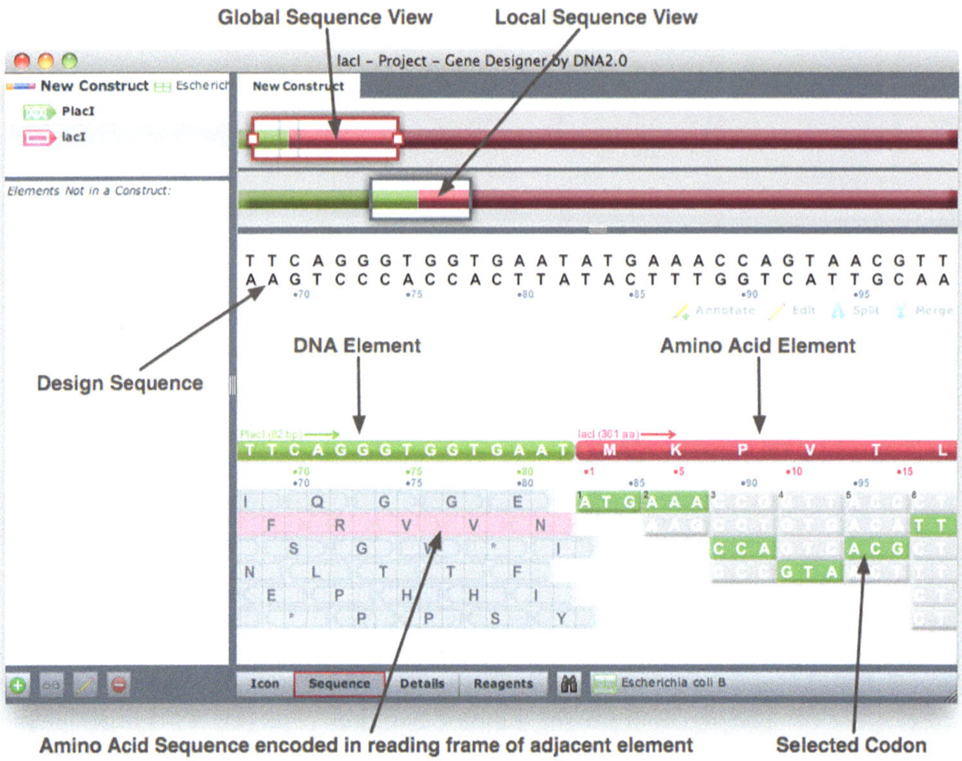


Fig. 3. Sequence View. The sequence of the construct can be viewed by selecting the Sequence View from the bottom of the Project Window.

annotation in the forward direction; selecting the bottom strand will add the annotation in reverse. The user can then choose between adding an oligonucleotide, using a melting temperature calculator, and using a general marker. Annotations have orientation and are marked with their start and end positions, and their lengths. In Fig. 4, a general annotation has been added with its ends at the center of the two lac operators. Here, the annotation indicates that the spacing is 91 bp, so an extra base is being added by editing within the Sequence View. This is done by selecting a portion of the sequence and selecting "Edit" from just below the Design Sequence.

### 3.2. Creating and Using Groups

A contiguous set of sequence elements can be combined into a functional unit by grouping them together. The two lac operators, the T5 promoter and the DNA element added to provide the correct spacing between the operators, are grouped together by selecting all of the elements (using the mouse to select the elements while holding down the shift key) and then selecting the "Group" command from the Edit menu on the Project Window (Fig. 5) and naming the group (in this case "inducible T5 promoter"). Once they

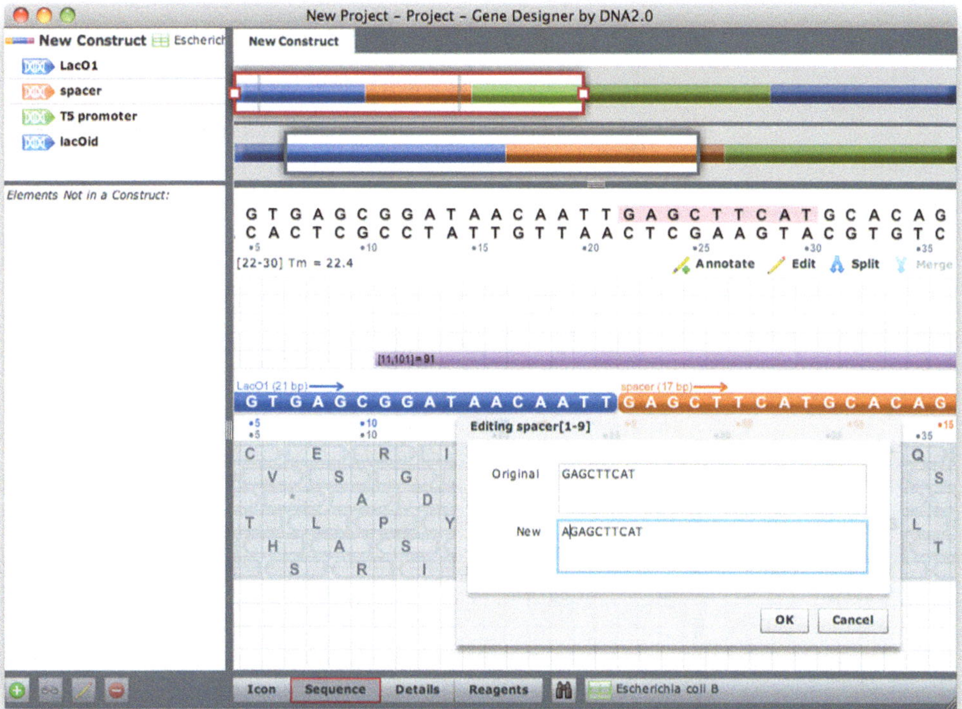


Fig. 4. Editing the sequence of a construct. Highlighting a section of the Design Sequence allows a user to edit the selected sequence in a dialog box.

are grouped together, they can be moved by dragging and dropping, just like any of the individual elements. Individual elements can also be dragged in or out of a group.

Binding of the lac repressor to the lac operator prevents RNA polymerase from binding to the promoter. Although the lac repressor is encoded on the *E. coli* chromosome, this may not produce enough repressor to turn off the promoter tightly, especially if the plasmid (and therefore also the lac operators) is present in hundreds of copies per cell. For the design of this expression plasmid, we therefore add the gene encoding the lac repressor and its own promoter to the plasmid. This is done by dragging these elements (lacI and PlacI) from the Library Explorer, as shown previously in Fig. 2, and then grouping them to create the Lac Repressor.

When elements are added to a construct, they are added in the forward direction. However, it is often desirable to reverse the direction of elements. In this case, we wish to ensure that no other promoters in the construct will inadvertently cause RNA polymerase to transcribe a gene that should be under the control of the inducible T5 promoter. We therefore reverse the orientation of the Lac Repressor by double clicking on the group name in the Navigator Pane and selecting "reverse" from the pop-up window.

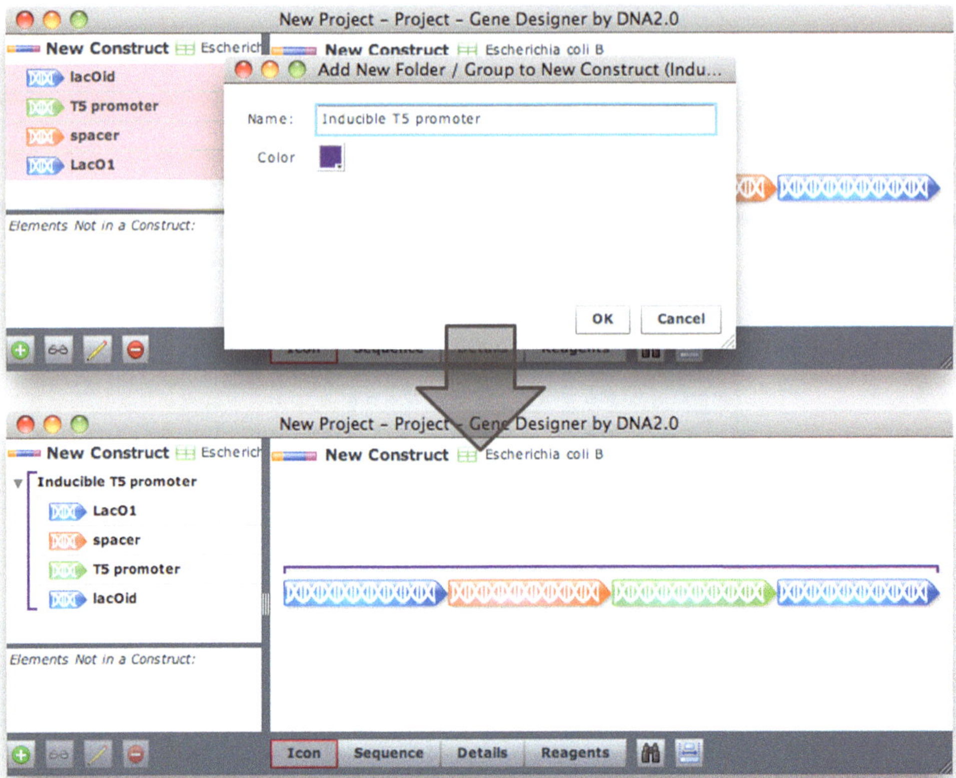


Fig. 5. Creating a group. Selecting a set of contiguous elements and choosing the "Group" option from the Edit menu allows the user to create higher order associations of sequence elements.

Reversing a group reverses each element within the group and reverses the order of each element, so that the Design Sequence for that group becomes its reverse complement.

Additional elements of the construct are added and positioned by drag and drop, and grouped into functional units. The remaining features needed for this expression vector are a Polylinker (comprising left and right cloning sites and a stuffer), a set of elements required for propagation of the construct in bacteria (an origin of replication and an antibiotic resistance marker), and a transcriptional terminator to prevent any transcription from the origin accidentally "reading through" the regulated promoter.

Although the vector has now been constructed as a linear arrangement of DNA and amino acid elements, the actual plasmid will be circular. This can be converted to a circular arrangement by opening the construct from the Navigation Pane. Double clicking on any element, group or construct in the Navigation Pane opens a pop-up window where the object's properties can be viewed or edited (Fig. 6). The Global Sequence Window displays circular constructs as circular, and the sequence can be viewed as continuous across the origin.

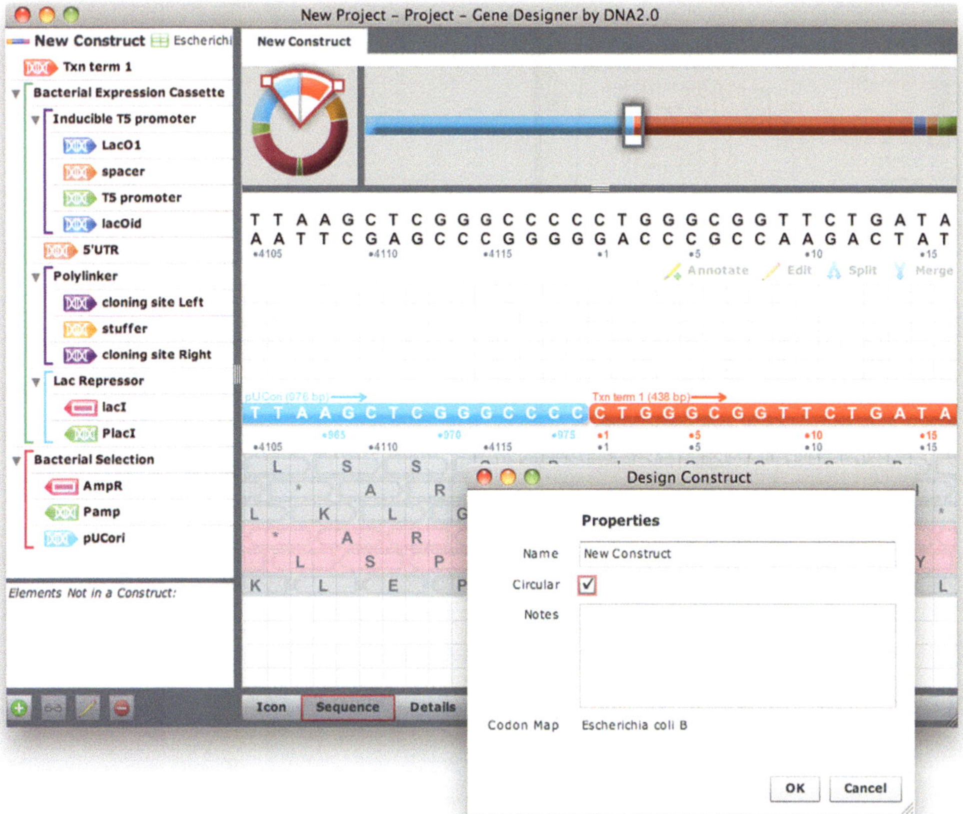


Fig. 6. Circular constructs. A construct can be circularized by double clicking on the construct name and checking the "*circular*" box. *Circular* constructs are treated as a single continuous molecule.

Elements, groups, and constructs can be copied into the Library Explorer or into a new construct by dragging them from the Navigation Pane. Dragging within a construct moves the element, while dragging from one construct to another copies the element. By creating and arranging folders within the Library Explorer, the user can build a customized library of sequence elements that will be available for future construct designs.

Sequences that are preloaded into Gene Designer cannot be overwritten. This is to allow automatic updates of preloaded library elements without overwriting any user's customization. In order to edit a preloaded sequence, it should be dragged into the Project Window, where it can be edited and renamed. This modified element can then be dragged back into the Library Explorer, where its name will now be shown in bold, as is the case for all user-entered sequences. This sequence will not be altered during library updates.

Some of the elements and groups from the bacterial expression vector can be reused to create an expression vector for use in

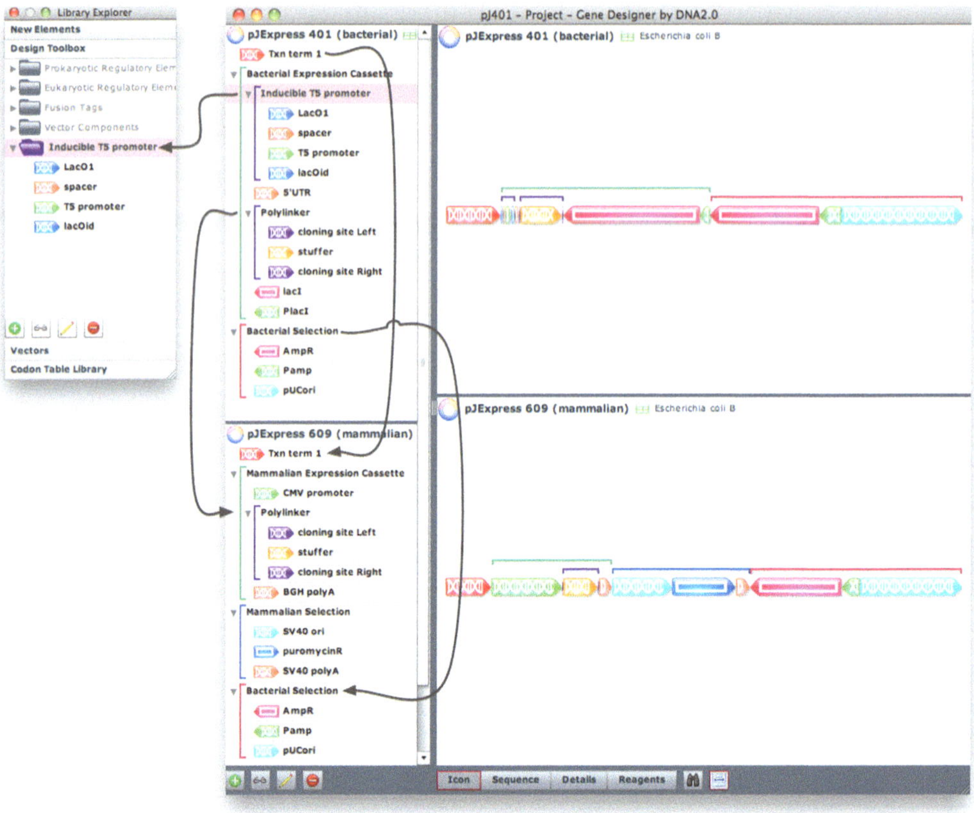

Fig. 7. Reuse of groups and elements. Elements can be moved by drag and drop into a new construct, or back into the Library Explorer for use in new projects.

mammalian cells (Fig. 7). A new construct is created, and the transcriptional terminator element and the Polylinker and Bacterial Selection groups are dragged into it. A Mammalian Expression Cassette is created by adding a CMV promoter before and a polyadenylation sequence after the Polylinker, then addition of a Mammalian Selection group comprising an SV40 origin/promoter, drug resistance marker, and polyadenylation sequence creates a mammalian expression vector.

### 3.3. The Cloning Tool

In silico design tools allow a user to specify the precise nucleotide sequence of a functional DNA molecule which can then be synthesized. However, in practice, most synthetic DNA is propagated in cloning vehicles such as the expression vectors designed in the preceding sections. In this case, it is useful for the design tool to perform in silico the same cloning processes that will be used to combine the gene with its vector in the physical world.

The cloning tool in Gene Designer is launched from the Tools menu in the Project Window (Fig. 8). The bacterial expression vector is first dragged into the Input Pane to serve as a fragment donor.

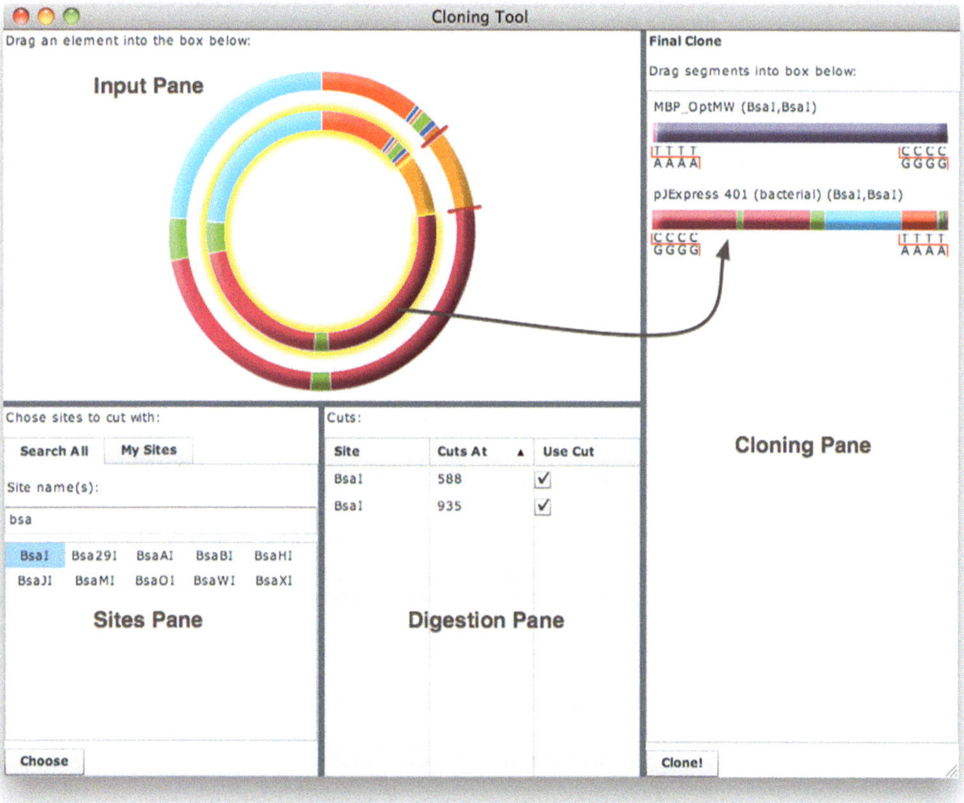


Fig. 8. Cloning tool. A construct can be digested in silico by one or more restriction enzymes, and any of the resulting fragments dragged to assemble into a new construct.

One or more restriction sites can then be selected within the Sites Pane, clicking the "Choose" button then displays the position of each of the sites within the construct, both as a sortable list in the Digestion Pane, and graphically on the image of the construct. Checking the box by a site digests the molecule at that position. The cloning sites in this vector were designed for digestion with BsaI to produce a CCCC overhang at one end and an AAAA overhang at the other. After in silico restriction digestion, the desired fragment is dragged out of the Input Pane and into the Cloning Pane.

In this demonstration project, we will take a protein sequence and clone it into the expression vector. To do this, we create a construct that comprises a ribosome binding site and the amino acid sequence of maltose-binding protein, flanked by DNA sequences containing BsaI restriction sites that produce ends complementary to those produced by the vector. This construct is then dragged into the cloning tool, digested with BsaI in silico, and the digested fragment dragged into the Cloning Pane. Clicking the "Clone" button then produces a molecule in which the two DNA fragments are ligated at their sticky ends.

### 3.4. Backtranslation of Polypeptide Sequence to an Encoding DNA

Because of the degeneracy of the genetic code, designing a DNA sequence from which a protein can be expressed requires choosing from an enormous number of possible sequences (11). Gene Designer backtranslates amino acid sequences by performing parallel optimizations to reduce secondary structure at the 5′ end of the mRNA (12–17) and to match the frequency of each codon within the DNA to a target frequency, preferably one determined experimentally to result in high expression levels in the desired host (11). Gene Designer selects from among sequences that meet the criteria for good expressibility, while simultaneously imposing constraints based upon other requirements: elimination or inclusion of specific restriction sites, elimination of DNA repeats that may compromise ease of synthesis or stable maintenance of the cloned DNA, and elimination of other potentially deleterious motifs such as cryptic ribosome binding sites or RNA splice sites.

Individual users frequently have the same set of requirements for many backtranslations, but one user's requirements can be very different from another's. Gene Designer therefore allows a user to create one or more backtranslation profiles in which all of the optimization parameters are defined; using this profile then applies the same optimization parameters to any backtranslation for which they are selected. Backtranslation Profiles are accessed from the Configure menu in the Project Window.

In backtranslating a sequence, it is often desirable to bias the use of synonymous codons. Codon tables describe the frequency with which each codon is used to encode its corresponding amino acid, generally within some real dataset (such as an organism's genome), or sometimes as some idealized target. A user selects a codon usage table to use within a Backtranslation Profile by dragging from the Codon Table Library in the Library Explorer into the Backtranslation Profile. If a desired table is not already present in the Library Explorer, it can be imported from DNA2.0's web service by selecting "Import" then "Codon Table" from the File menu in the Project Viewer.

In addition to the frequencies that are determined by the choice of Codon Table, the user also has the opportunity to select a usage frequency Threshold. This is used as a cutoff, where any codon below the Threshold frequency is treated as if its frequency were zero, so it is never selected in backtranslation. This allows a user to exclude rare codons.

During backtranslation, in some versions of the software, codons whose frequencies are above the Threshold will be chosen probabilistically at their relative frequencies in the Codon Table. Other versions of the software will iterate during backtranslation to approximate the frequencies defined in the Codon Table.

It is critical in backtranslation to avoid introducing sequence elements that might interfere with cloning (e.g., superfluous restriction sites) or expression of the gene (e.g., mRNA-processing motifs).

Sequence motifs to avoid can be selected from user-managed lists including "My Motifs" and "My Restriction Sites." Motifs to avoid can also include degenerate bases in their definitions.

Gene Designer will also seek to minimize repeated sequences. The user can set the size of repeats to be considered. In some versions of the software, there are also settings for avoiding alternate open reading frames.

Avoiding mRNA structure, particularly in the translational initiation region, can improve gene expressibility. The user can set parameters for avoidance of mRNA structure in the designs over user-defined windows. The algorithm seeks to minimize possible structure given helix and loop length constraints. In some versions of the software, structure minimization is confined to the region of bases −15 to +45 relative to the first base of the open reading frame to be back translated, where structure has been shown to inhibit translational initiation (13, 16, 17). This constraint is activated by checking the box for 5′ Structure Optimization.

In some cases, it may be advantageous for the design to be made dissimilar or similar to a homologous gene. For example, one may wish to avoid any unknown regulatory elements that may occur in a natural gene. One way of doing this could be to maximize dissimilarity to the wild-type sequence in the choice of codons throughout the gene. In another case, one might want to preserve similarity to a wild-type gene as much as possible while removing unwanted sequences, rare codons, and repeats. Within the backtranslation profile, at the Homologous DNA box, a weight can be set for the priority of similarity or dissimilarity to a user-defined sequence specified in the Amino Acid Element properties dialog box.

Backtranslation in Gene Designer uses a genetic algorithm to manage the multiple constraints with reduced risk of becoming trapped in a suboptimal local minimum in the optimization process. The algorithm evolves a population of genes using repeated rounds of selection, breeding, and mutation to find best solutions. The population size, selection criteria, and mutation rate can be altered by the user to optimize performance. For each constraint, the user may set a weight to bias the solution where two or more constraints may be in conflict. For example, the desired codon usage target may not be reachable given constraints on repeats and sequences to avoid. In a typical example, the user may choose to set a high weight for Sequences to Avoid to ensure these are not compromised, while using lower weights for Repeat Removal and Codon Usage. The result would be a gene where codon bias and repeat frequency are optimized under the constraint that sequences to avoid are maximally reduced.

Codons used for the initial gene population and for mutation at each generation of the genetic algorithm are chosen statistically according to the codon table settings. The degree to which the codon bias of a gene matches the bias in the table can be also constrained in some versions of the software. This will minimize deviation in bias due to statistical drift or conflict with other constraints applied.

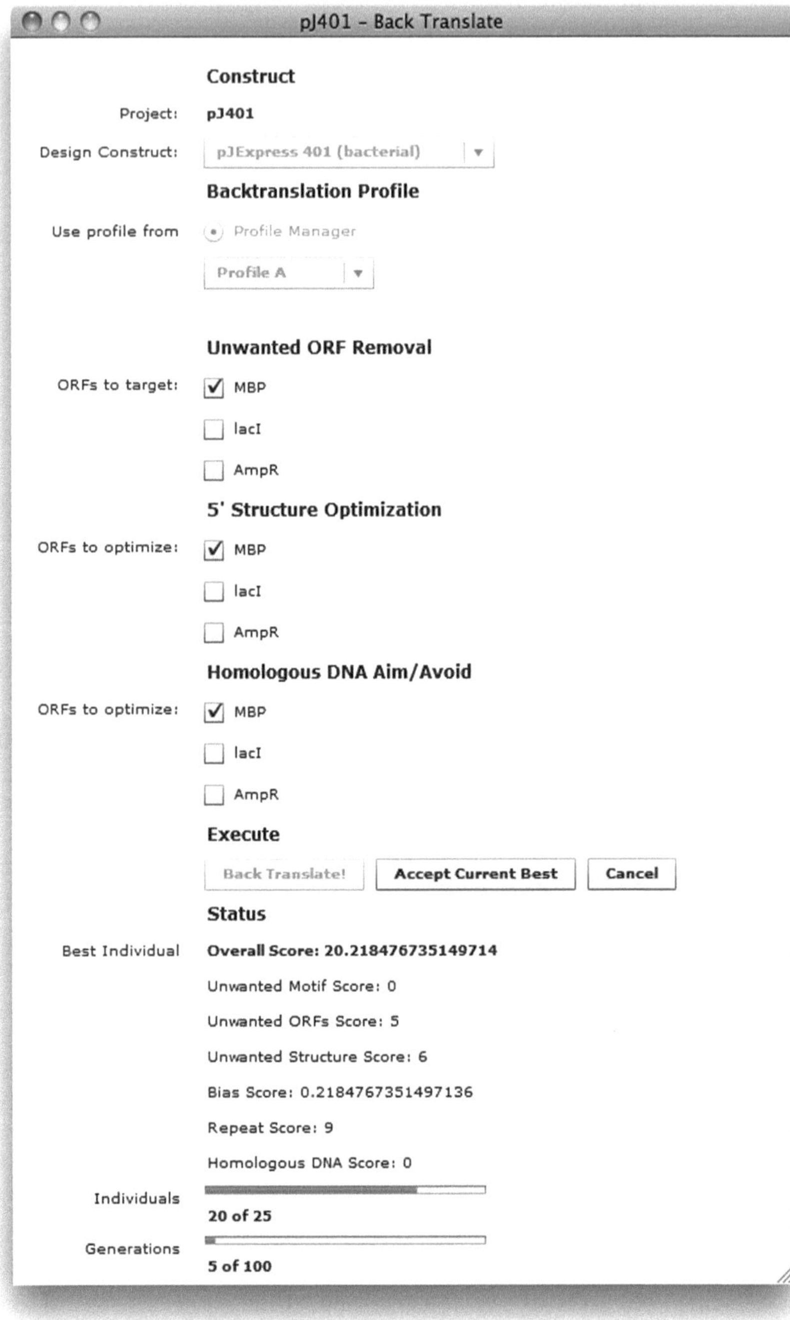


Fig. 9. Backtranslation. A user can impose design constraints on Gene Designer's choice of DNA sequences for encoding a selected polypeptide.

To backtranslate one or more open reading frames in a construct, the user chooses "Backtranslate" from the Tools menu. In the Backtranslate dialog (Fig. 9), the user chooses the construct to backtranslate and the profile to apply. The user may also specify whether 5′ Optimization, Homologous DNA, and/or Alternate ORF removal constraints should be applied to each open reading frame element in the construct. Once backtranslation is initiated, the progress of the genetic algorithm is displayed for each generation until the maximum number of evolutionary cycles is reached or the user terminates the process to accept the best current gene. If codons are manually selected in the open reading frame or if the sequence of an open reading frame element is set to fixed (colored green), no substitutions will be made throughout the backtranslation process, although fixed sequences will be included in assessing the overall gene fitness during the genetic algorithm.

The characteristics of the final gene may be analyzed using the Backtranslation Summary Report accessed from the Reports menu. Details on the final codon usage, mRNA structure, repeat sequences, and other features can be viewed.

## 4. Results and Conclusion

Gene Designer allows the user to readily assemble genetic elements into constructs (see Note 3). The vector designed here was completely synthesized de novo to yield a plasmid that propagated at high copy and conferred resistance to ampicillin. Cloning the backtranslated sequence for maltose-binding protein (MBP) into this vector, and growing cultures of the resulting construct in the presence of IPTG resulted in the strong expression of a 40.6-kDa protein, which was not expressed in the absence of IPTG (Fig. 10). Thus, all of the design elements and protocols used resulted in a construct with the desired function. To date, Gene Designer has been used successfully to construct and optimize thousands of genes for expression in a wide variety of hosts.

## 5. Notes

1. The chapter focuses on Gene Designer version 2.0.150, the current version at the time of this writing. See www.dna20.com/genedesigner2 for download instructions, online help, and tutorials.

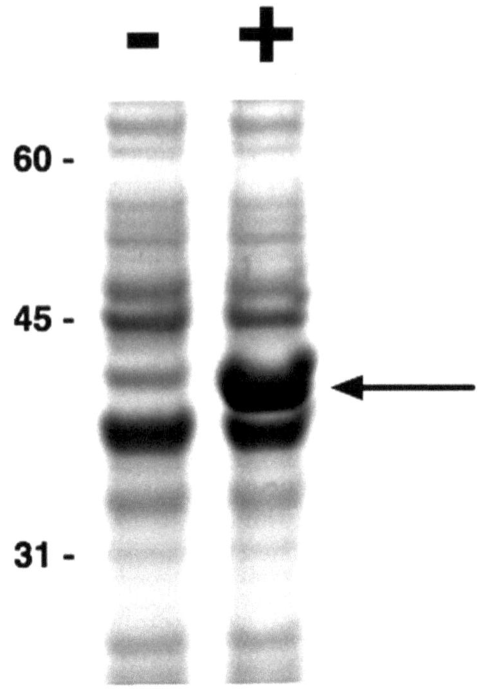


Fig. 10. Expression of an optimized MBP gene from pJ401 in BL21 *Escherichia coli*. Equivalent amounts of total cell culture were separated by polyacrylamide gel electrophoresis and stained with coomassie blue. (–) Noninduced culture. (+) Culture induced at mid-log phase with 1-mM IPTG. For protein expression, cells were grown at 37°C with shaking until they had reached an $A_{600}$ of 0.6. Expression was then induced by adding IPTG to 1 mM, and the cultures were incubated for a further 4 h. The equivalent of 2 μl of culture at an $A_{600}$ of 3.0 was run in each lane. The band in the noninduced lane that nearly comigrates with MBP is not plasmid-dependent. No detectable expression of MBP is seen under noninduced conditions.

2. See www.adobe.com/air for compatibility and system requirements.
3. Additional guidance in the use of Gene Designer is available from www.dna20.com/gdhelp. Angela Dixon at Utah State has created an excellent video guide, which can be found at https://telc.cngr.usu.edu/biophotonics_2010/Gene%20Designer/Gene%20Designer.html.

## References


1. Mitchell, M. (1998) An introduction to genetic algorithms, 1st MIT Press paperback ed., MIT Press, Cambridge, Mass.
2. Patil, K. R., Rocha, I., Forster, J., and Nielsen, J. (2005) Evolutionary programming as a platform for in silico metabolic engineering, *BMC Bioinformatics* **6**, 308.
3. Rocha, M., Maia, P., Mendes, R., Pinto, J. P., Ferreira, E. C., Patil, K., Nielsen, J., and Rocha, I. (2008) Natural computation meta-heuristics for the in silico optimization of microbial strains, *BMC Bioinformatics* **9**, 499.
4. Ukkonen, E. (1995) On-line construction of suffix-trees, *Algorithmica* **14**, 249–260.
5. Lanzer, M., and Bujard, H. (1988) Promoters largely determine the efficiency of repressor action, *Proc Natl Acad Sci USA* **85**, 8973–8977.

6. Bujard, H., Gentz, R., Lanzer, M., Stueber, D., Mueller, M., Ibrahimi, I., Haeuptle, M. T., and Dobberstein, B. (1987) A T5 promoter-based transcription-translation system for the analysis of proteins in vitro and in vivo, *Methods Enzymol* **155**, 416–433.
7. Gentz, R., Langner, A., Chang, A. C., Cohen, S. N., and Bujard, H. (1981) Cloning and analysis of strong promoters is made possible by the downstream placement of a RNA termination signal, *Proc Natl Acad Sci USA* **78**, 4936–4940.
8. Gentz, R., and Bujard, H. (1985) Promoters recognized by Escherichia coli RNA polymerase selected by function: highly efficient promoters from bacteriophage T5, *J Bacteriol* **164**, 70–77.
9. Sadler, J. R., Sasmor, H., and Betz, J. L. (1983) A perfectly symmetric lac operator binds the lac repressor very tightly, *Proc Natl Acad Sci USA* **80**, 6785–6789.
10. Oehler, S., Amouyal, M., Kolkhof, P., von Wilcken-Bergmann, B., and Muller-Hill, B. (1994) Quality and position of the three lac operators of E. coli define efficiency of repression, *EMBO J* **13**, 3348–3355.
11. Welch, M., Villalobos, A., Gustafsson, C., and Minshull, J. (2009) You're one in a googol: optimizing genes for protein expression, *J R Soc Interface*, 6 Suppl 4:S467–76.
12. Kozak, M. (1986) Influences of mRNA secondary structure on initiation by eukaryotic ribosomes, *Proc Natl Acad Sci USA* **83**, 2850–2854.
13. Kudla, G., Murray, A. W., Tollervey, D., and Plotkin, J. B. (2009) Coding-sequence determinants of gene expression in Escherichia coli, *Science* **324**, 255–258.
14. Salis, H., Tamsir, A., and Voigt, C. (2009) Engineering bacterial signals and sensors, *Contrib Microbiol* **16**, 194–225.
15. de Smit, M. H., and van Duin, J. (1990) Secondary structure of the ribosome binding site determines translational efficiency: a quantitative analysis, *Proc Natl Acad Sci USA* **87**, 7668–7672.
16. de Smit, M. H., and van Duin, J. (1994) Control of translation by mRNA secondary structure in Escherichia coli. A quantitative analysis of literature data, *J Mol Biol* **244**, 144–150.
17. Kozak, M. (2005) Regulation of translation via mRNA structure in prokaryotes and eukaryotes, *Gene* **361**, 13–37.

# Chapter 16

# Using DNAWorks in Designing Oligonucleotides for PCR-Based Gene Synthesis


## David Hoover



## Abstract

The availability of sequences of entire genomes has dramatically increased the number of protein targets, many of which will need to be overexpressed in cells other than where they have been identified originally. Gene synthesis often provides a fast and economically efficient approach. The synthetic gene can be optimized for expression and constructed for easy mutational manipulation without regard for the parent genome. Yet design and construction of synthetic genes, especially those coding for large proteins, can be a slow, difficult, and confusing process. DNAWorks automates the design of oligonucleotides for gene synthesis by PCR-based methods.



**Key words:** Gene synthesis, Codon optimization, Polymerase chain reaction, Cloning, Gene expression, Oligonucleotide design, Simulated annealing


## 1. Introduction

Gene synthesis is an indispensible method for creating DNA sequences, in cases where either there is no cDNA clone available or the gene is to be optimized for the purpose of gene expression or regulation (1). Current gene synthesis methods involve the chemical synthesis of short oligonucleotides (between 20 and 200 nucleotides in length), which can be assembled into larger DNA sequences (2–6). These assembly steps involve either DNA ligation or polymerization, or both. Because the assembly steps are often performed in a single tube, the oligonucleotides must be designed in such a way that they will spontaneously and correctly assemble under defined conditions.

To design such oligonucleotides manually is nearly impossible, not just because a manual approach is fundamentally error-prone, but also because of the sheer number of possible solutions to the problem. For a small protein, 100 amino acids in length, there are


Jean Peccoud (ed.), *Gene Synthesis: Methods and Protocols*, Methods in Molecular Biology, vol. 852,
DOI 10.1007/978-1-61779-564-0_16, © Springer Science+Business Media, LLC 2012

between $10^{30}$ and $10^{60}$ possible DNA sequences that will code for the same protein sequence. Additionally, the positioning of overlaps between the oligonucleotides to maintain a uniform assembly and minimization of repeated sequences must be addressed during the design process.

The program DNAWorks facilitates the design of a set of oligonucleotides that can be assembled to form a single DNA sequence (7). In the case of a given protein sequence, an initial gene is constructed by reverse translation. Codons are chosen from the set of codons that are possible for the given protein sequence based on a supplied codon frequency table. In the case of an input DNA sequence, no codons are required, and the sequence is taken as is. The set of resulting oligonucleotides is then optimized using a simulated annealing algorithm, which can find a global optimum without the need to evaluate all possible local optima, and in a greatly accelerated fashion.

## 2. Materials/Data Input

DNAWorks was originally developed as a stand-alone desktop application (7). It has since migrated to a web-based application (http://helixweb.nih.gov/dnaworks). Data input is through the Web form, and output is displayed to the screen, as well as echoed in an email to the user.

DNAWorks can generate synthetic genes for both protein and nucleotide sequences. Sequences can be entered as text or uploaded as files. DNAWorks can accept a wide variety of sequence formats. However, it cannot discriminate between a nucleotide and a protein sequence, so the user must indicate the type before the run (see below).

## 3. Methods

The work flow of DNAWorks is as follows. A set of oligonucleotides is scored based on a set of features that are critical in the gene synthesis procedure (Fig. 1) (see Note 1). The sequence features evaluated in the determination of a sequence score are annealing temperature, repeats and misprime potential, length of both overlap and oligonucleotide sequences, GC and AT content, codon frequency, and the presence of restriction sites or sequence patterns. These features are discovered, evaluated, and scored. The scores are applied to the sequence in such a way that the regions with unwanted features (i.e., potential misprime site, GC rich, Tm outside the desired range, etc.) have the highest local score, and the total score is the sum of all local scores. The scores for individual

**Parameters:**

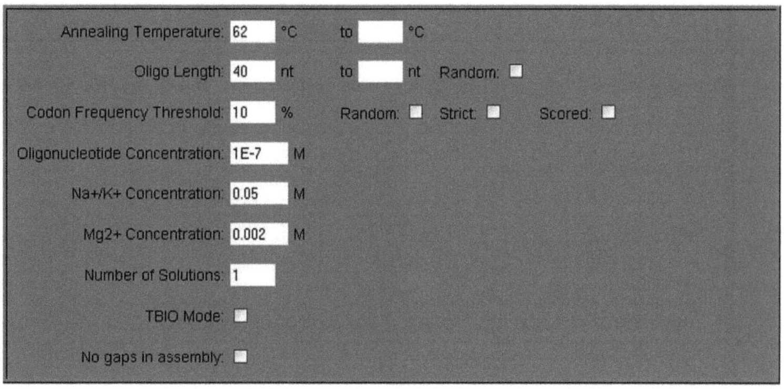


Fig. 1. Screenshot of the parameter input window of DNAWorks.

features are calculated so that the best possible score is zero (no violations present).

The gene is then silently mutated (codon swap) at a single position. The position is chosen based on its local score (low frequency, GC rich region, repeat, etc.), with some probability of a random choice as well. In this way, the mutation can be targeted, increasing the efficiency of the search, but allowing for some diversity. In the case of an input DNA sequence, the only changes that can be made are the positions of overlaps, and only the best set of oligos that result from the overlap generation step are shown in the output and the program ends.

After the mutation of the gene, its total score is recalculated. If the mutation lowers the score, or if the "temperature" (in simulated annealing) is sufficiently high enough, the mutation is kept. Otherwise, the program reverts back to the original codon. During every subsequent mutation round, another single silent mutation is generated and evaluated. When enough mutation rounds have been performed without further dropping the total score, the program exits, and the final set of oligonucleotides is printed.

Typically, a gene will optimize very quickly (within the first 500 rounds of mutation), but considerably smaller drops in the total score will continue afterward. Short simple sequences will drop to zero and will exit before 6,000 rounds of mutations are performed. Longer, more complicated sequences will drop in score more gradually and tend to trail for much longer before the cutoff number of rounds is reached (see Note 2).

Once the calculations for the final set of oligos are completed, the program displays the output of the results. If multiple solutions were requested, either by increasing the number of solutions or setting a range of annealing temperatures or lengths, the protein sequence is reverse translated as before, generating a new set of oligos, and the process is repeated for each solution. The results are printed to a plain text file that can be emailed to the user or accessed via the Web (Fig. 2).

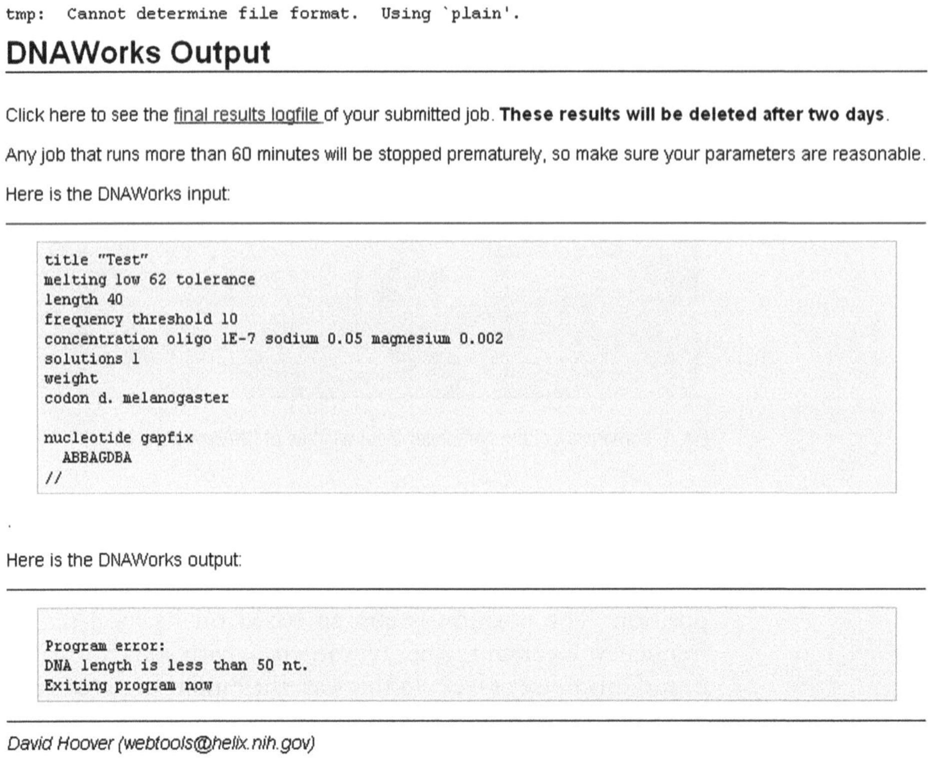


Fig. 2. Screenshot of an example output window.

### *3.1. General Run Procedure*

DNAWorks requires a job name. An email address is not required but is strongly encouraged, as many runs are not completed within a few minutes, and the user may lose the connection with the Web server during the run.

#### *3.1.1. Codon Frequency Table*

In order to optimize codon usage in a synthetic gene, the frequency of codon usage must be included with the run (Fig. 3). Codon frequency is determined using the method of Nakamura et al. (8). The codon frequency table also directs the reverse translation of amino acids to codons. The format is that of the GCG Wisconsin Package.

DNAWorks contains a small set of preloaded codon frequency tables for common organisms. A user can also manually type in codon frequencies or upload a file (Fig. 3). Nucleotide-only sequences do not require a codon frequency table.

#### *3.1.2. Parameters*

There are a number of parameters, for which values must be entered (see Fig. 1). The annealing temperature and the oligonucleotide length can either be a single value or include a range of values. Entering a range causes the program to try all values within the range, allowing the user to pick a solution with the best (lowest) score. The codon frequency threshold restricts the set of codons to those at or above a particular frequency. This allows rare

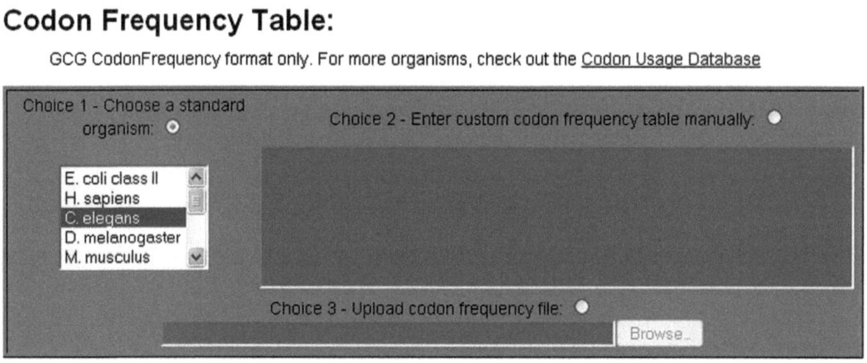


Fig. 3. Screenshot the codon frequency table selection.

codons to be avoided. The concentration of oligonucleotides and cations to be used in the subsequent PCRs can be adjusted. This is important as these values will affect the annealing temperature of the oligonucleotides. Finally, the number of solutions can be increased from 1 to 999, hopefully allowing a globally best solution to be found.

There are also parameters which toggle different modes of sequence optimization and scoring (see Fig. 1). Checking the "Random" box will cause the program to choose a random codon from those available for reverse translation in the first instance rather than the highest frequency codon. This can increase the range of possible sequences, perhaps finding better solutions, but with slightly worse codon frequencies. Opting for "Strict" will force the program to strictly use only those codons that are within the chosen codon frequency threshold, even if there is only one codon that satisfies this constraint. However, this parameter can limit the possible solutions. Opting for "Scored" will force the program to continuously evaluate the codon frequency score. This parameter may slightly increase the codon frequencies at the risk of slowing down the program to some extent. Checking the "TBIO" box will enable a thermodynamically balanced inside-out output, in which the first half of the oligonucleotides are all synthesized in the sense orientation, whereas the second half are synthesized as reverse complements in the antisense orientation of the gene. Clicking the "no gaps in assembly" checkbox will keep oligonucleotides as short as possible, with no gaps between the overlap regions. Restricting oligonucleotides to no gaps may slow down the optimization somewhat and may result in higher scores due to a higher probability of misprimes.

#### 3.1.3. Site Screen

Restriction sites can be excluded from the protein coding region of the synthetic gene (Fig. 4). A large set of preloaded restriction sites are available from a pull down menu. Custom sequences can also

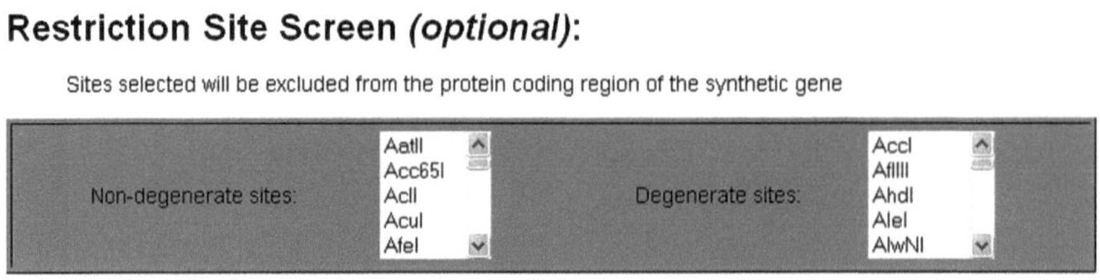


Fig. 4. Screenshot of the restriction site screen window, which allows the use to exclude specific restriction sites.

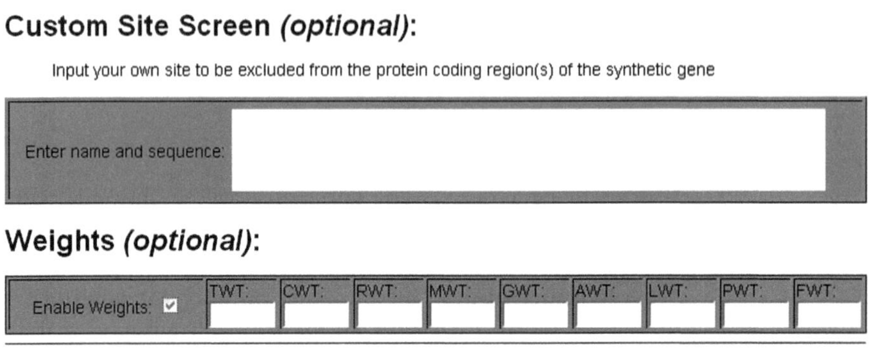


Fig. 5. Screenshot of the interface that allows the user to weigh individual features.

be excluded from the protein coding region. These sites can be represented in degenerate code to allow for multiple-specificity restriction endonucleases.

#### *3.1.4. Weights*

As mentioned above, DNAWorks optimizes a synthetic gene by evaluating the scores of a set of features: annealing temperature (T), codon frequency (C), repeat (R), misprime potential (M), GC (G) and AT (A) content, length (L), and pattern constraining (P). The default weights of each individual feature score are set to 1. By increasing the weight of an individual feature, the final output can be fine-tuned to favor one feature over the others (Fig. 5). For example, in cases in which potential synthetic genes for a set of sequences chronically suffer from a high number of repeats, increasing the weight of the repeat score (RWT) might decrease the final repeat score at the expense of the other feature scores.

However, please be aware when using this feature, as modulating the weights has not been fully tested. Remember that this merely skews the results toward one feature or another and might do more harm than good. Thus, in most cases, keeping the weights balanced is the best approach.

#### *3.1.5. Sequences*

Protein and DNA sequences can be inserted in separate blocks and this allows for more flexible designs (Fig. 6). The sequences can be entered as text or uploaded as files in multiple formats, including FASTA, GCG, GenBank, Swiss-Prot, and others. As many as 99

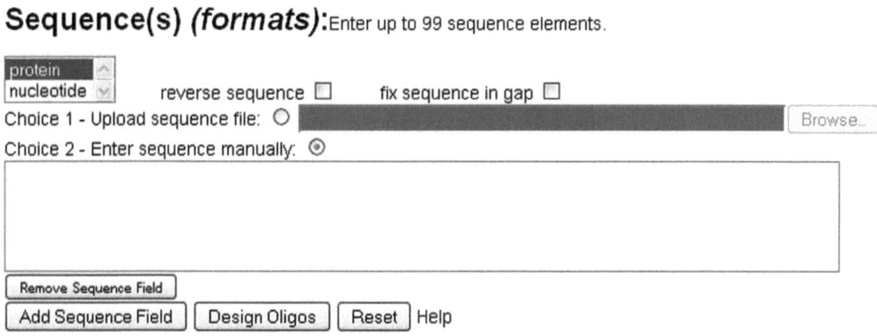


Fig. 6. Screenshot of the sequence input window.

individual sequence blocks can be inserted, with each block containing a minimum of a single nucleotide or amino acid.

Each sequence block can be reversed by clicking the "reverse sequence" box (Fig. 6). By checking the "fix sequence in gap," DNAWorks will attempt to keep the sequence block within a gap (the single-stranded region between overlaps). Gap-fixed sequences should be kept as small as the expected gaps, which depend on the melting temperature and length chosen. The lower the melting temperature and the longer the length, the larger the gaps will be, and the more room there is for gap-fixed sequences.

The typical reason for fixing a sequence in a gap is to allow the sequence to be swapped easily with a single oligonucleotide later on, for instance, in saturation mutagenesis experiments, and to eliminate problems that would occur with random mutagenesis.

### *3.2. Mutant Run*

Once a set of oligonucleotides has been designed, DNAWorks can evaluate a mutant sequence and design 1–3 new oligonucleotides that are needed to generate the mutant. Clicking on "mutant sequence" will change the Web form to allow uploading a previous DNAWorks logfile along with a sequence which has been mutated from the original sequence. The mutated sequence must be the same length as the original sequence, as the original overlap positions will be fixed. A trial number corresponding to the chosen trial from the previous DNAWorks run must be chosen. The parameters will be set to the same values as those of the trial number from the original logfile. Once all parameters have been entered, clicking "Design oligos" will generate the replacement oligos (Fig. 6). This will also generate an evaluation of scores for the mutated sequence. The mutation is printed in lowercase font, and it is highlighted in the oligonucleotide assembly.

### *3.3. Output*

After DNAWorks has finished calculating all solutions, a single output file is produced (see Fig. 2). The first section of the output displays the protein sequence(s) that has been entered, along with any input nucleotide sequences. The output file also contains the codon frequency table and active codons used during the optimization. Any restriction sites or user-defined sequence patterns that are to be excluded (or at least attempted to be excluded) are displayed in the sequence pattern section.

The next sections in the output file show the results of each solution. At the top of each section the parameters set for each solution are given. Next is the final, completed DNA sequence of the synthetic gene. Below this is the oligonucleotide assembly laid out in an alignment format. The alignment shows the numbering and direction of oligonucleotides and the translated amino acid sequence (for coding regions). Interspersed in the alignment are regions that are flagged as repeats or possible mispriming sites, as well as GC- or AT-rich regions, which the program was unable to eliminate. Toward the end of each solution are several short reports and histograms detailing the scores, frequencies of the codons used, lengths, annealing temperatures, and flagged sequence regions. At the very end of each section are the actual oligonucleotide sequences to be synthesized.

After all solutions are displayed, a final summary gives a table of the scores and statistics for each solution. Generally speaking, it is best to pick the solution which gives the lowest score. However, this should always be weighed against empirical problems, such as the presence of possible mispriming sites, very short overlap lengths (<12), or overly long oligonucleotide lengths (see Note 3).

## 4. Notes

1. DNAWorks was created for the purpose of automating the design of oligonucleotides for assembly into a synthetic gene. Alongside this was the need to increase the likelihood of success of the assembly process. However, the final score of a solution has little to do with the real physical properties of the gene, or of protein for which it codes. Often, the absolute value of a score has no association with the success of the actual assembly process. The score serves only to indicate which solutions are more likely to assemble correctly. A set of oligonucleotides which give a score of zero might generate multiple PCR artifacts. Alternatively, a set of oligonucleotides with a score of 500 may assemble very efficiently without experimental difficulties.
2. A major issue with optimization via simulated annealing methods is the possible lack of convergence toward a solution. The most obvious reason for this is sheer size; the longer a gene, the more individual scores need evaluating, and the more likely it is for the solution to be trapped in a local minimum. Another reason is the presence of problems that are inherent

to the sequence itself, such as multiple sequence repeats or highly biased amino acid preferences. It is highly advised that a full range of parameters is investigated before concluding that a sequence will never result in convergence to a reasonable solution. It is also not a bad idea to repeat a solution to make sure that an acceptable solution remains so after several addition trials.

3. DNAWorks is a tool for a well-informed user to simplify the task of synthesizing genes. It is not, however, an infallible predictor of success nor should the results be accepted blindly. Bad or flawed input will give bad or flawed results. As a rule of thumb, the longer the overlap length, and by definition the higher the annealing temperature of those overlaps, the more likely the oligonucleotides will anneal in a highly specific and unique way as predicted by DNAWorks (6). This rule must be balanced against the increased number of errors that are introduced for longer oligonucleotides.

## References


1. Carr PA, Church GM (2009) Genome engineering. Nature Biotech 27:1151–1162.
2. Prodromou C, Pearl LH (1992) Recursive PCR: a novel technique for total gene synthesis. Protein Eng 5:827–829.
3. Dillon PJ, Rosen CA (1990) A rapid method for the construction of synthetic genes using the polymerase chain reaction. Biotechniques 9:298–300.
4. Gao X, Yo P et al. (2003) Thermodynamically balanced inside-out (TBIO) PCR-based gene synthesis: a novel method of primer design for high-fidelity assembly of longer gene sequences. Nucleic Acids Res 31:e143.
5. Czar MJ, Anderson JC et al. (2009) Gene synthesis demystified. Trends Biotechnol 27: 63–72.
6. Xiong AS, Peng R et al. (2008) Chemical gene synthesis: strategies, softwares, error corrections, and applications. FEMS Microbiol Rev 32:522–540.
7. Hoover DM, Lubkowski J (2002) DNAWorks: an automated method for designing oligonucleotides for PCR-based gene synthesis. Nucleic Acids Res 30:e43.
8. Nakamura Y et al. (2000) Nucleic Acids Res 28:292.

# Chapter 17

# De Novo Gene Synthesis Design Using TmPrime Software

**Mo-Huang Li, Marcus Bode, Mo Chao Huang, Wai Chye Cheong, and Li Shi Lim**

## Abstract

This chapter presents TmPrime, a computer program to design oligonucleotide for both ligase chain reaction (LCR)- and polymerase chain reaction (PCR)-based de novo gene synthesis. The program divides a long input DNA sequence based on user-specified melting temperatures and assembly conditions, and dynamically optimizes the length of oligonucleotides to achieve homologous melting temperatures. The output reports the melting temperatures, oligonucleotide sequences, and potential formation of secondary structures in a PDF file, which will be sent to the user via e-mail. The program also provides functions on sequence pooling to separate long genes into smaller pieces for multipool assembly and codon optimization for expression based on the highest organism-specific codon frequency. This software has been successfully used in the design and synthesis of various genes with total length >20 kbp. This program is freely available at http://prime.ibn.a-star.edu.sg.

**Key words:** De novo gene synthesis, TmPrime, Bioinformatics, PCR, Ligase chain assembly, Melting temperature, Assembly efficiency

## 1. Introduction

The design and manufacture of custom genes is fast becoming an indispensable tool in synthetic biology (1) and protein engineering (2). Current de novo gene synthesis methods include ligase chain reaction (LCR) (3) and polymerase chain reaction (PCR) assembly (4). Both of them rely on the use of overlapped oligonucleotides (oligos) to construct genes. In LCR assembly, adjacent oligonucleotides with no gap between consecutive oligonucleotides are ligased, resulting in the extension of DNA, whereas PCR assembly utilizes DNA polymerase to extend the oligonucleotides. Regardless of whether LCR or PCR assembly is used, a successful synthesis requires appropriate oligonucleotide design to ensure that the oligonucleotides are highly specific for their targets and have uniform hybridization temperature to enhance the assembly efficiency.

Jean Peccoud (ed.), *Gene Synthesis: Methods and Protocols*, Methods in Molecular Biology, vol. 852,
DOI 10.1007/978-1-61779-564-0_17, © Springer Science+Business Media, LLC 2012

Programs have been developed for gene synthesis that include the design of oligonucleotides based on user-specific hybridization temperature and oligonucleotide length (5–9). The programs DNAWorks (5) or Gene2Oligo (6) provide fairly good synthesis results for DNA sizes below 1 kb. GeneDesign (8) and GeMS (9) have implemented a multipool function for the synthesis of multikilobase genes, in which long DNA sequences are split into smaller segments (~500 bp). These segments are first assembled in separated pools, before these intermediate segments are assembled into the full-length product in a final PCR step. DNAWorks provides an important and useful feature for predicting the potential for mishybridization and secondary structures among potential oligonucleotides.

This chapter presents a gene design program called TmPrime (10) that is capable to design oligonucleotides—and analyze their potential mishybridization and secondary structures—for up to 20 genes with very long gene sequences (≤40 kb) for use in LCR and gapless PCR assembly. This program allows to construct oligonucleotides with uniform melting temperatures ($\Delta T_m < 3°C$) which increases the yield of the assembled full-length DNA product by PCR gene assembly. These features are useful for de novo gene synthesis, especially for aspiring applications in genome synthesis and multiplex gene synthesis.

## 2. Materials

### 2.1. TmPrime Interface

TmPrime uses the equal-temperature (Equi-$T_m$) approach to design oligonucleotide sets. The program first divides the given sequence into fragments, from the beginning to the end of the DNA sequence, based on the user-specified melting temperature (Fig. 1). This process usually leaves a small DNA tail that has a melting temperature ($T_m$) that is lower than the user-specified $T_m$. The fragment boundaries are then shifted to accommodate this tail and to minimize melting temperature deviations of the fragments. Once the melting temperatures of the fragments are equilibrated, the oligonucleotides to be used in gapless PCR or LCR are constructed by connecting two adjacent fragments along both the sense and antisense strands. Each oligonucleotide overlaps with its complementary neighbors by exactly one fragment (see Note 1).

Figure 2 indicates all the parameters that are needed to generate the oligonucleotide sets using TmPrime. Most of the parameters are self-explanatory. The user is asked to provide gene information, gene assembly buffer condition, oligonucleotide and outer primer concentrations, optional parameters for long DNA

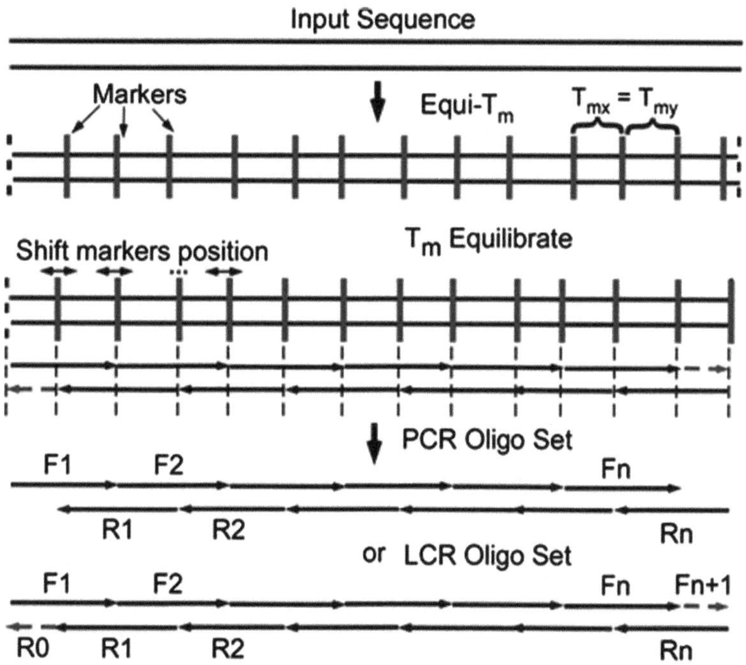


Fig. 1. An overview of the oligonucleotide design scheme. TmPrime first divides the input sequence into sections of approximately equal melting temperatures (Equi-$T_m$) using markers based on the user-specified melting temperature. The positions of the markers are iteratively shifted to globally minimize the deviation in melting temperature among the fragments ($T_m$ equilibrate). Two adjacent fragments are joined together to generate oligonucleotides for PCR gapless assembly. The two tail segments at the 3′ ends of the sense and antisense sequences are also included for LCR assembly (i.e., R0 and Fn + 1).

assembly, and parameters for mispriming analysis. The software will report melting temperatures, oligonucleotide sequences, primer sequences, potential formation of secondary structures, and statistical information of the oligonucleotide sets of each pool compiled in a PDF file.

***2.2. Reagents for LCR Gene Assembly***

1. 100 μM oligonucleotides.
2. T4 ligase and buffer.
3. Ampligase and buffer (Epicentre Biotechnology).
4. T4 polynucleotide kinase.
5. 100 μM outer primers.
6. 25 mM $MgSO_4$.
7. dNTP mixture (containing 25 mM dATP, 25 mM dGTP, 25 mM dCTP, and 25 mM dTTP).
8. High-fidelity KOD Hot Start DNA polymerase (1.0 U/μl) and 10× KOD buffer (Novagen).

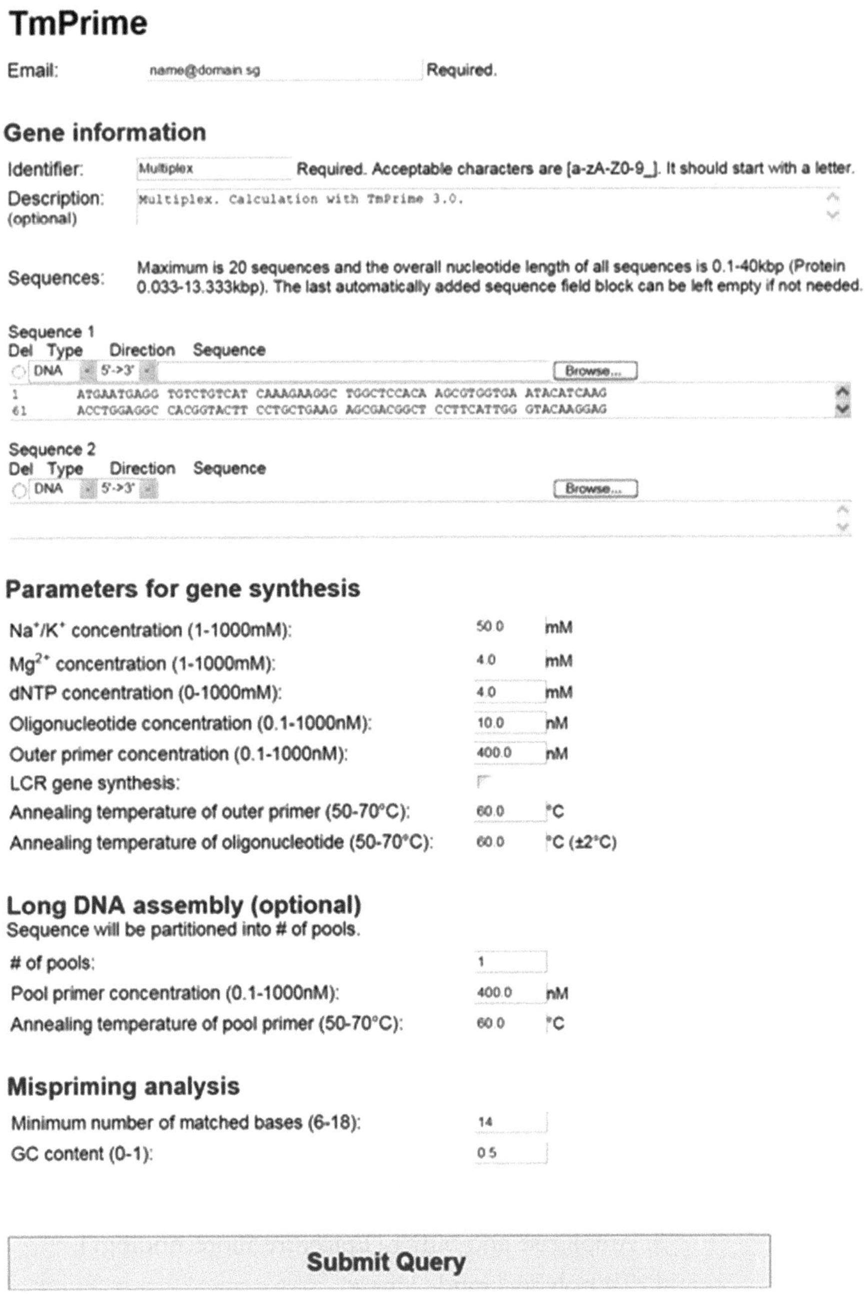


Fig. 2. Gene design web interface of TmPrime.

## 3. Methods

### 3.1. Calculation with Codon Optimization

TmPrime includes a codon optimizing feature. It implements global codon optimization that replaces each codon based on the organism-specific codon frequencies using the organism-specific codon data in the Codon Usage Database (http//www.kazusa.

or.jp/codon/). The user can select an organism for codon optimization from a list of organisms which exists in the Codon Usage Database (NCBI-GenBank Flat File Release 166.0).

### 3.2. Multiplex Gene Synthesis

TmPrime can handle up to 20 genes with a total DNA length of up to 40 kb (Fig. 2). This function is specifically useful for multiplexing gene synthesis, which allows users to screen the potential mishybridization among a set of multiple genes. When the parameter of "# of pools" is set to 1 (default), TmPrime automatically stitches the uploaded multiple gene sequences together into a single DNA sequence and conducts the oligonucleotide design and mishybridization analysis accordingly (see Note 2). The program performs mishybridization screening through a pairwise sequence alignment with a score based on the user-specified number of matched bases and G+C content (see "minimum number of matched bases" and "GC content" parameters in Fig. 2). The program connects adjacent potential mishybridization regions and reports the entire extended region. The oligonucleotides are displayed in alternating upper and lower case (Fig. 3). These features allow users to easily visualize and inspect any problematic DNA regions (see Note 3).

### 3.3. Single-Pool Assembly

TmPrime supports oligonucleotide design for conventional one-step and two-step PCR-based gene syntheses, "TopDown" one-step gene synthesis, and LCR-based gene synthesis. The software generates gapless oligo sets that have no gap between consecutive oligos, and reports the oligo set, which has the lowest melting temperature deviation, with the average melting temperature within ±2°C of the user-specified melting temperature. Oligonucleotides are displayed in alternating upper and lower case to make it easy for the user to find the boundaries with the prefix of oligonucleotide sets and primers defined in Fig. 4 (see Note 4).

Overlapping PCR assembly is a parallel process, by which the lengths of the overlapping oligonucleotides are extended after each PCR cycle. The theoretical minimum number of cycles ($x$) needed in order to construct a double-stranded (ds)DNA molecule of the length ($L$) from an uniform oligonucleotide length ($n$) and

**a**

```
  1 CAGTTAGGGGTACACATCCTATCTGCAAATTCCTTTCCAATGGTTCAAGAAAGAATGG
 59 GCCTGGATCATGTGGGAAGACAGGTAATGAAGCCCTGAGGGTAGA
104 GCCTGGATCATGTGGGAGGACAGGTCATGAAGCCCTGAGGGTAGA
149 GCCTGGATCACGTGGGAGGACAGGTCATGAAGTCCTCAGGGTAGA
194 GCCTGGATCATGTGGGAGGAAGGGTCATGAAGCCCTGAGGGTAGA
239 GCCTGGATCACGTGGGAGGACAGGTCATGAAGCCC
274 GGTAGGGTGCCACCCACTAAAGAATTCTAGTC
```

**b**

```
53 GAATGGGCCTGGATCATGTGGGAAGACAGGTAATGAAGCCCTGAgggtagagcctggat//GATCACGTGGGAGGACAGGTCATGAAGtcct 184
98 CCATCTCGGACCTAGTACACCctcctgtccagtacttcgggactcccatctcggaccta//CTagtacaccctccttcccagtacttcggga 229
```

Fig. 3. Mishybridization analysis of a 305-bp human minisatellite region. (**a**) DNA sequence. (**b**) Partial results of the mishybridization analysis.

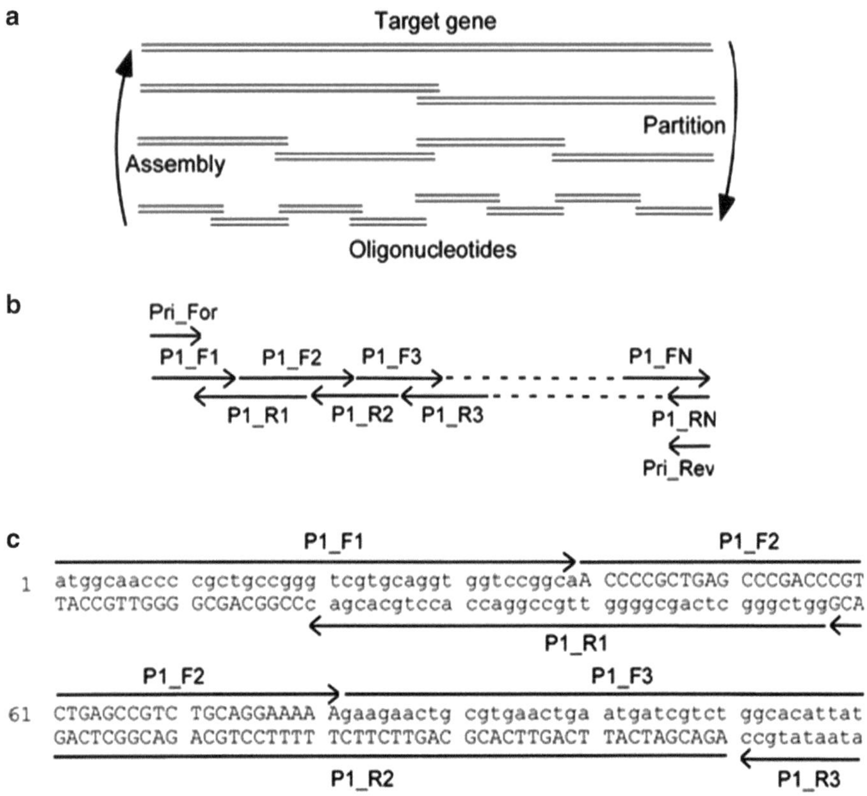


Fig. 4. (**a**) Schematic illustration of overlapping PCR assembly. (**b**) The prefix of oligonucleotide sets and primers in the output file. (**c**) Oligonucleotides are displayed in alternating upper and lower case for easy finding of the boundaries.

overlapping size (*s*), or from a pool of *m* oligonucleotides of various lengths can be calculated by Eqs. 1 and 2, respectively.

$$2^x n - (2^x - 1)s > L \tag{1}$$

$$x \geq \log_2(m) \tag{2}$$

Therefore, theoretically, six PCR cycles are sufficient for assembling a 1,000-bp DNA segment from a pool of oligonucleotides of 40 nucleotides (nt) in length with an overlap of 20 nt (see Note 5).

### 3.4. Multiple-Pool Assembly

For DNA with length of greater than 1.5 kb, we recommend splitting the gene into DNA segments and conducting the gene assembly in multiple steps (11). TmPrime automatically splits the gene into pools of shorter sequences of approximately equal length based on the user-specified number of pools, whereby the pool-pool overlap length is automatically adjusted according to the annealing temperature of the across-pool assembly of the outer primers. This function is implemented in the feature of "Long DNA Assembly" as shown in Fig. 2. Different annealing temperatures can be assigned for the assembly of outer primers across the

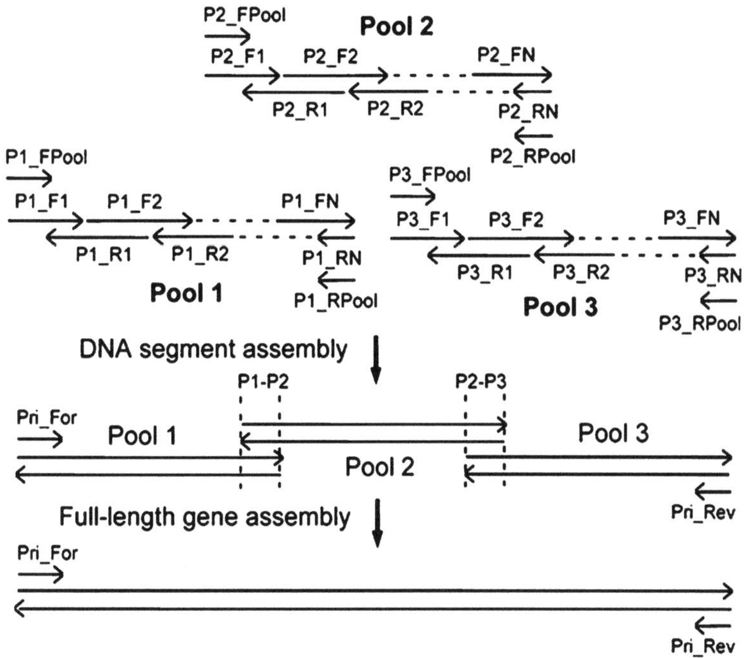


Fig. 5. Schematic illustration of multiple-pool assembly. A long gene is first split into DNA segments, and then, the DNA segments are further divided into individual oligonucleotide sets. The assembly process is conducted in two steps. DNA segments are first assembled from individual oligo sets in separate pools, followed by a final PCR to create the full-length gene. The lengths of pool overlapping regions (P1–P2 and P2–P3) are automatically defined based on the user-specified annealing temperature of pool primers.

pool (annealing temperature of outer primer), of outer primers within a pool (annealing temperature of pool primer), and of inner oligonucleotides (annealing temperature of oligonucleotide), thus providing flexibility for long gene construction. Oligonucleotides for each pool assembly are optimized at the same melting temperature to allow the parallel synthesis of different segments or different genes simultaneously in a single thermal cycler. The self-explanatory prefixes of outer primers, pool primers, and inner oligo sets are defined in Fig. 5 (see Note 6).

### 3.5. Comparison of Oligonucleotide Design Programs

The oligonucleotide design features of different synthetic gene design programs are summarized in Table 1, and Table 2 compares the performance of these programs for S100A4 (chr1:1503312036–1503311284), GFPuv (Genbank U62636; region of 261–1,020) and the entire genomes of poliovirus (Genbank FJ517648; 7,418 bp) and of øX174 bacteriophage (Genbank J02482; 5,386 bp). TmPrime offers the most homologous melting temperatures with $\Delta T_m < 3°C$, and a wider range of annealing temperatures (50–70°C) as compared to DNAWorks (58–70°C) and Assembly PCR Oligo Maker (50–60°C). GeneDesign cannot adjust the oligonucleotide

**Table 1**
**Comparisons of the oligonucleotide design features of gene synthesis programs**

| | $T_m$—optimization | Automatic pooling for long DNA | Mishybridization analysis | Codon optimization | Ultra-long DNA analysis[a] |
|---|---|---|---|---|---|
| TmPrime | Gapless PCR/LCR | Yes | Yes/improved | Yes | Yes |
| DNAWorks | Gapped/Gapless PCR | No | Yes | Yes | No |
| GeneDesign | Gapped PCR | Yes[b] | No | Yes | No |
| Gene2Oligo | Gapless PCR | No | No | No | No |
| Assembly PCR Oligo Maker | Gapped PCR | No | No | No | No |

[a]TmPrime is capable to handle DNA length up to 40 kbp, a unique feature for genome synthesis
[b]GeneDesign automatically searches unique restriction size and divides the long DNA sequence (> 600 bp) into chunks of approximately 500 bp. The user has no control on this function

**Table 2**
**Comparison of the oligonucleotide design performance of different gene synthesis programs**

| | S100A4 (752 bp) | | GFPuv (760 bp) | | |
|---|---|---|---|---|---|
| Annealing $T_m$ | 55°C | 65°C | 55°C | 65°C | Poliovirus, øX174 bacteriophage |
| TmPrime[a] | 2.49[b] | 1.56 | 1.55 | 0.86 | Yes |
| DNAWorks[c] | x[d] | 3.1 | x | 1.8 | Fail |
| GeneDesign[e] | Success | Success | Success | Fail | Fail |
| Gene2Oligo[f] | Fail | Fail | 7.32 | 5.84 | Fail |
| Assembly PCR Oligo Maker[g] | Success | x | Success | x | Fail |

[a]TmPrime: http://prime.ibn.a-star.edu.sg/
[b]Derivation of melting temperature (°C)
[c]DNAWorks: http://helixweb.nih.gov/dnaworks/
[d]Software does not support the oligonucleotide design at this annealing temperature
[e]GeneDesign: http://baderlab.bme.jhu.edu/gd/
[f]Gene2oligo: http://berry.engin.umich.edu/gene2oligo/
[g]Assembly PCR Oligo Maker: http://startrek.ccs.yorku.ca/~pjohnson/AssemblyPCRoligomaker.html

concentrations and PCR buffer conditions, and oligonucleotide design may fail when the sequence of consecutive oligonucleotides collides. Gene2Oligo has difficulty in designing S100A4 and fails to converge at specified annealing temperature. Only TmPrime can handle the poliovirus and øX174 bacteriophage genomes.

### 3.6. LCR Gene Assembly Protocol

1. LCR assembly
   The LCR assembly is carried out in a final volume of 50 μl containing 5 μl of 10× T4 ligase buffer, 5 μl of 10× Ampligase buffer, and 10–100 nM of TmPrime-optimized oligonucleotides that have been phosphorylated using 20 U of T4 polynucleotide kinase and 20 U of Ampligase.

   LCR assembly is conducted as follows: 37°C for 4 h, denatured at 95°C for 3 min, ramped to 60°C (matched with the average melting temperature of oligonucleotides) at 0.1°C/s for annealing, and incubated at 60°C for 2–8 h.
2. PCR amplification
   The full-length assembly product is amplified by a PCR containing 5 μl of the assembly mixture from step 1 above, 0.4 μM of outer primers, 1 μl of KOD Hot Start Master Mix, and 1× PCR buffer in a final volume of 25 μl.

   The PCR is conducted under the following conditions: 2 min of initial denaturation at 95°C—30 cycles of 95°C for 20 s, 55°C (matched with the melting temperature of primers) for 30 s, and 72°C for 30 s—followed by a final extension step of 72°C for 10 min.

## 4. Notes

1. The oligonucleotide sets designed for PCR gene synthesis cannot be directly utilized for LCR gene assembly as the two tail segments at 3′ end of the sense and antisense sequences are not included in the oligonucleotide sets. In addition, the average melting temperature of oligonucleotide sets decreases ~2.8°C with each order of magnitude decrease in the oligonucleotide concentration. The user should therefore adjust the annealing temperature of PCR and LCR assembly processes accordingly if the oligonucleotide concentration for oligonucleotide design and actual gene assembly are different.
2. The software skips the multigene mispriming analysis if the setting of "# of pools" is not 1. Under this condition, TmPrime assumes that the user will conduct multipool gene synthesis.
3. The potential mishybridization and secondary structures reported by TmPrime depend on the user-specified number of matched bases and the GC content. Users should adjust these parameters according to the GC content of target genes. We recommend starting the gene design with low value of GC content (such as 0.3). This would ensure capturing all potential mispriming and secondary structures even if the gene or portion of the gene has low GC content.

4. TmPrime generates oligonucleotides of various lengths, depending on the base composition profile of the gene sequence. Some genes may contain clusters of G+C or A+Tregions. The region with the high G+C content will generate shorter oligonucleotides than that with a high A+T content.
5. The assembly efficiency gradually decreases as the target gene length increases. For single-pool PCR gene synthesis, consistent and successful gene synthesis is obtained with DNA length below 1.5 kbp or from a pool of up to 60 oligonucleotides (12).
6. For DNA with high sequence repeats, PCR-based gene synthesis may not be the best choice. The LCR-based approach is more effective for these challenging DNA sequences as the LCR assembly inherently requires a more stringent assembly condition than that of the PCR process. Ligation only occurs when two adjacent oligonucleotides that do not have any gap are hybridized with an opposite pairing DNA. We recommend conducting the LCR gene assembly with a thermostable DNA ligase (such as Ampligase) and with an elevated annealing temperature to increase the annealing stringency of oligonucleotides and to minimize the potential mishybridization of oligos. We use the LCR gene assembly protocol described in Subheading 3.6.

## References


1. Cox JC, Lape J, Sayed MA and Hellinga HW (2007) Protein fabrication automation. *Protein Sci* 16:379–390.
2. Sprinzak D, and Elowitz MB (2005) Reconstruction of genetic circuits. *Nature* 438:443–448.
3. Au LC, Yang FY, Yang WJ, Lo SH and Kao CF (1998) Gene synthesis by a LCR-based approach: High-level production of leptin-L54 using synthetic gene in *Escherichia coli. Biochem Biophys Res Commun.* 248:200–203.
4. Prodromou C and Pearl L (1992) Recursive PCR: A novel technique for total gene synthesis. *Protein Eng* 5:827–829.
5. Hoover DM and Lubkowski J (2002) DNAWorks: An automated method for designing oligonucleotides for PCR-based gene synthesis. *Nucleic Acids Res* 30:e43.
6. Rouillard J-M, Lee W, Truan G, Gao X, Zhou X and Gulari E (2004) Gene2oligo: Oligonucleotide design for *in vitro* gene synthesis. *Nucleic Acids Res* 32:W176–W180.
7. Rydzanicz R, Zhao XS and Johnson PE (2005) Assembly PCR oligo maker: A tool for designing oligodeoxynucleotides for constructing long DNA molecules for RNA production. *Nucleic Acids Res* 33:W521–W525.
8. Richardson SM, Wheelan SJ, Yarrington RM and Boeke JD (2006) GeneDesign: Rapid, automated design of multikilobase synthetic genes. *Genome Res* 16:550–556.
9. Jayaraj S, Reid R and Santi DV (2005) GeMS: An advanced software package for designing synthetic genes. *Nucleic Acids Res* 33: 3011–3016.
10. Bode M, Khor S, Ye H, Li,M-H and Ying JY (2009) TmPrime: fast, flexible oligonucleotide design software for gene synthesis. *Nucleic Acids Res* 37:W214–W221.
11. Shevchuk NA, Bryksin AV, Nusinovich YA, Cabello FC, Sutherland, M and Ladisch S (2004) Construction of long DNA molecules using long PCR-based fusion of several fragments simultaneously. *Nucleic Acids Res* 32:e19.
12. Cheong WC, Lim LS, Huang MC, Bode M and Li M-H (2010) New Insights into the de novo Gene Synthesis Using the Automatic Kinetics Switch Approach. *Anal Biochem.*406: 51–60.

# Chapter 18

## Design-A-Gene with GeneDesign

**Sarah M. Richardson, Steffi Liu, Jef D. Boeke, and Joel S. Bader**

### Abstract

The manual design of synthetic genes is a tedious and error-prone process—even for very short genes—and it becomes completely infeasible when multiple synthetic genes are needed. GeneDesign is a set of modules that automate batch nucleotide manipulation. Here, we explain the installation, configuration, and use of GeneDesign as part of a synthetic design workflow.

**Key words:** Synthetic biology, Computer-assisted GeneDesign, Codon optimization, Synthetic genes, Synthetic biology software

## 1. Introduction

A synthetic gene is a DNA molecule that has been designed and built to the specification of a researcher. A synthetic gene is usually a close copy of a particular gene of interest, with changes to the nucleotide sequence affecting codon usage, base composition, and restriction enzyme recognition site placement. GeneDesign is software that automates the design of synthetic genes (1, 2). The current version of GeneDesign assumes that the protein sequence of a gene is of paramount importance and will not allow changes that disrupt it. GeneDesign accomplishes all manipulations with synonymous codon substitutions. The relative synonymous usage (RSCU) value for a codon is the ratio of how often that codon is seen over how often it would be seen given perfectly random usage (3). GeneDesign has a set of canonical RSCU values (4, 5) but also lets the user define its own. It uses these values to make decisions about which codons to change.

Jean Peccoud (ed.), *Gene Synthesis: Methods and Protocols*, Methods in Molecular Biology, vol. 852,
DOI 10.1007/978-1-61779-564-0_18, 

GeneDesign can be run as a web application or as a set of command line tools. We find the latter to be a good way to automate manipulations for large sets of genes.

## 2. Installation and Configuration

GeneDesign requires Perl 5 (6). A web server is required if you wish to use a graphical interface; we highly recommend Apache (7). Both Perl and Apache are widely distributed, well documented, and in some cases installed by operating system default, so we will not discuss their installation. The following directions assume that at least Perl 5.8 and Apache 2 are used.

### *2.1. Downloading GeneDesign Source Code*

The easiest way to obtain and update GeneDesign source is with the revision control software package git (8). All GeneDesign development is tracked in git, and the source code is hosted for public distribution through a git installation at http://www.github.com. Git makes it extremely easy to update GeneDesign, as well as to track—and potentially contribute—your own changes to the code. Once you have installed git, make a local copy of the GeneDesign repository in a working directory with the command: `git clone git://github.com/GeneDesign/GeneDesign.git /path_to/your_wd/GeneDesign.git`. If you want your repository to work as a web server, the working directory must be accessible by your server's user (see Note 1).

If you do not want to use git, you can download a snapshot of the GeneDesign repository at http://github.com/GeneDesign/GeneDesign/zipball/master. Unpack this zipped file into your working directory.

### *2.2. Configuring Perl*

GeneDesign requires a few additional Perl modules to run. Some may already be installed, depending on your Perl distribution. You can run the helper script `genedesign_mod_install.pl` as a privileged user to install these Perl modules for you (see Note 2).

### *2.3. Configuring Apache*

At this point, the command line scripts are ready to go, but if you want to serve GeneDesign to web browsers, there is more to do. If you are using git, you want to preserve version control and configure your Apache installation to serve GeneDesign from your repository by editing your web server's configuration file (see Note 3). Let `/path/to/your/GeneDesign.git` be the absolute path to your git repository. Make sure that the repository is accessible by the web server's user (see Note 1).

You must add a directory listing for the GeneDesign html files:

```
<Directory "/path/to/your/GeneDesign.git/documents/gd">
     Options Indexes MultiViews FollowSymLinks
     AllowOverride All
     Order allow,deny
     Allow from all
</Directory>
```

You must add another directory listing for the Perl scripts:

```
<Directory "/path/to/your/GeneDesign.git/cgi-bin/gd">
     AllowOverride None
     Options ExecCGI
     Order allow,deny
     Allow from all
</Directory>
```

You must add two aliases:

```
     ScriptAlias /cgi-bin/gd/
"/path/to/your/GeneDesign.git/cgi-bin/
     Alias /gd /path/to/your/documents/gd
```

If you are not using git, configuring a web server is much easier. Copy the `GeneDesign.git/cgi-bin/gd` folder to your server's CGI-Executables directory and the `GeneDesign.git/documents/gd` folder to your server's documents directory.

***2.4. Testing Your GeneDesign Installation***

To confirm that GeneDesign prerequisites have been installed and all necessary files have been downloaded and unpacked, run a test with the command `perl GeneDesign.git/bin/run_GD_tests.pl` in the folder (see Note 4).

To confirm that the GeneDesign web server is running properly, point a web browser to http://localhost/gd, where localhost may be replaced with your server's address. You should see a screen very similar to that in Fig. 1 (if not, see Note 5).

Fig. 1. Screenshot of the the GeneDesign main interface.

***2.5. Updating and Customizing Your GeneDesign Installation***

If you used git to download the GeneDesign source code, you can update your installation to the latest release by executing the following command in your local repository: `git pull origin master`. If you wish to make changes to your local installation while preserving git's ability to synchronize the code, it is recommended that you use git branching, which is a feature better covered elsewhere (9).

GeneDesign uses a mostly nonredundant set of restriction enzymes from REBASE (10) which are listed in the file `bs_enzymes.txt`. You can add or delete enzymes to this list—but make sure that your additions follow formatting rules described in the header.

There is a set of vector sequences, which you may edit in the directory `GeneDesign.git/cgi-bin/gd/vectors`; see methods in Subheading 3.6 for more information.

To add organisms and codon tables, you can edit the `gd_organisms.txt` file. Again, make sure that your additions follow the formatting rules.

## 3. Modules

GeneDesign modules are all designed to be independent so that you use as many or as few algorithms as you want. They all take FASTA formatted or plain text sequence as input, and they all offer the FASTA formatted or plain text as output. The simplest set of manipulations is a workflow we call the "Design-A-Gene Path," which begins with a protein sequence being entered in the Reverse Translation module, then goes through the Restriction Site Addition module, and ends with a list of oligonucleotides and building blocks in the Building Block Design module. You do not have to follow this path, but it works well as an example and is explicitly marked on the web pages.

This version of GeneDesign uses the standard genetic code; future versions will be more flexible about codon to residue assignment.

### 3.1. Web Server Modules Versus Command Line Modules

The difference between the Web Server modules and the Command Line modules is their user interface. The Web Server modules are used with a web browser, and the Command Line modules are used with a terminal. The difference in interface allows one or two Command Line modules to offer batch functions that the Web Server modules do not. Future versions of GeneDesign will offer more Command Line modules and have more robust batch processing interfaces for the Web Server modules.

### 3.2. Generate RSCU Values

GeneDesign offers a base set of RSCU values derived from the most highly expressed genes in *Saccharomyces cerevisiae*, *Escherichia coli*, *Bacillus subtilis*, *Drosophila melanogaster*, *Homo sapiens* (4), and *Caenorhabditis elegans* (5), but these will not be appropriate for every application. The Generate RSCU Values module allows you to define a set of highly used codons from a set of reference sequences. Paste a single nucleotide sequence or a FASTA formatted set of nucleotide sequences into the text box, and press the results button. You will receive two tables. The first is an abbreviated, 21-codon table, in which each residue is represented by the codon with the highest RSCU value, as culled from your input. This table is suitable for directly pasting into the Reverse Translation module. The second table is the full RSCU table, with the data for all codons.

This module is also available on the command line; run `perl GeneDesign.git/bin/Generate_RSCU_Table.pl --help` for additional documentation.

### 3.3. Reverse Translation

GeneDesign's Reverse Translation module takes protein sequences and returns synonymous nucleotide sequences using either a user-defined codon scheme or the most optimal codons for expression in a user-selected organism. To begin, paste in a single sequence or a FASTA formatted set of sequences. To define the codon usage, you may select one of the five common organisms from the radio buttons. Choosing an organism will cause the drop-down menus to automatically select the codon with the highest RSCU value in highly expressed genes in that organism (4). You may also paste a custom table into the field at right. Any input in that field will supersede all entries in the drop-down menus, so your table must be complete. The format for the table is residue, space, codon, space, and line break—the first line of a valid table may be "M ATG" and the second "L CTG." The easiest way to make a custom table is with the Generate RSCU Values module, as described in Subheading 3.2.

Once the codons have been defined to your satisfaction, press the "reverse translate" button to obtain your nucleotide sequences. If you used FASTA input, the identifiers will have changed to indicate that GeneDesign manipulated the sequence.

This module is also available on the command line, where it is capable of performing batch reverse translations using a variety of RSCU tables per sequence. Run `perl GeneDesign.git/bin/Reverse_Translate.pl --help` for usage details.

### 3.4. Codon Juggling

Reverse Translation only offers RSCU optimization because it takes a protein sequence as input. If an original nucleotide sequence is provided, GeneDesign can use several algorithms to change the profile of the synthetic nucleotide sequence relative to that of the original. The Codon Juggling module offers several algorithms to do just that. You can provide a custom RSCU table, just as you can for Reverse Translation, except that this time, you want a full table, not the abbreviated one (see methods in Subheading 3.2).

The Codon Juggling algorithms offer manipulations based on RSCU value and base composition, or some blend of the two. The "optimized" algorithm simply replaces every codon in the sequence with the codon that has the highest RSCU value, in an action analogous to the Reverse Translation function, while the "less optimized" algorithm replaces every codon with the highest RSCU-valued codon that is not the original codon. If the codon in question already has the highest RSCU value, it is replaced with the next most optimal codon. The "most different sequence" algorithm makes as many transversions as possible in an attempt to minimize sequence identity. The "least different RSCU" algorithm attempts to make as many changes as possible where the difference between the RSCU value of the old codon and that of the new codon is minimized but is no greater than 1. There is also a random codon-swapping algorithm that makes randomly chosen synonymous changes at every codon position. If no organism or RSCU table is supplied, you will only be offered the most different and random algorithms. Currently, the web version of Codon Juggling does not offer batch processing—but the command line interface does.

The Codon Juggling module also offers an alignment of all of the nucleotide sequences generated and a graph of the RSCU values for each sequence along the length of the input sequence. see Subheading 3.5 for more information on interpreting the graph.

This module is available on the command line, where it is capable of performing batch manipulations using a variety of RSCU tables per sequence. Run `perl GeneDesign.git/bin/Juggle_Codons.pl --help` for documentation.

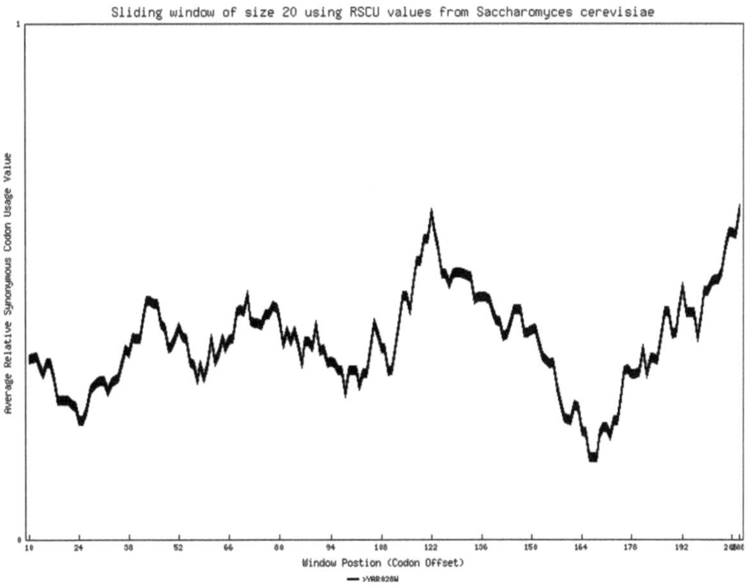


Fig. 2. Codon Bias Graphing module. Shown here is a screenshot of a plot of the change in average RSCU value along the length of a set of input sequences.

### 3.5. Codon Bias Graphing

It is sometimes helpful to visualize the change in RSCU values along a nucleotide sequence, as significant variations can correspond to conserved sequence necessary for expression or regulation (2). The Codon Bias Graphing module takes RSCU tables or reference organisms and plots the change in average RSCU value along the length of a set of input sequences (Fig. 2). You can change the window size (measured in codons) over which the average is determined to change the granularity of the graph.

### 3.6. Restriction Site Addition and Short Sequence Addition

We discuss these two modules together because both allow you to search a nucleotide sequence for positions where an arbitrary subsequence may be "silently" added—that is, the nucleotide sequence will be modified in place without altering its translation. The Restriction Site Addition is just a special case of the Short Sequence Addition where all the arbitrary subsequences are restriction enzyme recognition sites.

The Restriction Site Addition can be used to divide a large sequence into more manageable pieces by creating unique restriction enzyme recognition sites, called landmarks, at regular intervals. The module offers three different ways to create a list of restriction enzymes that will be suitable for this task. You may simply select the enzymes from a list, as the first box suggests. Select as many enzymes as you like from the left-hand list, and click the left arrow to move them to the right-hand list. Only enzymes in the right-hand list will be considered. To remove enzymes from consideration, highlight them, and click the right arrow to move them to the left-hand list (Fig. 3).

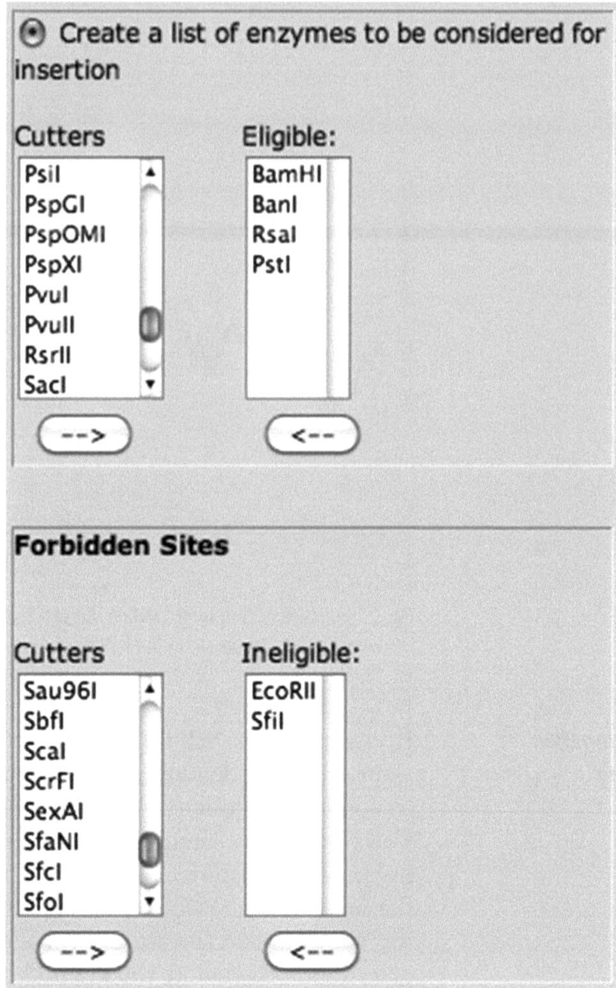


Fig. 3. Screenshot of the Restriction Site Addition module.

If you know the vector into which you will be cloning your gene, you can provide its sequence, and GeneDesign will automatically determine which enzyme recognition sites are absent from it. Only those enzymes will be considered for inclusion in your sequence. To modify which vectors appear in the drop-down menu, you can add or remove the FASTA files in the `GeneDesign.git/cgi-bin/gd/vectors/` directory. You can also just paste a vector sequence in at run time (Fig. 4).

You also have the option of specifying a list of "forbidden enzymes" that may not be considered, regardless of the absence of their recognition sites in a vector sequence (Fig. 3). You may define filtering criteria for enzymes as well—see Subheading 3.9 for a detailed description of enzyme filtering.

Once you have defined the restriction enzyme recognition sites to be sought, you can go to the next screen (Fig. 5). Here, you will

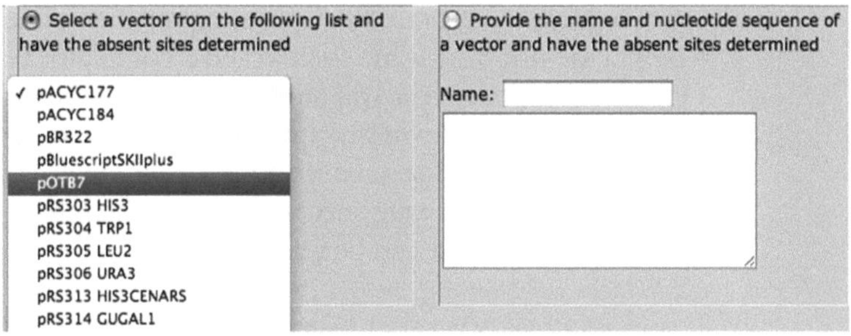


Fig. 4. Screenshot of the cloning vector selection tool.

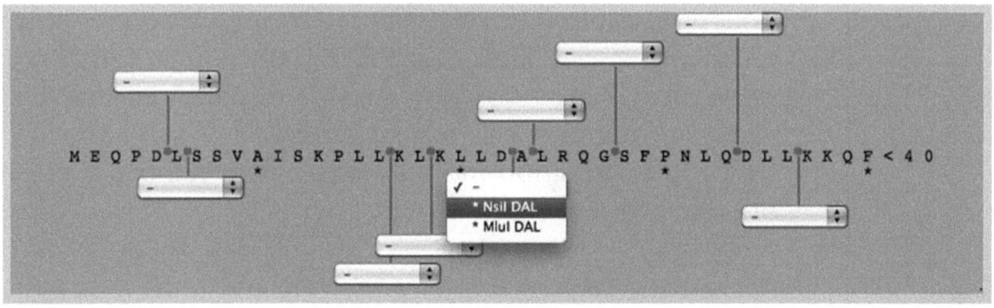


Fig. 5. Translated sequence window. The sequence is laid out in 40-column rows, with drop-down menus anchored between residues containing a list of the restriction enzymes whose recognition sites may be silently added to the sequence.

see your translated sequence laid out in 40-column rows, with drop-down menus anchored between residues. Every drop-down menu contains a list of the restriction enzymes whose recognition sites may be silently added to the sequence. If you do not see any drop-down menus, try changing your enzyme criteria—some genes and vectors are going to be incompatible. You may manually select your landmarks from the drop-down menus and choose to continue, or you can define an amino acid interval and have GeneDesign automatically select landmark enzymes for you. If you choose the latter, the view will be refreshed with the enzymes GeneDesign picks highlighted in blue, and you will have a chance to make manual changes or change the interval before you proceed.

At the next screen, the sequence modifications are actually made, and you will see a summary of the enzymes that were added. If there were any problems (for instance, two sites may be chosen so close to each other that one cannot be added without subtracting the other), you will be told. If you do not like the solution, you can reset the process entirely by, or you can mark the checkboxes by, enzymes in the solution to discard from consideration and return to the landmark selection screen to have GeneDesign reevaluate its selection by pressing the "reconsider" button. see Subheading 3.9 for a description of the columns in the summary table.

The Short Sequence Addition module offers a much more general form of this particular modification. You define a list of short sequences separated by spaces on the first page and indicate whether or not you want their reverse complements considered. On the next page, you will see the same 40-column amino acid layout with drop-down menus, and you must manually select where you want the sequences added. The final page will report which modifications were successful. There is no limit on the number of sequences that can be added this way.

### 3.7. Restriction Site Subtraction and Short Sequence Subtraction

Just as you may wish to add arbitrary subsequence to a gene, you may want to remove them. The Subtraction modules use a variant of the least different RSCU algorithm (see Subheading 3.4) to disrupt target sequences with as few synonymous codon substitutions as possible. The Restriction Site Subtraction module is just a specialized case of Short Sequence Subtraction, in which all arbitrary subsequences are restriction enzyme recognition sites. Unlike the Addition modules, where you can select which subsequences to add, the Subtraction modules attempt to remove every instance of target subsequence.

### 3.8. Building Block Design

A synthetic sequence is of very little use until it is synthesized. GeneDesign provides several algorithms to break each nucleotide sequence it is given into evenly sized "building blocks" composed of sets of overlapping oligos, which may be easily ordered from a vendor.

The restriction enzyme overlap algorithm designs building blocks that overlap on unique restriction enzyme recognition sites. An overlap parameter may be defined to determine how much adjacent building block sequences overlap, including the restriction enzyme recognition site itself. This algorithm makes use of existing sites only and does not add or modify sites. If there are not enough evenly spaced, unique restriction enzyme sites, the algorithm will fault; this algorithm is not suitable for dividing sequence over 12,000 base pairs long or sequences that are poor in unique restriction enzyme recognition sites.

The constant length overlap algorithm will design building blocks that overlap by a user-defined overlap length parameter that is agnostic about restriction sites. This is useful for assembly methods that exploit such overlaps (11, 12). Input sequences must be at least 1,000 base pairs long.

The USER overlap algorithm will design building blocks that overlap on A(N)xT sequences, so as to be compatible with a uracil-specific exicision reaction (USER) assembly protocol (13). The width of overlap is user definable, and a melting temperature may be defined for USER oligos. Input sequences must be at least 1,000 base pairs long.

All three algorithms will use the same oligonucleotide generation algorithm to divide building blocks into oligos. You may indicate whether oligos are to be gapped or ungapped, what the target length and maximum length of each oligo is, and what the average melting temperature of each gapped oligo overlap should be. Melting temperature is calculated using the thermodynamic nearest neighbor parameters for DNA/DNA duplexes (14, 15).

All three algorithms are available on the command line; run `perl GeneDesign.git/bin/Design_Building_Blocks.pl --help` for usage details.

### 3.9. Enzyme Property Filtering

GeneDesign comes with a list of about 250 nonredundant (both by recognition sequence and incubating temperature) isoschizomers. It can be hard to know which enzymes are suitable for inclusion or exclusion in a synthetic sequence, so we created a "filter" that lets you create a list of enzymes that fit your criteria (Fig. 6). You can filter enzymes by the biochemistry of the ends they leave in cleaved DNA, by their recognition site length, by the presence of ambiguous bases in their recognition sites, by the presence or absence of

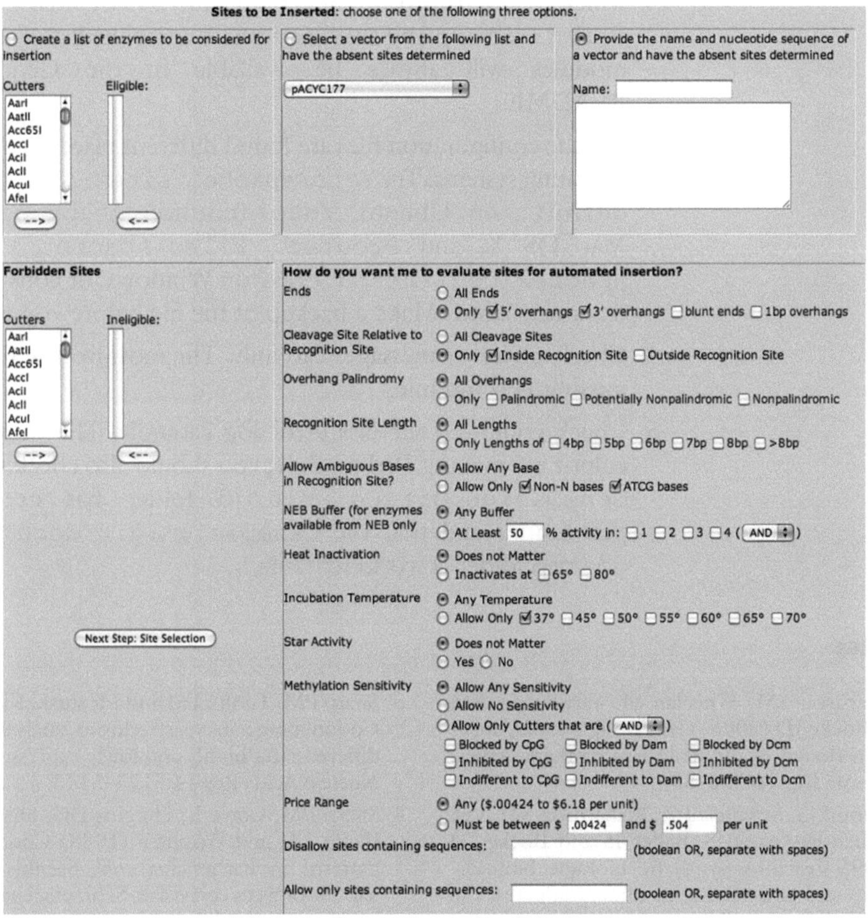


Fig. 6. Enzyme property filtering. Screenshot of the enzyme filtering window allowing the user to select enzymes with specific criteria.

specific subsequences in their recognition sites, and by whether or not the enzyme leaves palindromic or nonpalindromic overhangs, to name a few criteria. All enzymes are sorted in ascending order by an estimated cost per unit in US dollars so that the first results are the least expensive to order.

## 4. Notes

1. Apache usually runs as the user "www." We recommend placing the GeneDesign repository in a public folder of the user tree, as in `/Users/Shared/Public/` or `/Users/sarah/Public/` to ensure that the www user has access to it by default.
2. In the unlikely event that the helper script fails, you can use a package manager (ppm on Windows, cpan on Linux and MacOS X systems) to manually install the following bundles which are not usually installed by default: `List::MoreUtils`, `Perl6::Slurp` and `GD::Graph`. A complete list of required modules will always be available in the GeneDesign README.
3. Apache configuration files are found different places in different operating systems. Try `/etc/apache2/sites-available/default` on Ubuntu, `/etc/apache2/httpd.conf` on Mac OS X, and `Program Files\Apache Group\Apache2\conf\httpd.conf` on Windows, or consult your documentation. Make a backup of the file before you edit it.
4. Check any error messages carefully. The most likely culprit is a missing Perl module.
5. Check your web server's error log carefully. The most likely culprit is a missing Perl module; you should also check that the `GeneDesign.git/cgi-bin/gd` folder has executable permissions and that the `GeneDesign.git/documents/gd/tmp` folder is world writable.

### References


1. Richardson SM, Wheelan SJ, Yarrington RM and Boeke JD (2006) GeneDesign: rapid, automated design of multikilobase synthetic genes. Genome Res 16:550–556.
2. Dymond JS, Scheifele LZ, Richardson SM, Lee P, Chandrasegaran S, Bader JS and Boeke JD (2009) Teaching synthetic biology, bioinformatics and engineering to undergraduates: the interdisciplinary build-a-genome course. Genetics 181:13–21.
3. Sharp PM, Tuohy TM and Mosurski KR (1986) Codon usage in yeast: cluster analysis clearly differentiates highly and lowly expressed genes. Nucleic Acids Res 14:5125–5143.
4. Sharp PM, Cowe E, Higgins DG, Shields DC, Wolfe KH and Wright F (1988) Codon usage patterns in Escherichia coli, Bacillus subtilis, Saccharomyces cerevisiae, Schizosaccharomyces pombe, Drosophila melanogaster and Homo sapiens; a review of the considerable

within-species diversity. Nucleic Acids Res 16: 8207–8211.

5. Stenico M, Lloyd AT and Sharp PM (1994) Codon usage in Caenorhabditis elegans: delineation of translational selection and mutational biases. Nucleic Acids Res 22:2437–46.
6. http://www.perl.org/.
7. http://httpd.apache.org/.
8. http://git-scm.com/.
9. Chacon S (2009) Pro Git. Apress Publishers, ISBN 978–1430218333.
10. Roberts RJ, Vincze T, Posfai J and Macelis D (2010) Rebase- a database for DNA restriction and modification: enzymes, genes and genomes. Nucleic Acids Res 38:D234–6.
11. Gibson DG (2009) Synthesis of DNA fragments in yeast by one-step assembly of overlapping oligonucleotides. Nucleic Acids Res 37:6984–6990.
12. Gibson DG, Young L, Chuang R, Venter JC, Hutchison CA and Smith HO (2009) Enzymatic assembly of DNA molecules up to several hundred kilobases. Nat Methods 6:343–345.
13. Bitinaite J, Rubino M, Varma KH, Schildkraut I, Vaisvila R and Vaiskunaite R (2007) User friendly DNA engineering and cloning method by uracil excision. Nucleic Acids Res 35: 1992–2002.
14. Rychlik W, Spencer WJ and Rhoads RE (1990) Optimization of the annealing temperature for DNA amplification in vitro. Nucleic Acids Res 18:6409–6412.
15. Sugimoto N, Nakano S, Yoneyama M and Honda K. (1996) Improved thermodynamic parameters and helix initiation factor to predict stability of DNA duplexes. Nucleic Acids Res 24:4501–4505.

# Part IV

## Education and Security

# Chapter 19

## Leading a Successful iGEM Team

**Wayne Materi**

### Abstract

The International Genetically Engineered Machines (iGEM) competition allows undergraduate teams to develop projects in synthetic biology within the context of a large, international Jamboree. Organizing and managing a successful iGEM team is an exercise in advanced agile project development. While many of the principles applicable to such teams are derived from management of agile software teams, iGEM presents several unique challenges.

**Key words:** iGEM, Synthetic biology, iGEM Jamboree, International Genetically Engineered Machines, Teamwork

## 1. Introduction

### 1.1. What Is iGEM?

The International Genetically Engineered Machine competition (iGEM) is an annual undergraduate competition in synthetic biology, which culminates in a Jamboree in early November (see the 2010 iGEM site for more information—http://2010.igem.org/Main_Page). Since its inception in 2003, the Jamboree has been held at the Massachusetts Institute for Technology (MIT) in Cambridge, USA. The first competition was local to MIT but grew to involve 5 teams in 2004, 13 teams in 2005 (the first international year for the competition), 32 teams in 2006, 54 teams in 2007, 84 teams in 2008, over 100 teams in 2009, and more than 120 teams in 2010. Over time, projects have spanned a large range from the fun (e.g., bacteria that smell like bananas or wintergreen, buoyant bacteria) to the serious (e.g., a bacterial arsenic biosensor, a bacterial red blood cell substitute, bacteria flagellar display of *Helicobacter pylori* epitopes for vaccine development).

Many projects build upon previously developed Standard Biological Parts, known colloquially as BioBricks. Early in the iGEM cycle, teams receive a kit of biological parts from the Parts

Jean Peccoud (ed.), *Gene Synthesis: Methods and Protocols*, Methods in Molecular Biology, vol. 852, DOI 10.1007/978-1-61779-564-0_19, © Springer Science+Business Media, LLC 2012

Registry (http://partsregistry.org/Main_Page), which can be used, supplemented, or extended by their projects. Available parts include cloning and expression vectors, promoters, reporters, sensors, regulators, and genetic circuits, among others.

### *1.2. What Is Synthetic Biology?*

The goal of synthetic biology is the application of engineering principles to biological entities (1–3). The http://www.syntheticbiology.org web site has the following definitions: (1) the design and construction of new biological parts, devices, and systems and (2) the redesign of existing natural biological systems for useful purposes.

### *1.3. What Is Success?*

The goal of this document is to help your iGEM team to be successful. Our aim is to help teams achieve gold or silver medal status. If a team comes up with a great project, executes it well, and happens to catch the eyes of the judges, they might even make it into the finals and/or win one of the named awards. However, it is impossible to predict what may interest the judges in any particular year, and nothing can guarantee your team being selected as an iGEM finalist or winner. Nevertheless, it is our hope that the suggestions contained in this document might increase your team's chances.

In addition to doing well in the competition, other, perfectly valid measures of success might include the following:

- Getting publicity (locally or nationally).
- Continuing the research in a more highly funded way.
- Recruiting graduate students.

### *1.4. The iGEM Competition Cycle*

The iGEM cycle starts and ends with the Jamboree held in early November of each year. After reviewing the results of the Jamboree, you should start organizing for next year's team almost immediately. While it is possible to assemble teams just before the start of summer semesters (i.e., in April or May), this only gives enough time to pick a project, perform a bit of modeling, and maybe make a part or two. Successful iGEM teams actually accomplish much more than this, so starting early is a good idea. A useful first step is reviewing what the most recent winning teams have done and identifying the best characteristics to emulate.

## 2. Team, Advisors, Skills, and Project

### *2.1. Recruiting Team Members, Instructors, and Advisors*

By definition, iGEM teams are comprised mainly of undergraduates (this includes Masters students), although high-school students are also welcome. Ph.D. students, research associates, professors, and others are considered to be advisors or team instructors, and their role should be primarily instructional:

1. There are many ways to recruit iGEM team members. Some student groups self-organize after hearing about iGEM, while others are recruited through advertising.
2. If your university has courses on synthetic biology, this would be a natural group of students to recruit from.
3. The most successful iGEM teams contain members from a variety of disciplines, including (but not limited to) life sciences, biology, biochemistry, cell biology, microbiology, pharmacology, chemistry, chemical engineering, electrical engineering, computer engineering, computing science, web programming, mathematical biology, business, graphic arts, social science, and philosophy.
4. A broad advertising strategy is likely to be helpful in recruiting a well-rounded team.
5. An iGEM open house in mid-January, with a presentation by the previous year's team and instructor, followed by a mixer session provides a natural focus to a recruiting strategy.
6. Advertising should be directed toward having a good turnout at the open house as this provides an excellent place to describe synthetic biology, iGEM, the kind of people required for the team, and the general level of commitment required.
7. The optimal team size is probably between 8 and 12 members in total, but successful teams have had from 6 to over 30 members.
8. In the formative years of iGEM at your university, you may receive only a few applications and choose to invite all of them onto the team.
9. As the popularity of the iGEM competition increases, you will face the decision to either make team membership competitive or to split applicants into two or more teams. Many applicants may see this as an attractive research addition to their CVs.
10. Each additional team will require its own resources, so the number of teams one area can support may be limited. In addition to specific skill sets, successful iGEM team members are also required to exhibit considerable initiative, ingenuity, and innovation, so selecting for these personality traits may as important as good grades and experience.
11. Teams also need to have the right group dynamic or chemistry, so it might be wise to use the recruitment process to have a brainstorming session and watch how different groups of people work together.

12. You may want to use a brief questionnaire, like the following, to help in the selection process:

    *iGEM Team Application*

    Name (Last, First):

    E-mail address:

    Phone:

    Year/program:

    GPA from two most recent terms:

    Relevant course experience:

    Career/education aspirations:

    Why you want to participate in iGEM:

    Why you think you would make a good iGEM team member:

    What problems you would like to solve with this technology:

13. A critical element for successful iGEM teams is a high level of technical and scientific support. Teams require such support to assist with complex molecular biology and instrumentation. In addition, considerable training is required in the field of synthetic biology and in the many support activities an iGEM team engages in.
14. These support activities may include wiki construction (and HTML), giving presentations, making posters, proper scientific documentation, communication among team members, organizing teams, fund-raising, etc.
15. Thus, the iGEM organizers require a minimum of two team instructors to provide support and training activities. In addition, other scientific and technical advisors or instructors will greatly enrich the iGEM experience for everyone.
16. It is important to recruit additional instructors and advisors as appropriate for each team. Instructors are required to commit a fairly large portion of their time to the team in order to maximize training effectiveness, and many of them will want assurance that their considerable investment will be worthwhile. Instructors benefit greatly from their iGEM participation as it may help them to identify potential graduate students that demonstrate exceptional ambition and initiative. In addition, iGEM projects frequently can be extended and expanded into excellent projects for graduate students or postdocs as they raise many interesting scientific and engineering questions. Particularly strong projects could even bring in new grant funding or form the basis of commercial ventures.
17. Besides the active members for the current year, successful iGEM teams are always thinking about the future. Building upon the past experiences of iGEM team members can substantially increase the team's odds of success. You might

consider recruiting two or three second-year students to the team with a specific mandate to just learn about synthetic biology and iGEM and to help form the core for the team next year. Ideally, each team should have some senior members from the previous year, who will propose the majority of realistic ideas and perform much of the actual work on the current team, and some junior members whose primary responsibility is to learn.

#### 2.2. Full-Time or Part-Time, Volunteer, Course, or Paid

1. One important issue in formulating the iGEM team is whether it will include both full-time and part-time members. Full-time students working over the summer and/or fall semesters will be able to accomplish more in the lab than volunteer part-timers, so most teams include full-time people for at least part of their duration. The risk of mixed teams (both full- and part-time) is that it is possible to create two classes of students on the team, leading to clique formation and resentment. On the other hand, having only paid, full-time team members is beyond the reach of most institutions.
2. A balanced approach is most likely to be successful. However, all students should commit to an equal amount of volunteering time for the team, and if they are lucky enough to be able to work full-time on the project over the summer, that volunteering commitment should not be changed. In addition to time over the summer, students will need to spend time in the winter/spring learning about synthetic biology and iGEM and planning their project. Also, the fall semester leading up to the Jamboree requires a substantial time commitment so that presentations and posters can be prepared and perfected. Students' first priority should generally be to their course work, so incorporating a synthetic biology course, directed studies, or project course into the iGEM cycle can be an excellent way to encourage their participation and to reward them for their efforts.

#### 2.3. Team Agreement

The commitment from team members is sizable, though the potential rewards are substantial. Students participate in team-oriented, multidisciplinary research and have an opportunity to exhibit and develop scientific entrepreneurial skills. They participate in an international meeting with substantial public exposure and may even publish their results. Team members will also have expectations about what they will get from the experience, including gaining hands-on experience in molecular biology, bioinformatics, mathematical modeling, presentations, and public speaking, as well as learning about synthetic biology, in general. In order to avoid disappointment and possible recriminations, it is important to develop an agreement among team members, advisors, and instructors that outlines the commitment of each member to the others. We provide the following template:

*iGEM Team (Team or Project Name) Agreement*

Because participation in an iGEM team project is a privilege, an honor, a challenge, and a joyous celebration of science and engineering, we agree to the following guidelines:

Students:

- To commit our intellect and energies to the fulfillment of the team goals.
- To learn the principles of synthetic biology and the science behind our project.
- To conduct ourselves and our research to the highest scientific and ethical standards.
- To represent the ideals of synthetic biology and iGEM in a fair, balanced, and open manner to the general public.
- To work a minimum of xxx weekly volunteer hours during the project planning phase.
- To attend a full weekend Basic Molecular Biology course held <dates>.
- To work a minimum of xxx weekly volunteer hours during the project execution phase.
- To work a minimum of xxx weekly volunteer hours during the Jamboree preparation phase.
- To attend all group meetings and learning sessions or to notify coordinators if impossible.
- To travel to the Jamboree (and other local events) and participate joyfully.
- To work hard, learn lots, and have fun.

Instructors/Advisors:

- To commit our intellect and energies to the fulfillment of the team goals.
- To learn and teach the principles of synthetic biology and the science behind our project.
- To train students in all the skills and techniques required in the project or to find suitable instructors, where required.
- To conduct ourselves and our research to the highest scientific and ethical standards.
- To represent the ideals of synthetic biology and iGEM in a fair, balanced, and open manner to the general public.
- To work a minimum of xxx weekly volunteer hours throughout the project.
- To attend all group meetings and learning sessions or to notify coordinators if impossible.

- To travel to the Jamboree (and other local events) and participate joyfully.
- To work hard, teach/learn lots, and have fun.

Signed: Date:

### 2.4. Building Synthetic Biology and iGEM Background

Many consider a synthetic biology course as the best way to teach undergraduates about synthetic biology principles. For institutions lacking such a course, we recommend that the first few team meetings be used to teach some relevant basics. A large number of review articles are available that discuss principles of synthetic biology principles including those listed in the References section (1–6):

1. In addition, past proceedings of some synthetic biology conferences are available online. For example, webcasts from the international synthetic biology conferences SB1.0 to SB4.0 are available by following the conferences' links at http://www.syntheticbiology.org.
2. The best way to learn about iGEM is to participate, but the second best way is to review previous competitions. Fortunately, presentations, posters, and wikis are available online through the most recent iGEM.org web site (e.g., 2009.iGEM.org; 2010.iGEM.org). We recommend spending a few planning sessions in January or February to review past projects.

### 2.5. Getting Project Ideas

1. Generating iGEM project ideas is not necessarily all that difficult once some familiarization with synthetic biology has been achieved. Using a directed reading or synthetic biology course as a source of good project ideas is also likely to result in better conceived, scientifically more sound ideas, with more application potential.
2. Following some basic instruction on synthetic biology and past iGEM projects, your team should hold focused brainstorming sessions to generate some basic ideas.
3. Individual team members or groups of two or three should then elect to champion some of the ideas. This process should entail conducting deeper literature searches and developing proposals one- or two-page long.
4. Project champions can then make brief presentations to pitch their ideas to the rest of the team in subsequent meetings. Several meetings may be required before the team selects its favorite or best project.
5. A team should only have one project, although it may contain multiple subprojects. If a team cannot settle on just one project, consider splitting into two separate teams.
6. The most successful iGEM projects contain elements of mathematical modeling or simulation, molecular biology, assays of

results (perhaps, with instrumentation development), and thoughtful examination of EEELS (ethical, environmental, economic, legal, and social) issues.

### 2.6. Team Building

1. It is important that the iGEM team actually work like a team. Getting individuals to commit by signing the above agreement is only the first step.
2. Conducting brainstorming sessions during the winter/spring meetings is also a key element in building team spirit.
3. Consider holding some of these sessions in a more social environment (but one which permits some work to be done) to help build interactions and trust between the team members.
4. Team members also need to work in a rich communication environment, which can be difficult for young scientists and engineers.
5. Establish standards for documentation and encourage team members to share their results, problems, and thoughts among each other and with their advisors.
6. Operating a journal club where team members read and discuss a single paper can help facilitate this, depending on the time available.
7. Also, project milestones and deadlines can help heighten the sense of urgency and adventure, which will often help teams to coalesce.

### 2.7. Building Support Networks

1. In addition to the team instructors, successful iGEM teams reach out to the academic, commercial, or general community when they need to recruit additional expertise.
2. The more extensive and effective the network of specialists and consultants, the less likely the team is to become bogged down in problems, and the more they will feel part of something important.
3. Network building should be encouraged by having team members identify professors or companies that may have valuable relevant information, then contacting those people to ask for help or just to invite them for a chat.

### 2.8. Lab Space

1. Lab space may be contributed by team instructors or a sponsoring department/company.
2. In order to provide a workable environment, lab space should be available from the start of May to the start of September at minimum.
3. If at all possible, try to find a permanent lab and meeting space for the team, as this will permit year-round use. This will be especially important in the fall as the Jamboree approaches and lab work likely needs to be finished in a hurry.

#### 2.9. Funding

1. There are a large number of funding sources available to assist with the iGEM team and many ways to approach the funding question.
2. One fundamental question each team will need to answer is how much fund-raising to attempt and what sources to focus on.
3. Some projects are very suitable to approaching specific industries, such as biotech, energy, or pharmaceutical companies. Other teams will have easier access to more traditional forms of academic funding.
4. Some element of fund-raising should exist in all iGEM projects, as the entrepreneurial experience is an important element of iGEM. Cynical team members will be surprised at how receptive potential funding sources are to contributing to their projects.

#### 2.10. Publicity

Publicity can be a key element for teams as a recognition of their effort, promotion of their school or institution, attracting new team members, rewarding sponsors and sourcing new ones, developing new course programs, etc.

## 3. Planning

#### 3.1. Planning the Project

1. After acquiring some familiarity with synthetic biology, in general, and the iGEM competition, specifically, and after selecting a project, a detailed planning process should begin.
2. A successful iGEM project has an incredible number of elements, parts, and circuits to be designed and made, models to be written and tested, data to be collected and analyzed, presentations, posters, T-shirts, wiki pages, fund-raising, travel, etc. This would be overwhelming for any one person, so it is important to delegate (see below) and coordinate.
3. A dedicated project manager elected from the team might help this process and subsequent execution of the plan considerably.
4. Start with a broad plan. The general project should already be defined and the team should have a good idea of all the many things that need to be done to accomplish their goals. The plan can be fleshed out in more detail either by the whole team or by small working groups dedicated to particularly parts of the entire project.
5. Some parts of the project, such as the poster or presentation, will need to be planned at a later stage, once progress has been made and (perhaps) data collected.

6. Planning along with progress reporting should be a continuous process driven by the project manager.
7. The scientific portion of the project is most likely to constitute the greatest challenge for the team and will likely require considerable input from advisors and instructors. Effort expended at this point in the project will not only greatly enhance the chances of a successful conclusion but also reduce the amount of work required throughout the summer and fall.

### 3.2. iGEM Requirements

1. The iGEM organizing committee changes the requirements every year, so it is somewhat of a moving target. However, certain constants remain.
2. The basic requirements for a minimally successful project usually involve completing a team wiki, presenting a poster and talk at the Jamboree, and submitting a BioBrick part.
3. Higher levels of achievement require making and characterizing an existing or novel working part and contributing to the synthetic biology or larger community.
4. It is important to review the judging criteria each year on the iGEM homesite and to plan team activities to meet those criteria.
5. In past years, iGEM has awarded bronze, silver, and gold medals to teams based on their published judging criteria.
6. In addition to these, a number of named awards are presented (including the grand prize aluminum BioBrick) based on various criteria, such as best poster, best presentation, best new BioBrick, best model, etc.
7. These also change from year to year, so it is important to check the judging criteria for the current requirements. Generally, only silver and gold medalists are considered for named awards, though this is not a set rule.

### 3.3. Synthetic Biology Project

Planning the science and engineering that will comprise the project requires several members of the team to understand the project principles and the tools that will be used to execute the project. These may include cells, DNA, BioBrick parts, plasmids, enzymes, molecular biology, genetics, biochemical assays, microscopy, software, programming languages, etc. Where possible, instructors should include specific small courses or reference material that will help team members acquire the knowledge and skills they require. Clearly, though, the individual drive and initiative of team members will greatly determine their success in acquiring the necessary knowledge. To a large extent, this explains the importance of these characteristics even over background and knowledge in determining the success of the team.

### 3.4. Modeling

A number of modeling tools are available and may cover a large range from basic biochemistry texts to Mathematica or other simulation software. Having modeling expertise available is critical to the success of this portion of the project. Some good introductory articles are given in the References section (7–9).

### 3.5. Instrumentation

Characterizing a BioBrick part is an important component of any iGEM project. Frequently used assays for gene expression include lacZ, fluorescence-activated cell sorting (FACS) of GFP-expressing cells, Northern blots, Western blots, and other biochemical assays. A variety of instrumentation may be available to your team, but specialized instruction is often required to operate an instrument safely and reliably. Seek help from team instructors and advisors. Some iGEM projects may need to develop their own instrumentation. For example, digital cell tracking systems consisting of cameras and software have been developed by past teams.

### 3.6. Open Source

iGEM projects, BioBrick parts, and wiki documentation are all considered open source that belongs to the community at large. There is a debate as to whether wikis should be used as ongoing documentation tools or uploaded on the due date. We believe that secrecy, even for the sake of protecting your project from possible competitors, has little place within iGEM and should be actively discouraged. We, therefore, encourage teams to utilize their wikis as active, public documentation of their efforts.

### 3.7. Documentation

Good documentation serves multiple important purposes: it provides support for any intellectual property claims, it provides factual support when writing papers, it tells both you and other team members how to repeat an experiment, and it helps to organize your thinking and planning. Most importantly, documentation is the *public* property of the entire team, and it must be written for the entire team. Standards of documentation will reduce the amount of work that is required for one team member to understand the work and the thinking of another member. All work should be documented so that any team member can understand it easily:

1. The basic goal of good documentation is to communicate as efficiently as possible. Write everything that needs to be written; but nothing more. In particular, do not repeat what has been previously written, when a simple reference to a book and page number will do (plus a few notes remarking on what has been changed in the current experiment).
2. Use the appropriate lab book or wiki. While most documentation is kept in diary form, having the ability to organize it by project, subproject, or person is very useful for finding data quickly.
3. Document your thinking (rationale or purpose) along with the experimental protocol or procedure *before* you begin the

experiment. A hard copy lab book should always accompany you in the lab, except when performing the most mundane and repetitive of tasks.

4. All constructs should first be "assembled" electronically in silico. This greatly reduces the number of errors, as computers can easily check conflicting restriction sites and reading frames for fusion proteins. The BioBrick web site, among other available tools, provides construction capabilities.
5. As you perform an experiment (or make a construct), any changes to the expected protocol can be entered along with the results.
6. Standard or obvious steps need not be entered. However, standard protocols should be referenced, as should protocols that have been adapted from published work. These references greatly simplify the task of writing articles based on your work. There are far fewer "obvious" steps than one might think. Most of these are (or should be) in some standard protocol. Almost everything else should be written down.
7. Many steps in protocols involve the mixing of a number of reagents in a standard reaction, such as restriction enzyme digestion, PCR, ligation, etc. While it is not necessary (or even desirable) to write a detailed description of how each reagent was added to the tube (this is either "obvious" or left to personal preference), it is *critical* to always list all the reagents and amounts used in this particular experiment. As reagents are added, they should be checked off in the lab book so as not to lose one's place. This practice also helps focus the experimenter on their work. Other reaction conditions, including times and temperatures, should also be recorded. Running conditions, such as percentage of agarose used in a gel, voltage, and time should also be recorded.
8. Supporting documentation produced by lab equipment should usually be included in the documentation. Electronic images or scanned images may be uploaded to the wiki. Documents that are not uploaded due to space limitations should be filed carefully and cross-referenced in the wiki. Chromatograms may be held in binders, for example. Attached documentation should be annotated so that it is clearly related to the information in the wiki. For example, gel lanes and any bands that have been cut out should be marked. Electronic copies should be stored, and the file name and computer and folder (or directory) should be marked in the wiki.
9. Lab notebooks often reference material that is online in some computer file. In this case, the lab book and computer documentation should be cross-referenced and must be in agreement. At the very least, the same name must be used to describe the same construct in both sources.

10. Every experiment should end with some conclusion. Either something was made, verified, proven, disproven or inconclusive, or needs further work. In the latter case, problems and subsequent or alternate approaches should be discussed. At the very least, when no solid conclusion is possible, a link to the next page, on which the experiment is continued, should be included.
11. Most experiments (and *all* projects) take place over a number of days and may be interrupted by other work. The inclusion of "continued to" and "continued from" fields on each wiki page should assist in providing continuity, as should a table of contents.
12. Complete documentation standards should be developed on a continuous basis by the team.

Good labeling and storage of plasmids, glycerol stock, plates, intermediate constructs, etc., is *absolutely crucial* in enabling team members to find and identify the reagents they need. In addition to being correct and complete in their descriptions, you also need to ensure that using the labeling and filing systems does not become the major work activity; systems must be effective but efficient.

### 3.8. Presentation and Poster

Planning these is fairly straightforward. Review the efforts of successful past teams and try to emulate them. The standard for both presentations and posters at the iGEM Jamboree is very high. A thorough understanding of the subject material is only the starting point. iGEM teams are frequently more creative and have more fun with their presentations than what would normally be seen at most scientific conferences, so it is important to take this into account.

### 3.9. T-shirts and Memorabilia

Keeping with the theme of having fun at the Jamboree, team T-shirts and other memorabilia should be designed to be uniquely eye-catching and memorable. T-shirts are almost a required part of the Jamboree as it makes finding each other in the hubbub that much easier. Also team colors allow members to easily find their team in the traditional "picture from above." T-shirts have ranged from fairly standard forms, to soccer jerseys, to kimonos. Even other forms of clothing have made their appearance in some competitions, including hard hats. Other memorabilia include baseball caps, drink coasters, pens, pocket protectors, wristbands, and almost anything inexpensive enough to give away and small enough to transport to the Jamboree. Although not required, the memorabilia make a very nice secondary competition.

### 3.10. Raising Funds

The team needs to set funding goals and decide who will be approached for support. A short portfolio, describing the project, iGEM, synthetic biology, and the team, should be compiled by the funding focus group.

### 3.11. Publicity

At any early stage, the team should appoint a focus group to deal with publicity. Although it may seem premature to engage in public relations (PR) activities before anything has been accomplished, successful teams raise awareness (responsibly) at any early stage. Without overinflating expectations, a team should approach its institutional and student newspapers and tell their story. If teams are seeking more members, advisors, or instructors, or if they will be conducting public surveys, this is a good vehicle for raising awareness. Stress the general problem being addressed, the basics and purpose of synthetic biology and iGEM at this stage. As the team desires, local news agencies (and, especially, science news agencies) may also be contacted.

### 3.12. Delegating and Coordinating Work

Even in the planning stage, the energy level and coordination required to put together a successful iGEM team far surpasses almost any other undergraduate experience. Obviously one person cannot do it all, so work needs to be taken up by team members, either as individuals or small groups. This section has briefly described some of the activities that will need to be considered and planned. While overall planning should be done by the team as a whole, specific areas, such as modeling, lab work, fund-raising, presentations, posters, etc., are best performed by smaller working subgroups.

### 3.13. Team Meetings

1. Team meetings should be held on a regular basis; we recommend weekly meetings. During the planning phase, the team meetings will help everyone to share in the basic project ideas and to flesh out some details, as well as outline the other work required for the team. Team meetings will likely require 1–2 h, especially if instructional time is required. Team members who are enrolled in specific classes (e.g., synthetic biology or computational modeling) may be excused from specific instructional modules, but otherwise, everyone should attend. Instructors and advisors should attend the business portion of the meeting and may opt to attend the instructional sessions as well.
2. Instructional sessions may be held either first or last in a meeting. We would recommend carrying out instruction first then moving on to a brisker-paced business meeting afterward. We realize that the weekly meeting load, including the focus group meetings (below) may take 3–4 h/week at this stage, which is a fairly heavy workload on students. Our advice is that this will not only greatly enrich the iGEM experience for all team members but will also reduce the meeting time required over the summer. Obviously, if everyone on the iGEM team can enroll in a course (e.g., directed research option) then more time can be spent that is directly relevant to student members. The actual number of hours and, perhaps, a plan for the meetings at this stage can be part of the team agreement.

3. Meetings can be held at any mutually agreed-upon time and place. To a certain extent, one of the criteria for participating on an iGEM team should be the availability to attend weekly meetings. Because most iGEM teams contain 6–12 members, it can be difficult to arrange a convenient meeting time. We recommend a weeknight during regular semesters and the summer, with special weekend meetings during the fall to prepare for the Jamboree. A commitment to attend meetings is crucial to team spirit and to its eventual success. Nothing is more discouraging than team members who cannot bother to show up for a weekly meeting.
4. A wide variety of instructional sessions could be held during the planning period, constituting a minicourse in synthetic biology and iGEM. Instruction would preferably be for the entire team, with expanded discussion for focus groups. Some selected topics are suggested below:
   (a) Introduction to synthetic biology.
   (b) Introduction to iGEM.
   (c) Review of past iGEM competitions.
   (d) Literature searches and reading primary literature: PubMed, Google Scholar, patent literature.
   (e) Maintaining a literature database.
   (f) EEELS issues and studies.
   (g) Genetic circuits.
   (h) Protein engineering.
   (i) Metabolic engineering.
   (j) Molecular biology basics (digests, gels, ligation, transformation, sequencing).
   (k) Mathematical modeling.
   (l) Bioinformatics and support tools (Entrez, BLAST, Vector NTI, primer design).
   (m) BioBricks and the BioBrick Foundation.
   (n) Biochemical assays and analysis.
   (o) Instrumentation (FACS, microarrays, microscopy, etc.).
   (p) Advanced molecular biology (PCR, Northerns, Westerns, microarrays, etc.).
   (q) Fund-raising.
   (r) HTML and wikis.
   (s) Powerpoint and Photoshop.
   (t) Basics of presentations and posters.
   (u) Keeping a lab notebook.
   (v) Documentation of parts.

(w) Navigating the Registry of Biological Parts.
(x) Teamwork and leadership strategies.
(y) How to run a meeting.
(z) Lab safety.

### 3.14. Focus Group Meetings

Focus groups are subsets of the entire team with specific interests and/or skills that can meet separately to address specific subtasks. Smaller groups make for tighter working relations and more effective exchange of ideas. We recommend making focus groups to handle planning, execution, and management of most portions of the iGEM project. Focus groups can either meet following the general weekly meetings or at some other time convenient for the group.

### 3.15. Team Social Events

Even at the early planning stage, iGEM teams are already working hard to be successful. It is important that the team reward itself with some time for socialization. Pizza or snacks in the first half of the weekly meetings before getting down to work is recommended. After the weekly meeting, the team may want to get together for beer, coffee, tea, etc. Obviously, having the team members get along socially is almost as important as getting along intellectually if they are to work as an effective, dynamic team.

## 4. Extreme Execution

Planning is nice, but eventually, something real has to be produced. The most effective teams realize that communication at this stage is paramount. Appointing a full-time team manager to maintain a schedule of the many tasks to be done will go a long way toward maintaining everyone's sanity. Taking a pause once a month (or even more often) to ask everyone how the structure of the team and the division of tasks is working may help to identify and deal with problem areas. It is also a good idea to check how well team members' expectations of each other, of the project, and of the instructors are being met. This self-reflective exercise can highlight potential problems at an early stage. If the question is met with silence (rather than with overwhelming cheers of how great everything is), then the team is in real trouble and needs a serious review of its goals. Silence usually indicates that things are not going particularly well and that the team members do not trust each other enough to admit it.

### 4.1. In the Lab

A major goal of the iGEM project is to produce functional, well-characterized BioBrick parts. This will require wet lab work. Because iGEM members come from a variety of backgrounds and levels, experience with molecular biology and other required techniques will vary considerably. It is important to pair more-experienced members (or advisors) with less-experienced members so that transfer of skills and knowledge can take place. We find that it is usually best to hold a basic molecular biology course for all team members that will be conducting wet work. Because of the time required to conduct many basic molecular biology experiments (e.g., digests, gels, PCR, sequencing), it may be most convenient to run a basic molecular biology weekend course, while realizing that this might not be adequate for people to work successfully in the lab and they will still need assistance during their first few experiments.

Molecular biology remains a labor and time-intensive activity, though many procedures have seen considerable improvements in efficiency over the past decade. Full-time students will be capable of producing more results than part-time volunteers, so they confer an obvious advantage to any team. However, considerable work can be accomplished by a committed volunteer team with good ideas and good support. The team with the best of both worlds has a core of full-time summer students supporting a larger group of volunteers. This may require full-timers to work some evenings and weekends during the summer months. Such a commitment should be spelled out in the team agreement, remembering that being a full-timer does not remove the obligation to carry out volunteer activities as well.

### 4.2. Maintaining Focus and Energy

Eventually, all lab work falls into a rut because either the work itself becomes easy but repetitive as experiments have to be repeated multiple times with different samples, nothing is working and the researcher is frustrated, or large stretches of mundane work are required to verify reliability. iGEM projects are usually so short and intense that there is little danger of this happening until well into the summer. One important thing to remember is that scientists and engineers are just people. We get discouraged by failure, we fall into patterned modes of thinking, and we enjoy staying within our comfort zone. Lab work can be reinvigorated by shuffling tasks among team members and by cross-training. Although this may reduce overall efficiency, it makes for happier team members.

### 4.3. Surpassing Failure

Psychological security is important during execution as well as in the planning stages. Science is hard, and many things will not work out the first time, or second (or third, etc.). Plans and schedules are not intended simply to make it easier to blame the responsible person when things go wrong. The early recognition of mistakes and failures should be encouraged and congratulated as this will enable the team to get back on track most quickly. Admitting error is much less costly than trying to hide it.

#### 4.4. Nearing Completion

After the summer, when students return to classes, the project will experience a lull of 2 or 3 weeks. It is important to continue to hold weekly meetings to help get past this point and to settle the team in for the finishing kick. This lull may be even a bit longer if full-time students decide to take some summer vacation before heading back to classes. Instructors need to remember that a student's first priority is to their educational program and to their intended career path, so some time for reinvigoration is very important.

#### 4.5. Course or Volunteer

At the end of the summer, it is easy to think the project is done, but really, it is only getting started. Usually, there will be lab work to complete, documentation to finish, and the Jamboree to prepare for. The most effective way to maintain student interest in the project is to make it worth their while. This will usually mean some course credit for their iGEM work, either through a specific iGEM course or through a directed studies or project course. Encourage your students to register in such a course and encourage their home departments to recognize this effort.

#### 4.6. Document to Win

1. Apart from the poster and presentation, iGEM projects must be fully documented on the wiki and BioBricks parts submitted. Wiki pages from past competitions are available on the web and reflect an amazing amount of talent and creativity. Because formatting and imaging as well as more interactive features are limited with the standard wiki formatting, teams may want to enhance their wikis with advanced HTML scripts.
2. Winning wikis contain well-formatted pages with many interesting images. The main page should briefly describe the project, the institution, the home city, sponsors, etc. Other pages can include more detailed descriptions and photos of the team and its members. It is good to include actual photos of the team at work and at play here; the team manager may also want to take on the role of documenting (and blogging about?) team activities. The project details page documents the ideas, relevant references, and explains basic concepts of the project. A modeling page can include formulae and modeling results along with source code for simulation. A parts page could include an overview of BioBrick parts for the project, including their design, construction, and characterization. Full parts descriptions should be documented in BioBricks though judges may only look at the "favorite" parts. Colorful, well-designed images which clearly convey the important information are the goal for these pages.
3. Daily wet lab progress is to be documented in the notebook pages. This may contain detailed information and act as an electronic lab notebook, or it may be diary-like summaries of

lab work. While we prefer a more detailed approach, it is not clear if this is important to the judges. The standard wiki comes with a calendar-like notebook, which is minimally useful. Many groups replace it with summary pages, but others enhance it to provide better browsing capabilities (e.g., day-by-day flipping, project or researcher cross-references). Some groups include scans or gels and other such machine-generated raw data, while others do not. Above all, the wiki must be clear, attractively formatted, easy to navigate, and complete.

### 4.7. Organizing the Presentations

Presentation teams are generally fairly small, usually three to five team members when all members speak English. Presenting your research is an important part of the scientific and engineering enterprise, so being on the presentation team is a valuable enhancement to an individual's experience. However, it will require considerable extra work. Not only are presenters required to know the project (or at least their part) thoroughly, they must be able to communicate it clearly and speak with authority. This is one of the most visible aspects of the entire competition, and it is easy to be judged harshly. A typical presenter will require 12 h of team practice and at least as much individual practice in order to competently present their portion. As presentations are only 20 min, there is no time to hesitate or stumble. Speaking quickly but clearly and correctly with confidence and authority requires a lot of rehearsal; longer talks are actually easier, in general.

The presentation should include the following:

- Present the team and institution (maybe city and country).
- Outline the reason the project is important.
- Discuss the basic background science.
- Describe the approach to solving the problem.
- Present the model and its predictions.
- Describe BioBrick parts made, sequenced, submitted, and characterized.
- Describe assays and results.
- Draw conclusions of what worked and what did not.
- Talk about future plans.
- Thank advisors and instructors.
- Thank financial supporters.
- All in about 20–30 slides (at 1 min or less per slide).

### 4.8. Poster

Posters are also an important part of the competition, and previous posters should be examined for ideas on producing successful posters.

### 4.9. Team Meetings and Focus Group Meetings

As students return to classes, it can be very difficult to maintain regular meeting schedules. Most teams begin planning and preparing their presentations and posters during this time period. This is unfortunate as many students do not have enough free time to easily contribute to the project during the fall. As mentioned previously, having students register in a class to allow them to get credit for their iGEM work will ensure they have some available time to continue working on completing the project.

## 5. The iGEM Jamboree (and After)

### 5.1. Publicity

Before leaving for the Jamboree, inform your various news agencies that your team has been working hard and is ready to compete. Not only will this raise local interest and team spirits, judges will be impressed by efforts of teams to promote synthetic biology and iGEM.

### 5.2. Organizing: Working the Program

As with any conference, download and read the program and any abstracts before you get to the Jamboree. As a team, you should plan to look in on other teams' presentations and posters to use the Jamboree as a learning experience. Preparation for attending the Jamboree begins a few weeks prior, when wikis are frozen. At this point, the team should meet for a few hours to review what other teams have accomplished and to determine whether any last minute changes to the presentation or poster need to be incorporated.

### 5.3. Practice Talk

Upon arriving at the Jamboree, practice times are arranged for the night preceding the next day's competition. This will give team members a chance to practice in a room similar to the one they will actually use and, most importantly, at a place outside their comfort zone. Try to keep the tone of the presentation relaxed but serious. Minimize anxiety by being polite to other teams, who may be running a little late. Be kind and considerate. Feel free to sit in on other teams' practice talks but be appreciative and respectful that they are likely as anxious as you are. Your team should try to convey that they are ambassadors of goodwill and interested in others' success as much as their own. Be supportive.

### 5.4. Attending Other Talks

Team members should try to take in a fairly wide variety of other talks, both within and outside their stream, to enhance their iGEM learning experience. Certainly attend as many of the "big school" talks as possible, but try to take in a few of the lesser-known schools' efforts as well. Try to think of questions to ask the presenters. Culture your curiosity about other teams' work and frame your questions from curiosity rather than from a challenging perspective. Do not try to make others look bad.

### 5.5. Viewing Other Posters

Everybody likes to have others express interest in what they are doing. So the team should make an organized effort to visit a large number of posters and talk to team members there about their project and their iGEM experience. In addition to enjoying the Jamboree more, this will enhance the learning experience.

### 5.6. Team Presentation

Arrive at the presentation room on time or early and be prepared to set up for your presentation quickly. Each team needs to bring its own computer and remote presentation device or laser pointer, but the Jamboree provides the projector and sound, along with technical support. The key technical consideration in public presentations is to minimize surprises. Run the presentation on known hardware with known software, whenever possible. Bring more than one copy of the presentation (and possibly more than one presentation computer) to the talk. Be prepared for any equipment failure.

### 5.7. Finals

If you are fortunate enough to make it into the finals presentations, the first thing to remember is DON'T PANIC! You will be presenting your talk in front of a very large group of 600–1,000 people, and you will be judged by everyone present, so some nervousness is to be expected. Stick to your training and remember that you rehearsed for this, so just give your presentation as always and you will be fine. Beyond that, the selection of the final winners is a mystery, so do not worry about it.

### 5.8. Debriefing Initial Impressions

Immediately after the Jamboree, begin to collect impressions from your team about how it went. Either on the plane ride home, in the airport, or over breakfast the next day, try to collect some initial ideas of what worked and what did not and what can be improved next time. Collecting these ideas while they are fresh is crucial to annual improvement. Do not just talk and listen though, write them down and include them in your final report.

## 6. Follow-Up

### 6.1. Post-Jamboree

So the Jamboree is over, and you have returned home. You may think your work is done for a few months, but successful iGEM teams get to work almost immediately on the next year. As soon as possible, write a final report on your achievements for your sponsors, thanking them and discussing next year's plans. Immediately contact your news agencies to report on your success and strike up interest in the next team. Plan a celebratory gathering and the kickoff open house to help recruit the next team.

### 6.2. Detailed Debriefing

Team members should review all finalists' (and many gold medal winners') projects from the iGEM results page in detail. Ideally, a number of team members will conduct this review, though the team instructors may have to do the majority of the work as final exam time will be fast approaching. Review all team wikis, their presentations, and posters, and try to determine the winning criteria. This will be very difficult but informative. If the team has proposed hypotheses as to why some teams did very well, try to collect objective statistics to evaluate these ideas. Although arduous and time-consuming, this will help the next year's team considerably.

### 6.3. Celebration and Reinitiation

After all the work is completed, the Jamboree is over, and final reports have been written, it is important for the team to take a moment to celebrate their achievement and reflect on the entire experience. One excellent way to do this is to hold an open house, inviting advisors and supporters of the current team. A small presentation of the project can be made, and serving food always helps people to mingle. The open house should be coordinated with advertising and publicity in advance so that recruiting for the next iGEM team is part of this event. And so, the iGEM cycle begins again, setting your team up for another successful year.

## References

1. Endy D (2005) Foundations for engineering biology. Nature 438:449–453. 11.
2. Andrianantoandro E, Basu S, Karig DK and Weiss R (2006) Synthetic biology: new engineering rules for an emerging discipline. Mol Syst Biol 2:2006 0028.
3. Heinemann M and Panke S (2006) Synthetic biology – putting engineering into biology. Bioinformatics 22:2790–2799.
4. Alon U (2007) *Introduction to Systems Biology: Design Principles of Biological Circuits.* Boca Raton: Chapman & Hall/CRC Press.
5. Drubin D A, Way JC and Silver PA (2007) Designing biological systems. Genes Dev 21:242–54.
6. Marguet P, Balagadde F, Tan C and You L (2007) Biology by design: reduction and synthesis of cellular components and behaviour. J R Soc Interface 4:607–23.
7. Wilkinson DJ (2006) *Stochastic Modelling for Systems Biology.* London: Chapman & Hall/CRC Press.
8. Bolouri H (2008) *Computational Modeling of Gene Regulatory Networks – A Primer.* London: Imperial College Press.
9. Demin O and Goryanin I (2009) *Kinetic Modelling in Systems Biology.* Boca Raton: Chapman & Hall/CRC Press.

# Chapter 20

## The Build-a-Genome Course

**Eric M. Cooper, Helöise Müller, Srinivasan Chandrasegaran, Joel S. Bader, and Jef D. Boeke**

### Abstract

Build-a-Genome is an intensive laboratory course at Johns Hopkins University that introduces undergraduates to the burgeoning field of synthetic biology. In addition to lectures that provide a comprehensive overview of the field, the course contains a unique laboratory component in which the students contribute to an actual, ongoing project to construct the first synthetic eukaryotic cell, a yeast cell composed of man-made parts. In doing so, the students acquire basic molecular biology skills and gain a truly "graduate student-like experience" in which they take ownership of their projects, troubleshoot their own experiments, present at frequent laboratory meetings, and are given 24-h access to the laboratory, albeit with all the guidance they will need to complete their projects during the semester. In this chapter, we describe the organization of the course and provide advice for anyone interested in starting a similar course at their own institution.

**Key words:** Build-a-Genome, Undergraduate course, Synthetic biology, Synthetic yeast, CloneQC, GeneDesign

## 1. Introduction

For the past several years at Johns Hopkins University, we have offered an undergraduate course called Build-a-Genome (1), which provides a comprehensive introduction to the field of synthetic biology. The course consists of lectures on various aspects of synthetic biology, including methods of gene synthesis, design and implementation of genetic devices, and the ethical considerations of constructing synthetic genomes. But what makes the course truly unique is the intensive laboratory component in which the students, rather than performing "cookbook" exercises, make significant contributions to an actual, ongoing research project aimed at building the world's first eukaryotic cell, specified by a yeast genome made entirely of synthetic parts.

Jean Peccoud (ed.), *Gene Synthesis: Methods and Protocols*, Methods in Molecular Biology, vol. 852, DOI 10.1007/978-1-61779-564-0_20, © Springer Science+Business Media, LLC 2012

The aims of the project are severalfold: to provide a greater understanding of the "rules" of genome organization and structure, to ask whether activities such as splicing are essential for "life," and to engineer a controllable rearrangement system in yeast to explore relationships between optimal genome configurations and different environmental conditions. This recombination system may also allow genome minimization and might be suitable as a host for engineered gene networks that could impart useful properties to the cell, perhaps for biofuel or pharmaceutical production (2).

The students in our course contribute to this ambitious undertaking by assembling commercially synthesized 60-mer oligonucleotides into ~750-bp fragments we call "building blocks (or BBs)." These are the smallest assembled units of synthetic genome, which are later strung together by methods described in detail in previous chapters (3, 4). But most importantly, the course offers our students a truly authentic laboratory experience in which they learn basic molecular biology techniques such as PCR and gel electrophoresis, design and troubleshoot their own experiments, present data to each other at frequent lab meetings, and are responsible for organizing their time to complete experiments for their own projects.

In this chapter, we describe how we organize the Build-a-Genome (also known as BAG) course, provide detailed protocols on the gene synthesis procedures, and offer tips and advice for anyone interested in starting a similar course at their own institution.

## 2. Methods

### 2.1. Overview of Molecular Biology Boot Camp

The BAG course begins with several lectures that introduce the synthetic yeast project to the students. The lectures cover topics such as why yeast is a useful model organism for studying cellular biology, how the synthetic genome is designed and assembled, the costs involved in doing such a project, and what we hope to learn from it. Although additional lectures are sprinkled throughout the semester, we strive to familiarize the students with the experimental workflow and relevant laboratory techniques as quickly as possible. This is accomplished during an eight-session training period, which we refer to as molecular biology "boot camp," in which we guide the class, as a group, stepwise through all the procedures used to assemble, clone, and sequence the ~750-bp "building blocks" (or "BBs") (Figs. 1 and 2). After boot camp, our students are assigned their own 10,000 bp segments of synthetic genome to synthesize, and they are given keys to the laboratory so they can work on their own schedules, albeit with all the help and guidance they will need, as explained later.

During boot camp, our students are assigned to groups of two or three in which each group is given the necessary PCR and cloning

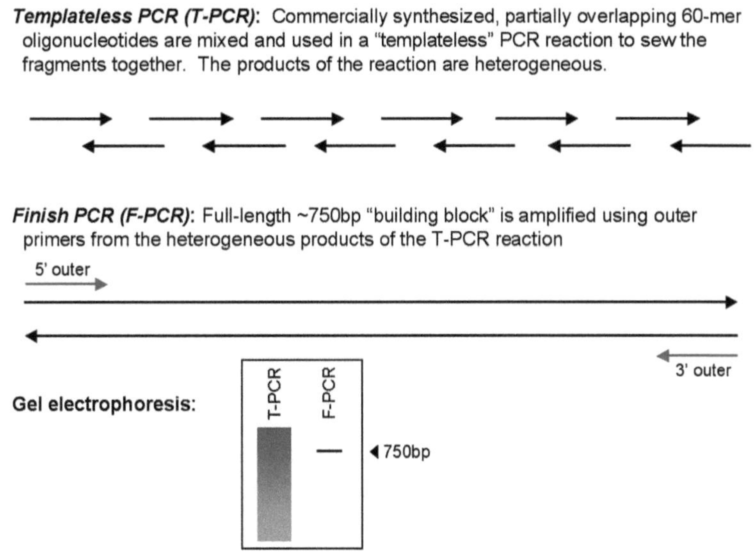


Fig. 1. Build-a-Genome student workflow I—PCR and gel electrophoresis.

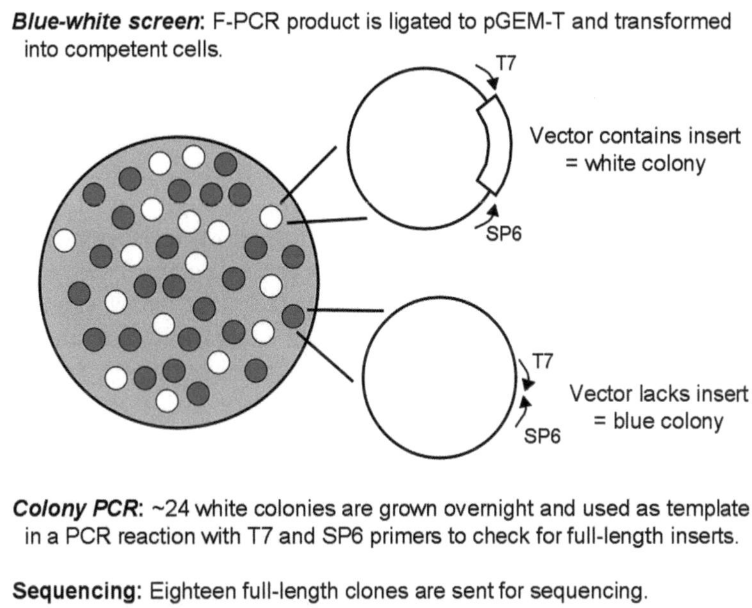


Fig. 2. Build-a-Genome student workflow II—cloning and sequencing.

reagents (for details, see below), and sets of oligos that will ultimately be assembled into BBs. Both members of the group use the same set of reagents to assemble the identical sets of BBs. Besides allowing our students to get acquainted and allowing them to help each other with the experimental procedures, assigning our students to groups has an additional advantage. Since each group member works on identical building blocks with identical reagents, we can "control" for the technique of each individual. If one member of a group successfully assembles a particular building block, but the other does not, we can work individually with that student to identify and correct the problem. Should any step in the workflow fail for a student during boot camp, he/she is required to repeat it as soon as possible, side-by-side with the instructor, so that the student catches up with the rest of the class. Most often, errors in assembling the PCR cocktail or poor mixing of the reaction are the culprits in boot camp PCR failures, but our students occasionally discover new and interesting ways to undermine their experiments.

As with any undergraduate course, we try to minimize this by paying special attention to the clarity of our directions and by adhering to the philosophy that the more we can *show* our students through hands-on approaches, the better. This was dramatically illustrated by our observation that *describing* how to adjust the settings on the pipettors correlated with a very low PCR success rate. However, when we gathered our students around a table with their pipettors, led them through a detailed lesson in pipettor anatomy, and tested their skill and accuracy by having them dispense fixed volumes of water onto a balance to measure the mass of pipetted liquid, the class PCR success rate improved remarkably. We can never provide too much detail when leading our students through the laboratory procedures and feel like the hands-on learning approach is highly beneficial to the students.

We also assume that our students possess no prior molecular biology experience, although many of the Hopkins undergraduates have used pipettors in other laboratory courses and are generally familiar with the theory behind PCR. That said, nonbiology majors who completely lack laboratory experience prior to taking the course have both successfully completed it and made important and unexpected contributions. For example, one of our students was a computer science major who developed a sequence analysis software program called CloneQC (see below) that has become an essential part of the course, and which was the subject of a recent first-author publication (3). Another student designed and built a robot for the course in an attempt to automate some of the tedious pipetting work. An attractive aspect of synthetic biology is that it is, by its nature, interdisciplinary. Biologists, chemists, mathematicians, computer scientists, ethicists, and engineers all have their place, and we encourage interested students from diverse academic disciplines to participate in the course.

### 2.2. Molecular Biology Boot Camp Workflow

The boot camp period is designed to guide students through the experimental workflow as a group to introduce the basic molecular biology skills they will need to complete their projects. Each group of two receives a set of the required reagents that have previously aliquotted into small volumes by the instructor and teaching assistants. The reagents include Taq polymerase, 10× Taq polymerase buffer, dNTPs, T7 and SP6 primers (to check clones for inserts by colony PCR, see below), a positive control consisting of a pGEM-T plasmid containing a cloned BB, and the appropriate oligonucleotides for assembling their assigned BBs.

We have pared down the assembly, cloning, and sequencing procedures such that any handling of dangerous or toxic substances is kept to a minimum. For instance, when plating cells onto agar plates, our students do not use a Bunsen burner flame to sterilize a spreader to distribute the cells on the plate. Instead, cells are pipetted onto the agar along with several sterile 5 mm diameter glass balls. The plate is then shaken such that the glass balls spread the cells evenly upon the agar surface. The only potential hazard is the ethidium bromide (EtBr) used in agarose gels, but we try to minimize their contact with it, and, perhaps more importantly, the boiling agarose itself. Instead of having the students directly pipette concentrated EtBr themselves, we premake gel-ready agarose. The instructor or TA autoclaves several bottles containing 300 mL of 1% agarose for 15 min to melt large batches of agarose simultaneously, and EtBr is added once the agarose has cooled. The agarose is then allowed to harden in the bottles. When students need to pour gels, they melt the agarose/EtBr mixture in the microwave and then proceed with their gel casting. We train the students very carefully to avoid boilovers and alert them to the danger of molten agarose burns, which we view as a much more significant potential hazard than EtBr.

The BB synthesis protocols have been optimized such that ~75% yield the desired full-length products at first pass. Our students are encouraged to push ahead to the cloning steps for successfully assembled BBs so they will get results to analyze as soon as possible and simultaneously devise methods for troubleshooting any BBs that failed during the initial attempts. In this section, we outline the experimental procedure depicted in Figs. 1 and 2, although detailed protocols appear in other chapters of the book.

### 2.3. Templateless PCR

The first step of the assembly process, referred to as the "templateless" PCR (T-PCR), involves "sewing" together the ~18 overlapping oligos that comprise the ~750-bp BB. In fact, the primers in these reactions serve as both the template and the primer for the reaction. The products of this reaction are heterogeneous and appear as a smear when examined by agarose gel electrophoresis (Fig. 1). We like to point out to the students that in the early cycles of this PCR, the molecules increase in length but not in number. This is different from a conventional PCR in which exponential increases in number occur starting in the first cycle.

### 2.4. Finish PCR

The desired full-length ~750-bp product is then amplified from the heterogeneous T-PCR product in a second reaction, called the "finish" PCR (F-PCR) (Fig. 1). The T-PCR product is diluted and then used as the template in this reaction, and the outermost

5′ and 3′ oligos from the particular BB are used as the primers. For BBs that fail at first pass (i.e., do not give the expected, full-length product, or do so in a very low yield), we ask students to think about why the PCR did not work. It is sometimes due to the particular sequence of the BB itself, which may contain repetitive elements, be unusually A-T rich, or possess other unusual sequences.

Based on their analysis of sequences, our students generate a plan for changing the PCR conditions to increase the probability of success and retry the BB assembly reactions using several different conditions. Problematic BBs can usually be assembled using a "touchdown" PCR protocol in the F-PCR step, in which successively lower annealing temperatures are used during each cycle in the early phase. We have also had a lot of success by lowering the extension temperature of the reaction from 72° to 68°, which helps amplify through A-T rich templates that are less stable at the higher temperature.

### 2.5. Cloning and Checking for Inserts

Each full-length ~750-bp BB is ligated to the pGEM-T vector (Invitrogen) in a T-A cloning step that allows for blue-white screening (Fig. 2). Ligations are typically performed at 4°C overnight, since this is convenient for the students, but the standard 1-h incubation at 16°C works fine as well. The ligated products are then transformed into either homemade or commercially available competent cells. We emphasize that the cells must be thawed on ice!

Not only is blue-white screening an effective method for identifying plasmid clones that contain inserts, but using the colorimetric screen, in which the students can directly see the results, is also a great way to teach about plasmids and ligations, as well as active genes, inactivation of ORFs, the probability of hitting a stop codon in random sequence, etc. Our students pick around 24 white colonies, which usually contain inserts, and perform colony PCRs using the T7 and SP6 priming sites available within the pGEM-T vector. The products of these PCRs are then run on agarose gels, and the data are immediately uploaded to the course Web site (details in Subheading 2.6). Students aim to send 18 full-length clones of each BB for sequencing to maximize the chances that at least one will have the desired sequence (details in Subheading 2.9). We send the bacteria in 96-well plates to an external Sanger sequence provider (Beckman-Coulter, formerly Agencourt).

### 2.6. Course Organization

We only utilize free or open-source software tools in our course, including ClustalW (5) (http://www.ebi.ac.uk/clustalw) for performing sequence alignments, FinchTV (http://www.geospiza.com/finchtv/) for viewing sequence peak data, and GeneDesign (6) and CloneQC (7), our in-house generated tools for designing oligos and analyzing sequence data, respectively. For course management, we have been using Moodle (http://moodle.org), which has become the repository for all course-related information.

This includes the syllabus, lecture notes, protocols, a Google calendar containing a PCR machine sign-up list, and a drop box for submitting assignments. But the Moodle site has also been modified to serve as universal "lab notebook" for the students. There is a gel database where students upload and label pictures of their agarose gels (and can easily access them during lab meeting presentations), and sections for them to describe their attempts to assemble each BB. These may be as simple as "Worked at first pass using standard PCR conditions!" or may include detailed troubleshooting protocols about problematic BBs that required specific PCR conditions for amplifying full-length products. Moodle is useful because as students come and go from semester to semester, we have one constant source housing all the data in a consistent and searchable format.

### 2.7. After Boot Camp

After completing the boot camp period and submitting the required milestone assignments, students are each given a key to the laboratory so they can complete their work on their own schedules. Although they have 24/7 access to the lab, they do not have 24-h access to the reagents; instead, an instructor or teaching assistant is available during the evening 6 days/week to dispense reagents and help students who have need help troubleshooting, etc.

During this part of the course, some students find it challenging to manage their schedule in a way that allows them to complete their BB assemblies. Of course, different students have very different work habits, learning styles, and commitments to additional activities. Some students prefer doing the bulk of their work nocturnally, whereas others are more comfortable working while the TAs or instructors are present. Moreover, some students methodically march through the assembly of their assigned BBs through steady and consistent work, whereas others complete their experiments in productive bursts.

However, we put several mechanisms in place to keep the students on track. First, an instructor or TA is available at the laboratory 6 evenings/week for questions and dispensing reagents. Second, students are expected to be present at the laboratory during the normal class time (although they will need plenty of out-of-class time to complete their work). Third, we hold mandatory lab meetings every other week where the students present their data and receive feedback from the group. Fourth, they must submit occasional "milestone" assignments, which are exercises based on each of the step of the workflow to assure that they are both progressing in their work and understanding the methods they are using in their experiments. These include uploading gels, calculating transformation efficiencies, and completing a sequence alignment and analysis exercise, all of which that assures that they are familiar with all the tools they will need to complete their work. So, despite the differences in student work styles, there is enough structure to the course such that all of our students have successfully assembled the bulk of their assigned sequences in plenty of time to send up to 18 clones of each BB for sequencing.

### 2.8. Designing the Oligos

We use an open-source, in-house generated software package called GeneDesign (4), which takes a chunk of synthetic yeast genomic sequence and divides it first into ~750 base pair BBs. GeneDesign then breaks down the BBs into ~60 base pair oligos that overlap adjacent sequences by ~20 nucleotides. The length of the oligos varies somewhat as GeneDesign attempts to generate oligos with similar melting temperatures to facilitate assembly. Conveniently, GeneDesign also generates properly formatted order sheets that can be sent directly to our commercial oligo provider and generates "maps" indicating the positions of oligos that will be delivered in a 96-well format. Our students learn how to use this software after completing boot camp, and they use it to design the BBs and oligos from their assigned synthetic sequence.

### 2.9. Sequence Analysis

Because of the intrinsic error rate in chemical DNA synthesis (which can be between 0.2 and 0.5%), we expect that there are, on average, between 1 and 5 mutations per 750-bp building block. In addition, the intrinsic error rate of Taq polymerase (~0.0001%), although much lower than that of oligo synthesis, may add significantly to the given amount of DNA that we are synthesizing and the two PCRs (at least) per BB. To deal with the error problem, we send 18 clones of each BB for sequencing, which is usually sufficient for getting at least one perfect clone that completely matches the desired synthetic sequence. At first, the sequence analysis was done "manually" by our students, who would align each clone sequence one-by-one to the desired BB sequence.

This sort of analysis is tedious and prone to human error, but a very talented former Build-a-Genome student wrote a program called CloneQC to automate the sequence analysis, and it has become an essential component of the project. CloneQC first takes the data from the forward and reverse sequencing runs for each clone, uses BLAST to identify the desired synthetic sequence to compare the sequences to, and then aligns the three sequences (the forward and reverse clone sequences, and the desired sequence) using ClustalW (4). CloneQC then scores each clone as being either a "Pass," indicating that it completely matches the desired sequence, a "Fail," indicating that there are errors in the sequence, or a "Check" indicating that the clone is likely a perfect match but that some of the sequence is of low quality and warrants a visual confirmation by the student. In addition, CloneQC lists the type (missense, deletion, etc.) and position of any mutations in a particular clone, so one can easily skip to that area of the sequence to confirm the presence of any mutations. CloneQC is publicly available, and additional details about accessing the program and about its design are available at http://cloneqc.thruhere.net (3).

By using CloneQC, the number of clones that need to be checked by eye is reduced dramatically, saving a lot of work for the students. However, we do require that all submitted clones be confirmed visually by the students. Moreover, they present their sequence analyses at lab meetings, which provides an additional check. During the process, our students all become experts at aligning sequences using ClustalW, interpreting the peaks in sequencing runs, and explaining why there are differences in sequence quality between the beginning and the end of a sequencing run. By explaining to the group how they performed their experiments and sequence analyses, they must master all the steps in their gene synthesis projects.

### 2.10. Submitting Clones for Sequencing

Because it is more efficient and cost-effective to send completely full 96-well plates for sequencing, we have set up a system on the course Moodle site that organizes student clone submissions. Once students have accumulated 18 full-length clones of a particular BB to send for sequencing, they sign up for wells (in a 96-well dish) using a wiki that is linked to the course Moodle site. The wiki simply has 96-well template spreadsheets that can be accessed by the students. Student #1 may sign up for wells A1–B6 for his 18 clones, and student #2 may sign up for wells B7–C12 for her clones. Once they sign up for wells, they put the remaining bacterial cultures (1 μL of which was used in the colony PCR) into the corresponding well of their own 96-well dishes and place them in the refrigerator. So, student #1 would submit a 96-well plate that has cultures only in wells A1–B6, and student #2 would submit a 96-well plate that has cultures only in wells B7–C12.

Once all the wells on a wiki have been signed up for by the students, the instructor consolidates all the bacterial cultures from their individual plates and makes two copies which are grown freshly overnight in LB + carbenicillin. One copy will be submitted for sequencing, whereas the other is kept at −80°C for the archives.

### 2.11. Lab Meetings and Grading

Once the students have completed boot camp and are working individually on completing their assigned building blocks, we convene as a group every other week for lab meetings, whose purpose is severalfold. First, it is a forum for questions and discussion where each student is given the opportunity to present his/her data to the group for feedback. Second, having regularly scheduled lab meetings keeps the students on schedule to complete their work. Third, since we have never included traditional exams in the course, we rely on their lab meeting presentations and their ability to answer related questions to assess their understanding of the material. By the end of the course, they should be able to explain (with drawings on the chalkboard) the details and theory behind all the experimental procedures they have performed in the lab, from PCR and DNA structure, to how gel electrophoresis, cloning, and sequencing work. For most students, their mastery of the material improves markedly over the course of the semester, and they become more comfortable presenting data in front of an audience. In addition, student grades depend on their ability to deliver perfect BBs for as many of their assigned building blocks as possible. On average, students complete about 10 kb of perfect synthetic genome by the end of the semester.

## 3. Discussion

In this chapter, we presented the organization of the Johns Hopkins Build-a-Genome course, an intensive laboratory course in which our students are introduced to the field of synthetic biology and learn molecular biology techniques, all while contributing to the synthesis of the world's first eukaryotic genome. What is unique about our course is that after a several-week training period, our students are given full access to the laboratory where they have the freedom to work on their own projects, albeit with enough structure that we keep their progress moving forward.

We have also introduced a related course called Build-a-Genome Mentor, in which students that have taken the first course, and are now experts in the methodology of the course, work on additional projects that we help them to develop. These projects have included optimizing the standard protocols for the templateless and finish PCR steps, devising methods to increase the frequency of "perfect" clones using affinity-based methods to remove mismatched duplex DNA from the finish PCR product, testing ligase chain reaction as an alternative to PCR, and working on improved methods to string together BBs into larger and larger pieces for introduction into yeast.

One additional benefit of the course is that it allows undergraduates to work alongside graduate students and postdoctoral fellows, which helps them get a sense of what a career in laboratory research is like. Of course, this is also a great teaching experience for the teaching assistants and instructors. We find that running the course in this way gives students a sense of ownership toward their work that makes them committed to the project's success. Also, by having them manage their own time, troubleshoot their own experiments, and present their data to the group, they truly master the theory and practice of the techniques they use and gain an authentic laboratory experience.

## References

1. Dymond JS, Scheifele LZ, Richardson S, Lee P, Chandrasegaran S, Bader JS, Boeke JD (2009) Teaching synthetic biology, bioinformatics and engineering to undergraduates: the interdisciplinary Build-a-Genome course. *Genetics*, **1**, 13–21.
2. Dymond JS, Richardson SM, Coombes CE, Muller H, Annaluru N, Blake WJ, Schwerzmann JW, Dai J, Lindstrom DL, Boeke AC, Gottschling D, Chandrasegaran S, Bader JS, Boeke JD (2011) Synthetic chromosome arms function in yeast and generate phenotypic diversity by design. *Nature*, **477**, 471–6.
3. Annaluru A, Muller H, Ramalingam S, Kandavelou K, London V, Richardson SM, Dymond JS, Cooper EM, Bader JS, Boeke JD, Chandrasegaran S (2011) Assembling DNA Fragments by USER Fusion, Methods in Molecular Biology (this volume).
4. Muller H, Narayana Annalur N, Schwerzmann JW, Richardson SM, Dymond JS, Cooper EM, Bader JS, Boeke JD, Chandrasegaran S (2011) Assembling large DNA segments in yeast Methods in Molecular Biology (this volume).
5. Higgins DG, Thompson JD, Gibson TJ (1996) Using CLUSTAL for multiple sequence alignments. *Methods in Enzymology*, **266**, 383–402.
6. Richardson SM, Nunley PW, Yarrington RM, Boeke JD, Bader JS (2010) GeneDesign 3.0 is an updated synthetic biology toolkit. *Nucl. Acids Res.* **9**, 2603–06.
7. Lee PA, Dymond JS, Scheifele LZ, Richardson SM, Foelber KJ, Boeke JD, Bader JS (2010) CLONEQC: lightweight sequence verification for synthetic biology. *Nucl. Acids Res.* **8**, 2617–23.

# Chapter 21

# DNA Synthesis Security

## Ali Nouri and Christopher F. Chyba

## Abstract

It is generally assumed that genetic engineering advances will, inevitably, facilitate the misapplication of biotechnology toward the production of biological weapons. Unexpectedly, however, some of these very advances in the areas of DNA synthesis and sequencing may enable the implementation of automated and nonintrusive safeguards to avert the illicit applications of biotechnology. In the case of DNA synthesis, automated DNA screening tools could be built into DNA synthesizers in order to block the synthesis of hazardous agents. In addition, a comprehensive safety and security regime for dual-use genetic engineering research could include nonintrusive monitoring of DNA sequencing. This is increasingly feasible as laboratories outsource this service to just a few centralized sequencing factories. The adoption of automated, nonintrusive monitoring and surveillance of the DNA synthesis and sequencing pipelines may avert many risks associated with dual-use biotechnology. Here, we describe the historical background and current challenges associated with dual-use biotechnologies and propose strategies to address these challenges.

**Key words:** DNA engineering, DNA synthesis, Biosecurity, Biotechnology, Pathogen, Toxin

## 1. Introduction

The discovery of DNA structure by James Watson and Francis Crick (1), and Rosalind Franklin (2), and the subsequent elucidation of the genetic code at the Cavendish Laboratories (3) ultimately afforded scientists the ability to modify DNA and change the genetic makeup of living systems. Since then, areas of biotechnological power—in particular DNA sequencing and synthesis—have grown at a rapid pace. Some have compared such growth to "Moore's Law"—the exponential growth in computing power (4).

This growth in biotechnological power is coupled with the rapid diffusion of technologies. Just as Moore's Law led to a world of personal computers that are more powerful today than the most advanced computers were only decades ago, powerful biotechnologies are now increasingly accessible, affording to technically

Jean Peccoud (ed.), *Gene Synthesis: Methods and Protocols*, Methods in Molecular Biology, vol. 852,
DOI 10.1007/978-1-61779-564-0_21, 

competent groups and even the individual user the sophisticated synthesis and manipulation of biological systems.

These advances have had a profoundly positive impact on human understanding of biological processes, and they have consequently improved health and well-being: genetically manipulating a cell's signal transduction pathway has led to the identification of drug targets, antibiotics, and other therapies; food security has been enhanced through the genetic modification of crops to produce varieties that are higher in yield than their wild-type counterparts or even ones that are resistant to disease and to drought; DNA manipulation has enabled the development of novel vaccines and rendered microbes that for centuries plagued humankind obsolete to the vaccinated.

But these same tools that have fundamentally improved human health and agriculture can also be used for ill: the same molecular biology methodologies that are employed to alter wild-type viruses into attenuated vaccine strains, for instance, could also be used to enhance their infectivity and pathogenicity for purposes of biological terrorism or warfare.

The challenge associated with biotechnology—and particularly with DNA synthesis technologies—is to ensure that they are only used for the advancement of humankind, rather than to its detriment. Striking this balance in a practical and efficient manner has thus far eluded policy makers because biotechnologies are extremely diffuse and exist in tens of thousands of academic and commercial laboratories worldwide (5).

Recent trends in the life sciences, however, lend themselves to the adoption of safeguards that were previously untenable. That is because biotechnology has become increasingly high-throughput and therefore increasingly automated. Automation of technologies such as DNA synthesizers and sequencers may enable the incorporation of automated safeguards either into machines or into a centralized clearinghouse to reduce the possibility of misusing such technologies for nefarious purposes (e.g., the creation of genetic material for bioterrorist purposes). During the development of novel technologies, there may be a small window of opportunity that scientists can capitalize upon in order to curb risks to ensure that their discipline is used only for legitimate purposes (6).

## 2. Dual-Use Biotechnology

The problem of biological weapons can be traced back centuries; whether it was catapulting bubonic plague-infested bodies over city walls by invading armies (7) or delivering smallpox virus-contaminated blankets to Native American tribes, humans resorted to biological warfare long before modern biology took hold.

With advances in biological research also came more advanced biological warfare programs. During WWII, for instance, the United States, United Kingdom, Soviet Union, and Germany all developed extensive chemical and biological warfare programs. These countries refrained from using them, but Japan attacked civilian populations extensively, including with fleas infected with *Yersinia pestis*—the causative agent of the bubonic plague (8). The US biological weapons program was ceased by executive order of President Richard Nixon. Subsequently, the Biological Weapons Convention—a multilateral treaty that prohibits the development, production, and stockpiling of biological weapons—was signed by the president and ratified by the US Senate. The Soviet Union, despite being a signatory to the treaty, continued a secret but elaborate and extensive offensive biological warfare program that apparently came to an end only with the disintegration of the USSR.

The ease with which biological organisms and toxins can be harnessed, amplified, and stockpiled has made them not just a threat that emanates from nation states but also from nonstate groups and even individuals. In 1984, the Rajneeshee, an Oregon-based cult, acquired Salmonella and contaminated restaurant salad bars, causing food poisoning in over 700 individuals. In 1995, the Japanese Aum Shinrikyo cult, which had previously attempted but failed in attacks on civilians with biological weapons, carried out attacks using the chemical weapon sarin, killing 13 and injuring dozens more in the Tokyo subway system. Many questions still remain with respect to the culprit behind the 2001 *Bacillus anthracis* "anthrax" attacks in the United States (9). Although he took his own life prior to the filing of formal charges by the FBI, that attack may have been carried out by Bruce Ivins, a biodefense researcher at the United States Army Medical Research Institute of Infectious Diseases. Even though the anthrax attacks resulted in only five deaths, they caused insecurity and substantial economic damage throughout the country (9).

One concern is that as biotechnologies diffuse throughout the world, and become increasingly user-friendly, the likelihood and potential damage of bioterrorism and biological warfare simply grow. There is already a list of well-known experiments that illustrate the capability and possible dangers intrinsic to biotechnologies, which are of dual-use: these technologies can be used for good or for ill. These experiments include genetic manipulation of mousepox—a cousin of smallpox—to the extent that the modified virus overcomes natural host immunity (10) (although humans are not susceptible to mousepox, the study's findings provide a potential blueprint to developing a vaccine-resistant smallpox virus); the creation of polio virus from scratch through the purchase of synthetic DNA molecules (11)—so that even if the World Health Organization (WHO) repeats its success with eradicating smallpox with that of eradicating polio worldwide, the virus could be created, de novo, in laboratories around the world; and the laboratory resynthesis of the extinct Spanish influenza virus—the agent that killed tens of millions of people worldwide in the pandemic that began in 1918 (12). Although these experiments help elucidate the biology of disease-causing pathogens, the underlying tools—if misused—could result in catastrophic public health consequences. These dangers have been broadly recognized by committees of the US National Academies and the British Royal Society (5, 14, 15), but solutions that do not do more harm than good remain elusive. More recently, DNA synthesis capabilities integral to the emerging field of "synthetic biology," whose aims are to allow practitioners to fabricate small "biological devices" and ultimately new types of microbes (16) have further elevated previous security concerns. The synthesis of a 1.08-mega-base pair *Mycoplasma mycoides* genome and its transplantation into a *Mycoplasma capricolum* recipient cell to create new *M. mycoides* cells (17), for instance, raised concerns at the highest political levels and led President Obama to task his commission on bioethics to consider synthetic biology risks (18).

## 3. Challenges in Biological Security

Traditionally, there have been severe challenges to regulatory schemes that address risks for the field of biology (19) because (a) there is a mismatch between the rapid pace with which biotechnology advances, and the comparative sluggishness of creating and updating a regulatory regime; and (b) because monitoring and inspection of molecular biology laboratories is difficult, given that technologies are small scale and widespread. In the case of DNA synthesis technologies used for large, gene-length DNA synthesis, however, these challenges have been moderated in part because technologies are currently confined to relatively few facilities. This centralization makes a monitoring regime possible and has enabled much of the DNA synthesis industry to adopt safeguard strategies. Security concerns over this industry were heightened shortly after the technology was employed at the Stony Brook Laboratories in 2002 to synthesize the poliovirus DNA (11). To address risks associated with the fledgling field of DNA synthesis, Harvard biologist (and biotechnology developer and entrepreneur) George Church proposed a safeguard strategy to ensure that the technology will not be used for the illegitimate synthesis of potentially harmful genomes (20). In 2010, the National Institutes of Health published rules urging all companies engaged in large-molecule DNA synthesis to screen incoming requests for DNA above 200 nucleotides in length to ensure that sequences belonging to particular pathogens and toxins are not commercially provided to certain individuals and states. Sequence comparison software is used to "read" DNA sequences of incoming customer orders and compare them to genes of toxins and genomes of a list of known pathogens, so that hazardous material is not produced for those that might have illicit intent (21).

The support of the synthesis industry for and adoption of these procedures is in part due to the nonintrusive nature of the proposal; rather than requiring oversight and cumbersome regulatory structures to which industry and many scientists might be opposed, the screening tool allows the DNA synthesis pipeline to go about business as usual while computer software engages in the invisible detective work. A major challenge going forward will be to harmonize the approach among all DNA providers worldwide since any noncompliant entity providing harmful material to unauthorized users undermines the framework. Possible strategies to address this deficiency include international guidelines, agreement, or even licensing for all DNA providers to follow a screening protocol, or establishing a centralized international clearinghouse that receives and screens all customer orders, clearing them for synthesis.

## 4. Synthesizer Proliferation and a New Security Paradigm

This risk-management framework embraced by industry may help prevent the misuse of commercially provided DNA molecules, but it will only be effective so long as the underlying tools remain confined to a relatively small number of facilities (22). Meanwhile, the increasing demand for synthetic DNA has made large-DNA synthesis lucrative, leading to the development of novel platforms that have potential for automation. One possible outcome could be the diffusion of advanced synthesizers—those capable of constructing large DNA molecules—to individual users around the world. Just as powerful computers once confined to the most advanced nations are now accessible to those who can afford them, one can envision a world in which graduate, undergraduate, perhaps even some high school biology students are provided with personal advanced DNA synthesizers as a standard laboratory bench-top research device. The diffusion of the polymerase chain reaction (PCR) machine to tens of thousands of laboratories worldwide demonstrates this possibility. This outcome would undermine, or even render irrelevant, the current risk-management framework adopted by the DNA synthesis industry. Therefore, alternative security strategies ought to be explored.

## 5. Conventional Biosecurity Strategies

Past attempts to safeguard the life sciences from misuse have proved challenging. This is because strategies that rely on restriction and classification of sensitive research, or on the curbing of scientific communication, limiting tacit knowledge, and restricting know-how, tend to be counterproductive for the life sciences, whose central mission is to improve human health and well-being. Life science research is the cornerstone of modern medicine. Despite its risks, for example, it is virology research that keeps lethal pathogens such as HIV at bay. And without information and material sharing regarding threatening pathogens such as methicilin-resistant *Staphylococcus aureus* (MRSA) and many others, there will be no treatment or cure. Any risk-management proposal that substantially hinders these medical-relevant efforts is and impractical—particularly given that life science research is already a thriving global endeavor.

The difficulty in devising concrete and unintrusive risk-management proposals for biotechnology has left the greater biology community with softer measures to guard against the possibility of misuse. Domestically, through the National Science Advisory Board for Biosecurity (NSABB), and internationally, through annual Biological Weapons Convention meetings, scientists, policy makers, and governments are largely relying on measures that include awareness-raising initiatives and ethics training as primary mechanisms by which biotechnology risks are addressed. Although these promote norms and encourage good behavior among well-intentioned individuals, they ought to be complemented with stronger measures that hinder the acquisition of biological material by those with illicit intent. Efforts to establish a monitoring regime for DNA synthesis companies represent an important approach. In addition, technologies, which are inherently resistant to being misused for illicit purposes, could further fill this gap.

## 6. Misuse-Resistant Technologies

As high-throughput large-scale research efforts gained traction in the life sciences, technologies that were only recently manual became increasingly automated. The absence of other measures, automation, and the consequent user-friendly nature of these technologies lowers the required expertise and increases the potential for these technologies to be misused for nefarious purposes. But if appropriately safeguarded, automation also provides opportunities to build safeguards into these technologies, so that only illicit applications are hindered. In the case of DNA synthesis, for instance, safeguards could include built-in software or hardware that screens DNA sequence inputs and compares them to the sequences of pathogens and toxins of concern (22). Sequences could then be vetted as "legitimate" or "potentially nefarious."

An analogous concept is the so-called V-chip, a feature that can block the display of television programs of a particular rating. The V-chip is intended only to exert parental control over television viewing and can easily be reprogrammed; in the case of dual-use DNA synthesizers, such security measures would have to be more effective and robust.

For researchers registered to perform experiments with the genetic material of biological agents of concern, however, a software update or a modified computer chip permitting the user to bypass the synthesizer's restrictions could be utilized. Just as US regulations prohibit the transfer of hazardous biological agents to nonlicensed users, any software or hardware updates that permit bypassing of regulations would similarly only be available to appropriately vetted users. Of course, "hacking" these protections would have to be anticipated and countered.

Whereas these approaches would safeguard stand-alone synthesizers, alternatively, a "network security" approach could be employed, in which a central server that is in communication with remote synthesizers serves as the security focal point. Under this model, after users request DNA sequences online, the sequence filters through the server, which functions as a virtual DNA clearinghouse. Automated DNA screening at the server determines the identity of the DNA, as well as the identity of the user. While "legitimate" requests would simply filter through the clearinghouse and make their way to the user's synthesizer, any illegitimate order, such as a hazardous agent that is requested by an unauthorized user, would be terminated and the "job" transmitted to the appropriate national or international oversight agency.

One major challenge going forward is determining exactly what sequences would be regulated. Initially, a regulatory framework that already applies to the possession of pathogens and toxins of concern could be extended to their DNA sequences. In the USA, the possession of such "select agents" requires licensing by the Centers for Disease Control and Prevention or the Department of Agriculture. The Select Agent Regulations target the most dangerous pathogens. In order to minimize the regulatory burdens on biological research, these are currently limited to a small subset of dangerous organisms and toxins, despite the existence of numerous disease-causing agents that occur in nature. The National Academy of Sciences has now recommended that the extension of the select agent framework from biological agents to their genetic material be considered (23). This would facilitate the monitoring approaches the commercial DNA synthesis industry has adopted. Sequencing discrimination would be more challenging for genetic material that is very similar, but not identical, to select agents. One can imagine, for instance, that an experienced researcher with illicit motives could simply request sequences that deviate from the original wild-type strain only slightly and yet still encode for pathogenic biologically active products. In the future, more comprehensive monitoring approaches will be needed to infer the pathogenicity of DNA sequences, despite human-induced changes in the sequence. More sophisticated approaches, however, such as the prediction of pathogenicity from novel sequences, are currently not available. Moreover, the Academies recently concluded that this approach is unlikely to be feasible for the purposes of a regulatory framework in the foreseeable future (23).

## 7. Oligonucleotide Synthesis

Thus far, concerns over DNA synthesis have involved only technologies that are capable of synthesizing large DNA molecules. Guidance from the National Institutes of Health, for instance, only urges companies to screen orders for DNA that are 200 nucleotides or longer. The majority of the synthetic DNA market, however, revolves around the production of oligonucleotides (oligos), which are DNA fragments that range from only a few to tens of nucleotides in length. Lack of a risk-management framework for this industry indeed enables a person to obtain a series of oligos that can be stitched together through standard biology techniques to make up longer genes.

Screening at the oligo level is challenging because such sequences are too small to provide "unique" features, making it impossible for sequence comparison software to assign a particular sequence to the organism of origin. If a number of these short sequences are pooled together, however, unique sequences could be revealed (Nouri, Goudarzi, and Chyba, unpublished). A security protocol for the oligonucleotide industry might be feasible if a centralized clearinghouse was established to pool a customer's disparate oligo requests—even if placed over separate time frames—so that the identity of the pooled sequences and thus the legitimacy of the order could be assessed.

Rather than providing a false sense of comfort by only safeguarding gene-size DNA manufacturers, a security framework that encompasses the much larger oligonucleotide industry ought to be explored.

## 8. Extending Screening from the DNA Synthesis to the DNA Sequencing Industry

While DNA screening at the synthesis bottleneck could be one effective tool for dealing with commercial DNA, it will not alleviate all biotechnology risks, which go well beyond DNA synthesis and include, for instance, genetic engineering experiments to enhance pathogens and toxins. We therefore set out to identify any bottlenecks associated with genetic engineering experiments that could be appropriately safeguarded. To uncover these, we analyzed scientific publications that utilize molecular biology and genetic engineering techniques and found that a necessary step in these experiments is DNA sequencing: almost invariably, researchers rely on DNA sequencing to gauge the success of molecular biology experiments such as constructing or altering a fragment of DNA or even the genome of an entire organism. Other potentially dual-use experiments, such as modifying hazardous bacterial and viral genomes, or construction of gene-encoding toxins, also require this verification step. The monitoring of the DNA sequencing phase of these experiments may provide clues as to the legitimacy of the underlying genetic engineering research and captures a host of potentially dangerous experiments that synthesis monitoring does not. While this proposal would have been much too intrusive and costly to implement in the days of manual DNA sequencing, advances in the field are opening new possibilities.

## 9. Outsourcing DNA Sequencing

DNA sequencing is among the fastest advancing biotechnologies; it used to be performed in-house using manual methodologies but is now increasingly automated. Centralized facilities that employ high-throughput technologies have sprung up to accommodate DNA sequencing needs of the life science community. This outsourcing is less error prone than traditional techniques and relieves researchers of the cumbersome task of manual sequencing. By 1999, 40% of US life science researchers engaged in DNA sequencing were outsourcing all sequencing work to only few facilities that served large numbers of users. By 2005, this figure had jumped to 80% (24). This centralization of DNA sequencing permits the adoption of monitoring and surveillance at these few locations. As customers submit DNA material for sequencing, samples can automatically be cross-referenced against the select agent database. Instances in which researchers who are not authorized to work with select agents submit select-agent DNA could result in the notification of appropriate authorities. This is similar to the screening proposal already operating at some DNA synthesis factories where companies screen DNA orders to ensure that pathogenic genes and genomes are not provided to unauthorized customers. By adapting this protocol to the sequencing industry, a broad spectrum of risks that encompasses virtually all of recombinant DNA technologies could be partly mitigated.

Simultaneous with this outsourcing of DNA sequencing, however, there are major efforts on the part of the biotechnology industry to develop affordable, user-friendly, advanced personal sequencers. The diffusion of these to an increasing number of users would render a more centralized monitoring and surveillance strategy insufficient. To cope with this outcome, we also suggest that, in addition to the monitoring of centralized facilities, advanced personal DNA sequencers incorporate safeguards that recognize "illicit" sequences. Similar to the misuse-resistant synthesizers discussed above, sequencers could also be fitted with security software or built-in security chips that recognize select agent sequences and prohibit the sequencer from carrying out its function.

## 10. Paths Forward

Safeguarding the dual-use life sciences has been elusive, particularly because they are extremely widespread. Increasing automation in the synthesis and sequencing areas, however, provide opportunities to monitor DNA sequences and, thus, provide an opportunity to safeguard dual-use genetic engineering research. Those who wish to evade these safeguards could continue to perform in-house sequencing and synthesis using traditional (manual) tools. These methods, however, are more error prone, costly, and time-consuming, thereby reducing the likelihood and pace of success. Moreover, market forces will increasingly contribute to the eventual replacement of such manual technologies by advanced technologies amenable to safeguards. In any event, no safeguards regime can aspire to the total elimination of risk. The objective, rather, must be to mitigate risk without causing more harm than good.

These strategies can serve as focal points to safeguard the life sciences. The DNA synthesis industry has already begun implementing some of these approaches to deal with the narrow field of commercial DNA synthesis. If other biotechnology areas follow suit, a wide range of risks encompassing genetic engineering could be mitigated. Moreover, these preventative measures could be an important supplement to the current strategy of attempting to explore countermeasures against potential biological agents of concern.

Security strategies discussed above should be prioritized during early stages of technology development. Since the 2001 mail anthrax attacks, the federal government has spent over $50 billion just on civilian biodefense projects that include developing vaccines, drugs, and disease surveillance systems (25). In comparison, little attention is paid to developing biotechnologies and systems that are intrinsically more secure. Designing and deploying these would help to prevent misuse of the technology, thus relieving some of the need to develop measures aimed at neutralizing laboratory-generated pathogens. If such technological-based security systems are prioritized, gradually improved automated technologies that are also safeguard-friendly will replace the older, less efficient, and difficult-to-safeguard tools.

Concurrent with the development of such technologies, it is also essential for a proper regulatory framework to be similarly advanced. As suggested by the Academies, this could include appropriate modification (26) and possible extension of rules that exist for the possession of particular organisms and toxins to their genetic sequences. For novel sequences, predicting pathogenicity based on sequence alone may not be possible in the foreseeable future, but better monitoring tools can be developed to detect select agent genome variants. Another important challenge will be to advance an international biotechnology security framework. Many countries lack national frameworks for dealing with the agents themselves, let alone their genetic material. And for those that have regulations in place, perceived biological threats vary greatly, leaving many hurdles to the creation of a harmonized global framework. The UN Secretary-General has recognized these challenges and called a global forum to address biotechnology risks. A methodical development of policies, nationally and internationally, together with the development and deployment of biotechnologies and bioservices that are intrinsically more secure will help ensure that the revolution in synthetic biology will only be used to benefit society.

## References

1. Watson JD, Crick FH (1953) Molecular structure of nucleic acids; a structure for deoxyribose nucleic acid. Nature. 171:737–8.
2. Franklin RE and Gosling RG (1953) Molecular configuration in sodium thymonucleate. Nature 171, 740–741 (1953).
3. Crick FH, Barnett L, Brenner S, Watts-Tobin RJ (1961) General nature of the genetic code for proteins Nature 192:1227–32.
4. Moore, G (1965) Cramming More Components onto Integrated Circuits. Electronics pp114–7.
5. National Research Council (2006) Committee on Advances in Technology and the Prevention of their Application to Next Generation Biowarfare Threats; Globalization, Biosecurity, and the Future of the Life Sciences. Washington, DC: National Academies Press.
6. Nouri A and Chyba CF (2008) Biotechnology and biosecurity. In: Bostrom N and Cirkovic MM (eds) Global Catastrophic Risks. Oxford University Press.
7. Wheelis, M (2002) Biological Warfare at the 1346 Siege of Caffa. Emerg Infect Dis 8:971–5.
8. Unit 731 Criminal Evidence Museum (2005). Unit 731: Japanese Germ Warfare Unit in China. China Intercontinental Press.

9. Shane S (2010). F.B.I., Laying Out Evidence, Closes Anthrax Case. The New York Times, February 19.
10. Jackson RJ et al (2001) Expression of mouse interleukin-4 by a recombinant ectromelia virus suppresses cytolytic lymphocyte responses and overcomes genetic resistance to mousepox. J Virol 75:1205–10.
11. Cello JP, Paul AV and Wimmer E (2002) Chemical synthesis of poliovirus cDNA: generation of infectious virus in the absence of natural template. Science 297:1016–8.
12. Tumpey TM et al (2005) Characterization of the reconstructed 1918 Spanish influenza pandemic virus. Science 310:77–80.
13. National Academy of Sciences (2003) Biotechnology research in an age of terrorism: Confronting the 'dual use' dilemma.5th ed National Academies Press, Washington DC.
14. Committee on Advances in Technology and the Prevention of their Application to Next Generation Biowarfare Threats, Globalization, Biosecurity, and the Future of the Life Sciences. National Academies Press Washington DC.
15. The Royal Society (2009) New Approaches to Biological Risk Assessment. Royal Society Policy Document, United Kingdom.
16. Fu P (2006) A perspective of synthetic biology: Assembling building blocks for novel functions. Biotechnol J 1:690–9.
17. Gibson DG et al (2010) Creation of a bacterial cell controlled by a chemically synthesized genome. Science 329:52–6.
18. US President Obama's letter to his bioethics committee. www.bioethics.gov/documents/Letter-from-President-Obama-05.20.10.pdf.
19. Chyba CF. (2006) Biotechnology and the Challenge to Arms Control. Arms Control Today. http://www.armscontrol.org/act/2006_10/BioTechFeature.asp.
20. Church G (2005) Let us go forth and safely multiply Nature 438:423.
21. Bugl, H et al (2007) DNA synthesis and biological security Nat. Biotechnol 25:627–629 (2007).
22. Nouri A and Chyba CF (2009). Proliferation-resistant biotechnology: an approach to improve biological security. Nat Biotechnol 27: 234–236.
23. National Research Council (2010) Committee on Scientific Milestones for the Development of a Gene-Sequence-Based Classification System for the Oversight of Select Agents, Sequence-Based Classification of Select Agents: A Brighter Line. National Academies Press Washington, DC.
24. U.S. MSPPSA report on DNA Sequencing Market Analysis (2005/2006). http://www.phortech.com/2005seq.htm.
25. Franco C (2008). Billions for Biodefense. Biosecurity and Bioterrorism 6:131–146.
26. National Research Council (2009) Committee on Laboratory Security and Personnel Reliability Assurance Systems for Laboratories Conducting Research on Biological Select Agents and Toxins; Responsible Research with Select Agents and Toxins. National Academies Press Washington DC.

# Index



Jean Peccoud (ed.), *Gene Synthesis: Methods and Protocols*, Methods in Molecular Biology, vol. 852, DOI 10.1007/978-1-61779-564-0, 

MIX
Papier aus verantwortungsvollen Quellen
Paper from responsible sources
FSC® C105338

If you have any concerns about our products,
you can contact us on
**ProductSafety@springernature.com**

In case Publisher is established outside the EU,
the EU authorized representative is:
**Springer Nature Customer Service Center GmbH**
**Europaplatz 3, 69115 Heidelberg, Germany**

Printed by Libri Plureos GmbH
in Hamburg, Germany